Lecture Notes in Physics

The Editorial Policy for Proceedings

The series Lecture Notes in Physics reports new developments in physical research and teaching – quickly, informally, and at a high level. The proceedings to be considered for publication in this series should be limited to only a few areas of research, and these should be closely related to each other. The contributions should be of a high standard and should avoid lengthy redraftings of papers already published or about to be published elsewhere. As a whole, the proceedings should aim for a balanced presentation of the theme of the conference including a description of the techniques used and enough motivation for a broad readership. It should not be assumed that the published proceedings must reflect the conference in its entirety. (A listing or abstracts of papers presented at the meeting but not included in the proceedings could be added as an appendix.)
When applying for publication in the series Lecture Notes in Physics the volume's editor(s) should submit sufficient material to enable the series editors and their referees to make a fairly accurate evaluation (e.g. a complete list of speakers and titles of papers to be presented and abstracts). If, based on this information, the proceedings are (tentatively) accepted, the volume's editor(s), whose name(s) will appear on the title pages, should select the papers suitable for publication and have them refereed (as for a journal) when appropriate. As a rule discussions will not be accepted. The series editors and Springer-Verlag will normally not interfere with the detailed editing except in fairly obvious cases or on technical matters.
Final acceptance is expressed by the series editor in charge, in consultation with Springer-Verlag only after receiving the complete manuscript. It might help to send a copy of the authors' manuscripts in advance to the editor in charge to discuss possible revisions with him. As a general rule, the series editor will confirm his tentative acceptance if the final manuscript corresponds to the original concept discussed, if the quality of the contribution meets the requirements of the series, and if the final size of the manuscript does not greatly exceed the number of pages originally agreed upon. The manuscript should be forwarded to Springer-Verlag shortly after the meeting. In cases of extreme delay (more than six months after the conference) the series editors will check once more the timeliness of the papers. Therefore, the volume's editor(s) should establish strict deadlines, or collect the articles during the conference and have them revised on the spot. If a delay is unavoidable, one should encourage the authors to update their contributions if appropriate. The editors of proceedings are strongly advised to inform contributors about these points at an early stage.
The final manuscript should contain a table of contents and an informative introduction accessible also to readers not particularly familiar with the topic of the conference. The contributions should be in English. The volume's editor(s) should check the contributions for the correct use of language. At Springer-Verlag only the prefaces will be checked by a copy-editor for language and style. Grave linguistic or technical shortcomings may lead to the rejection of contributions by the series editors. A conference report should not exceed a total of 500 pages. Keeping the size within this bound should be achieved by a stricter selection of articles and not by imposing an upper limit to the length of the individual papers. Editors receive jointly 30 complimentary copies of their book. They are entitled to purchase further copies of their book at a reduced rate. As a rule no reprints of individual contributions can be supplied. No royalty is paid on Lecture Notes in Physics volumes. Commitment to publish is made by letter of interest rather than by signing a formal contract. Springer-Verlag secures the copyright for each volume.

The Production Process

The books are hardbound, and the publisher will select quality paper appropriate to the needs of the author(s). Publication time is about ten weeks. More than twenty years of experience guarantee authors the best possible service. To reach the goal of rapid publication at a low price the technique of photographic reproduction from a camera-ready manuscript was chosen. This process shifts the main responsibility for the technical quality considerably from the publisher to the authors. We therefore urge all authors and editors of proceedings to observe very carefully the essentials for the preparation of camera-ready manuscripts, which we will supply on request. This applies especially to the quality of figures and halftones submitted for publication. In addition, it might be useful to look at some of the volumes already published. As a special service, we offer free of charge LATEX and TEX macro packages to format the text according to Springer-Verlag's quality requirements. We strongly recommend that you make use of this offer, since the result will be a book of considerably improved technical quality. To avoid mistakes and time-consuming correspondence during the production period the conference editors should request special instructions from the publisher well before the beginning of the conference. Manuscripts not meeting the technical standard of the series will have to be returned for improvement.

For further information please contact Springer-Verlag, Physics Editorial Department II, Tiergartenstrasse 17, D-69121 Heidelberg, FRG

T. P. Ray S. V. W. Beckwith (Eds.)

Star Formation and Techniques in Infrared and mm-Wave Astronomy

Lectures Held at the Predoctoral Astrophysics School V

Organized by the European Astrophysics Doctoral Network (EADN) in Berlin, Germany, 21 September - 2 October 1992

Springer-Verlag
Berlin Heidelberg New York
London Paris Tokyo
Hong Kong Barcelona
Budapest

Editors

T.P. Ray
Dublin Institute for Advanced Studies
5 Merrion Square
Dublin 2, Ireland

S.V. W. Beckwith
MPI für Astronomie
Königstuhl
D-69117 Heidelberg, Germany

ISBN 3-540-58196-0 Springer-Verlag Berlin Heidelberg New York
ISBN 0-387-58196-0 Springer-Verlag New York Berlin Heidelberg

CIP data applied for

Printed in Germany

This book was processed using the TEX/LATEX macro packages from Springer-Verlag.
SPIN: 10080361 55/3140-543210 - Printed on acid-free paper

Preface

The European Astrophysics Doctoral Network (EADN) is an affiliation of 27 universities whose aim is to encourage the mobility of astrophysics students within Europe. This it does through mobility grants funded by the EU ERASMUS programme.

In addition the EADN organizes a summer school aimed at students in either the first or second year of their doctoral studies. The first such school was held in Les Houches (France) in September 1988 and subsequent schools have been held in Ponte de Lima, Dublin and Graz. As a rule, each school proposes two closely related themes: one being astrophysical and the other more methodological, e.g. in the field of technology or in numerical studies. The content of the lectures, although advanced, is aimed at a broad audience. The presentations should, in general, be understandable to students who are still beginners in their field.

The past decade has seen an enormous growth in our understanding of star formation due largely to developments in infrared and mm technology. The choice of themes for the Fifth EADN Summer School in Berlin, i.e. Star Formation and Techniques in Infrared and mm-Wave Astronomy, was therefore an easy one. The School itself was held at the Technische Universität Berlin from 21st September to October 2nd 1992.

In Part I of this volume, sites of star formation, i.e. molecular clouds, the characteristics of low and high mass young stellar objects (YSOs) and their interaction with their environment, are amongst the topics considered in depth. One of the most important findings in recent years has been the discovery that stellar birth is associated not only with the accretion of matter but with the outflow of material as well. These outflows in turn seem to be related to another ubiquitous phenomenon amongst young stars, namely the presence of disks. Both disks and outflows are examined in Part I.

Because stars form in the dusty environments of molecular clouds, our understanding of the processes governing their formation, particularly at the earliest stages, has relied heavily on infrared and mm observing techniques. Far-infrared astronomy has to be carried out from space (for example in the past by satellites like IRAS and in the future by ESA missions like ISO) but it is possible to observe from the ground at sub-millimeter and millimeter wavelengths. Observing methods in the far-infrared and sub-mm are explored in Part II. The widespread availability of infrared arrays has revolutionized the field of infrared astronomy. Such arrays have rapidly replaced traditional detectors and these are discussed in Part II along with other techniques for studying star formation at near-infrared wavelengths. Part II ends with a detailed review of the rapidly developing field of high resolution infrared studies including a discussion of adaptive optics. The only lecture course given at the Berlin School but not included in this volume was on Molecular Line Emission. An important part of the EADN school programme, is that students present their

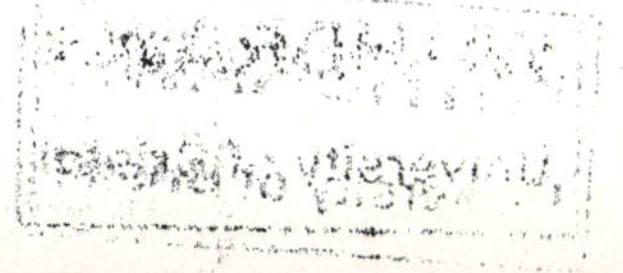

research to their fellow students. Abstracts of the student presentations at the Berlin School are given in Part III.

As in previous years, participants had time off to enjoy their surroundings and it must be said that in this respect Berlin was not found wanting! Organized events included a welcoming reception by the Technische Universität Berlin who hosted the School and a visit to the Potsdam Observatory.

Seven lecturers from various parts of Europe, delivered approximately seventy lectures over the ten working days of the School. The Scientific Directors would like to thank them for their hard work which was appreciated not only by us but quite clearly by the students themselves. We would also like to express our sincere gratitude to the Technische Universität Berlin and to the local organizers, Dr. J.P. Kaufmann and Prof. Dr. E. Sedlmayr, not only for their help but for their hospitality as well. Finally thanks are also due to the School secretaries, A. Grace and I. Birambaux, who made sure that the administrative "wheels" of the School ran very smoothly.

Dublin/Heidelberg
April 1994

T.P. Ray
S.V.W. Beckwith

Contents

Part I. Star Formation

Part III. Student Presentations

Participants

Lecturers:
Steve Beckwith, MPI für Astronomie, Heidelberg
Claude Bertout, Observatoire de Grenoble
Sylvie Cabrit, Observatoire de Grenoble
John Dyson, University of Manchester
Jim Emerson, Queen Mary and Westfield College
Christoph Leinert, MPI für Astronomie, Heidelberg
Richard Hills, Mullard Radio Astronomy Observatory, Cambridge

Local Organizers:
Jens P. Kaufmann, Technische Universität Berlin
Prof. Dr. E. Sedlmayr, Technische Universität Berlin

Scientific Directors:
Tom Ray, Dublin Institute for Advanced Studies
Steve Beckwith, MPI für Astronomie, Heidelberg

Students:
Nancy Ageorges, Observatoire de Grenoble
Juan Manuel Alcala, Landessternwarte Heidelberg
Bernhard Aringer, Institut für Astronomie der Universtität Wien
Philippe Audinos, Observatoire de Grenoble
David Barrado y Navascues, Universidad Complutense de Madrid
Jean Philippe Beaulieu, Institut d'Astrophysique de Paris
Susana Biro, University of Manchester
Vitor Bonifácio, Centro de Astrofisica, Porto
Marie Helene Boulard, Centre d'Etude Spatiale des Rayonnements, Toulouse
Anthony Brown, Leiden Observatory
Lars Flemming Brunke, Astromomisk Observatorium, Københavns Universitet
Nicolas Cardiel Lopez, Universidad Complutense de Madrid
Sophie Casanova, Service d'Astrophysique, Centre d'Études de Saclay
Myles Corcoran, Dublin Institute for Advanced Studies
Uwe Corneliussen, Physikalisches Institut Universität zu Köln
Vincent Coudé du Foresto, DESPA, Observatoire de Paris-Meudon
Elvira Covino, Osservatorio Astronomico di Capodimonte Napoli
Francesco Damiani, Osservatorio Astronomico di Palermo
Isabel Enríquez de Salamanca, Observatoire de Meudon
Daniel Folha, Centro de Astrofisica, Porto
Oriol Fuentes, Instituto de Astrofísica de Canarias
Carme Gallart, Instituto de Astrofísica de Canarias
Jose Girart, Universitat de Barcelona
Erik Gullbring, Stockholms Observatorium
Päivi Harjunpää, Helsinki University Observatory
Frank Helmich, Leiden Observatory
Gerhard Hirth, MPI für Astronomie, Heidelberg
Cathy Horellou, DEMIRM, Observatoire de Meudon

Jean Marc Hure, DAEC, Observatoire de Paris
Kostas Kleidis, Aristotle University of Thessaloniki
Kimmo Lehtinen, Helsinki University Observatory
Natércia Lima, Centro de Astrofisica, Porto
Fabien Malbet, Observatoire de Grenoble
Alessandro Marconi, Department of Astronomy, University of Florence
Oscar Morata, Universitat de Barcelona
Christophe Morisset, DAEC, Observatoire de Meudon
Konstantinos Pavlakis, University of Crete
Didier Raboud, University of Lausanne
Cecile Rastoin, Service d'Astrophysique, Centre d'Études de Saclay
Leonardo Sánchez, Département d'Astrophysique, Université de Nice
Mathias Schultheis, Institut für Astronomie der Universität Wien
Inma Sepulveda, Universitat de Barcelona
Johannes Staguhn, Universität zu Köln
Teresa Teixeira, Centro de Astrofisica, Porto
Caroline Terquem, Observatoire de Grenoble
Leonardo Testi, Dipartimento di Astronomia, Universita di Firenze
Viktor Toth, Department of Astronomy, Lora'nd Eotvös University
Clara Viegas, Centro de Astrofisica, Porto
Sophie Warin, Observatoire de Grenoble
Peter Wintoft, Lund Observatory

Molecular Clouds and Star Formation

Sylvie Cabrit

Observatoire de Grenoble
BP 53 X
38041 Grenoble Cedex, France

1 Introduction

Molecular clouds are the coldest, densest component of the interstellar medium (ISM). Only there will self-gravity eventually overcome thermal, turbulent, and magnetic pressures, and ultimately lead to the birth of new stars. Comprehensive reviews and recent results about molecular clouds in the context of star formation can be found for example in the proceedings of the NATO ASI school on "The Physics of Star Formation and Early Stellar Evolution" (1991), and the IAU Symposium No. 147 on "Fragmentation of Molecular Clouds and Star Formation" (1991). Rather than presenting a complete summary of this extremely rich topic, the present lectures concentrate on simple aspects which are particularly important in the context of star formation, with an emphasis on the basic physical assumptions involved. §2 deals with molecular cloud structure and equilibrium, §3 summarizes our observational and theoretical understanding of where, how, and why stars form in molecular clouds, and §4 reviews the phenomenon of molecular outflows and their role in cloud support and star formation.

2 Overall Properties of Molecular Clouds

2.1 The Molecular Component of our Galaxy

When a large atomic cloud in the interstellar medium (with typically $T_{\rm k} \sim$100 K, $n \sim 30$ cm^{-3}) assembles and slowly contracts under its own self-gravity, inner regions become sufficiently shielded by dust from the ambient ultraviolet (UV) field that they become mostly molecular. Self-shielding of the H_2 molecules makes the transition very sharp, at $A_V \sim N_H/2 \times 10^{21}$ cm^{-2} $\sim 0.5-1$ mag (see e.g. Elmegreen 1989).

Models of interstellar chemistry predict, and observations show, that a variety of complex molecules will also be formed, with abundances varying both in space and time (e.g. Blake 1988). The most ubiquitous molecules include CO, NH_3, H_2CO, and CS, with average fractional abundances relative to H_2 of approximately 10^{-4}, 10^{-8}, 10^{-8}, and 3×10^{-9} respectively in dark clouds (Irvine *et al.* 1987). While H_2 has no allowed electric-dipole transitions at all in the radio and infrared (IR) that could be excited at low temperature, these other molecules can radiate in a variety of rotational and ro-vibrational lines of low energy, thus providing extremely useful tracers of conditions in quiescent molecular gas (see e.g. Genzel 1991; Beckwith, this volume). Radiative cooling in CO lines establishes a low equilibrium temperature $T_k \sim$ 8-15 K inside molecular clouds, and thereby densities $n_{H_2} \sim 10^2$-10^4 cm^{-3} to ensure thermal pressure equilibrium with the hotter atomic ISM (Goldsmith 1988). Temperatures as high as 50-100 K are reached in the photo-dissociation layers exposed to strong radiative and shock heating by massive stars (Tielens & Hollenback 1985; Genzel 1991).

Our most complete picture of the cold molecular component in our Galaxy comes from large scale surveys in the J=1-0 rotational line of CO at λ= 2.6mm (e.g. Dame *et al.* 1987; Combes 1991). Molecular clouds are found predominantly in the galactic plane, with the highest concentration in a "ring" at galactic radii $4 < R < 8$ kpc. They represent 50% of the total mass of the ISM, i.e. $\sim 2 \times 10^9\ M_\odot$. Most of the mass lies within self-gravitating complexes called Giant Molecular Clouds (GMCs), of mass $10^4 - 10^6$ M$_\odot$ and size 10-100 pc, that are typically surrounded by extended envelopes of atomic hydrogen with masses up to $10^6\ M_\odot$. An extensive review of observational properties of GMCs can be found in e.g. Blitz (1991). GMCs are concentrated in spiral arms and toward the Galactic center, whereas smaller clouds are spread throughout the whole disk. Nearby molecular clouds within 800 pc of the Sun appear associated with an expanding ring system, the Gould's Belt (cf. Blaauw 1991). The process assembling diffuse interstellar medium into clouds is not yet identified. Possible scenarios for GMC formation and evolution are reviewed e.g. by Elmegreen (1991).

Molecular clouds are the sites of star formation in the Galaxy. All known GMCs within 1 kpc from the Sun, and all but one within 3 kpc, are currently forming stars, as evidenced by H II regions, infrared sources, heated dust, and outflows. Isolated dark clouds of only a few tens of solar masses (the "Bok globules") also form low-mass stars: many harbor optical jets and Hα emission stars (Reipurth & Petterson 1993), and 23% contain embedded infrared sources (Yun & Clemens 1990). Quiescent, cold GMCs with no apparent star formation also exist but their number apparently does not exceed that of active star-forming complexes. These statistics suggest that star formation is initiated quite rapidly after the formation of a molecular cloud, and that it lasts for a significant fraction of the total cloud lifetime (cf. Blitz 1991). Understanding the structure of molecular clouds is therefore an important step for building a global theory of star formation.

2.2 Structure, Kinematics, and Magnetic Fields

Over the last decade, it has become increasingly apparent that molecular clouds are highly inhomogeneous and turbulent, and that they possess magnetic fields which are dynamically important. The following subsections describe these general properties of molecular clouds, in particular their hierarchical structure.

The results will be combined in §2.4 to discuss the equilibrium of various scales, based on the virial theorem (§2.3).

Spatial structure of molecular clouds. The inhomogeneity of molecular clouds is apparent in a variety of ways. Molecular maps of line emission, e.g. the ^{13}CO map of the L1495 region in the Taurus-Auriga complex by Duvert *et al.* (1986) in Fig. 1a, reveal concentrations of mass in filaments and "clumps". Figure 1b shows a map of the same region, where column density is traced by the visual extinction A_V, measured through star counts (Cernicharo *et al.* 1985). A clumpy, hierarchical structure is again seen down to the resolution limit of 0.1pc; each resolved structure has on average half of its mass distributed in 2 to 5 smaller and more opaque fragments (Cernicharo 1991). Figure 2 shows an IRAS map of 100μm emission from dust in the same region, with a similar linear resolution. The filamentary and clumpy appearance, reminiscent of terrestrial clouds, is even more striking, probably because dust emission is detected over a larger dynamical range.

Since the 1980s, maps of GMCs in the mostly optically thin lines of ^{13}CO have allowed detailed quantitative studies of this structure. In particular, the velocity information present in line maps can be used as a "third dimension" to separate distinct clumps superposed on the same line of sight. For each of them, one can then determine:

1. the mean observed clump radius $R = \sqrt{R_{min} R_{max}}$ (measured at half-power or down to a given intensity threshold)
2. the average column density $\bar{N}_{\mathrm{H}_2}$, which is proportional to the average integrated over velocity line intensity, if the line is optically thin (see Bertoldi & McKee 1992, Appendix D).
3. the total mass $M = \pi R^2 m \bar{N}_{\mathrm{H}_2}$, and the average number density $\bar{n}_{\mathrm{H}_2} = 3\bar{N}_{\mathrm{H}_2}/4R$ (here $m = 1.4 m_{\mathrm{H}_2}$ to include Helium). H_2 densities can also be derived using line ratios, though source structure introduces additional uncertainties (e.g. Cernicharo 1991).

These studies reveal that at least 50% of the mass of a GMC is in dense ^{13}CO "clumps" ($n_{H_2} \sim 10^3$ cm^{-3}, R $\sim$ 0.3-2 pc) — defined as the largest coherent structures in velocity-position space — which occupy only a small fraction of the total volume, and are embedded in tenuous molecular and/or atomic gas (Blitz & Shu 1980; Pérault *et al.* 1985).

Observations in molecules requiring higher densities for detection ($C^{18}O$, CS, NH_3) reveal increasingly smaller mass concentrations inside clumps, down to the so-called molecular "cores", of size $\sim$ 0.05-0.5 pc and densities $\sim 10^4 - 10^5$ cm^{-3}(e.g. Myers *et al.* 1991; Dutrey *et al.* 1993). Those seen in NH_3 appear to be the direct progenitors of stars or stellar clusters (see §3). The mass spectrum for both ^{13}CO clumps and the denser cores obeys a similar power-law $dN/dM \propto M^{-1.5\pm0.2}$ from about 1-10M$_\odot$ to 3000 M$_\odot$, compatible with a simple coagulation-fragmentation scenario (see e.g. Scalo 1985; Blitz 1991; Elmegreen 1991).

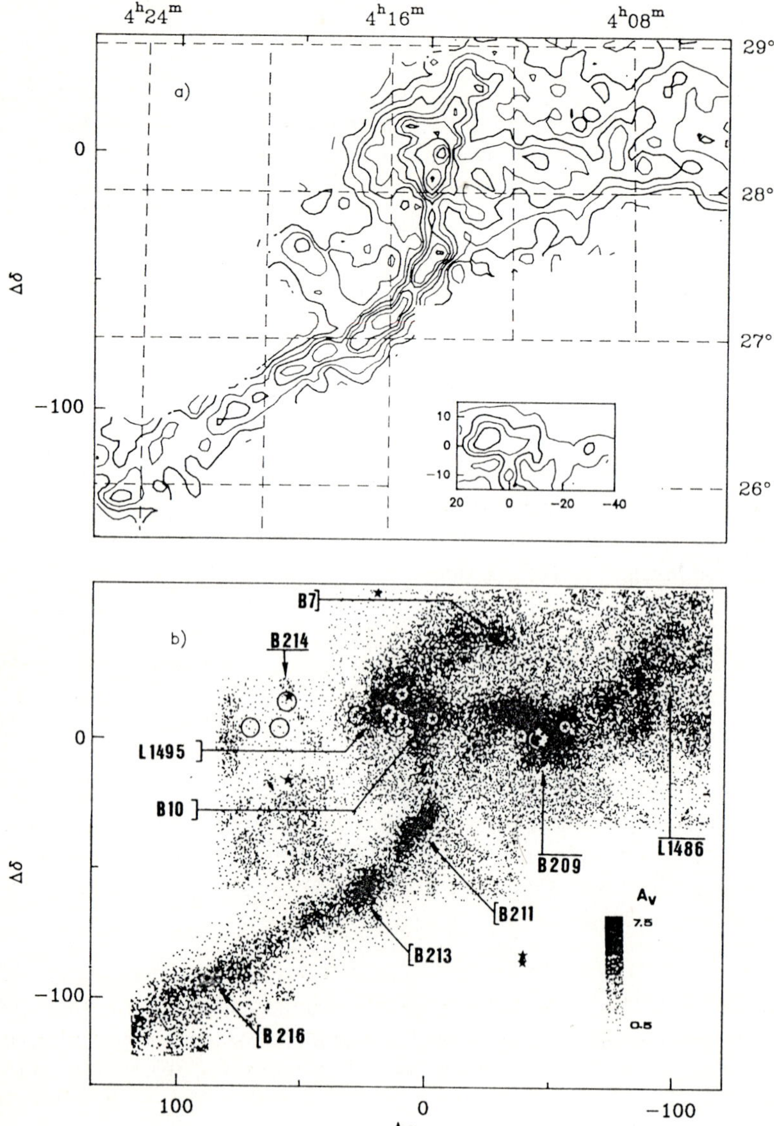

Fig. 1. a Map of integrated ^{13}CO(J=1-0) emission towards the L1495 dark cloud in Taurus-Auriga, with a resolution of 5'. Lowest contour and contour interval: 1 K kms^{-1}. **b** Greyscale map of visible extinction A_V derived from star counts (2'.5 pixels). T Tauri stars are shown as star symbols and reflection nebulae as open circles (from Duvert *et al.* 1986).

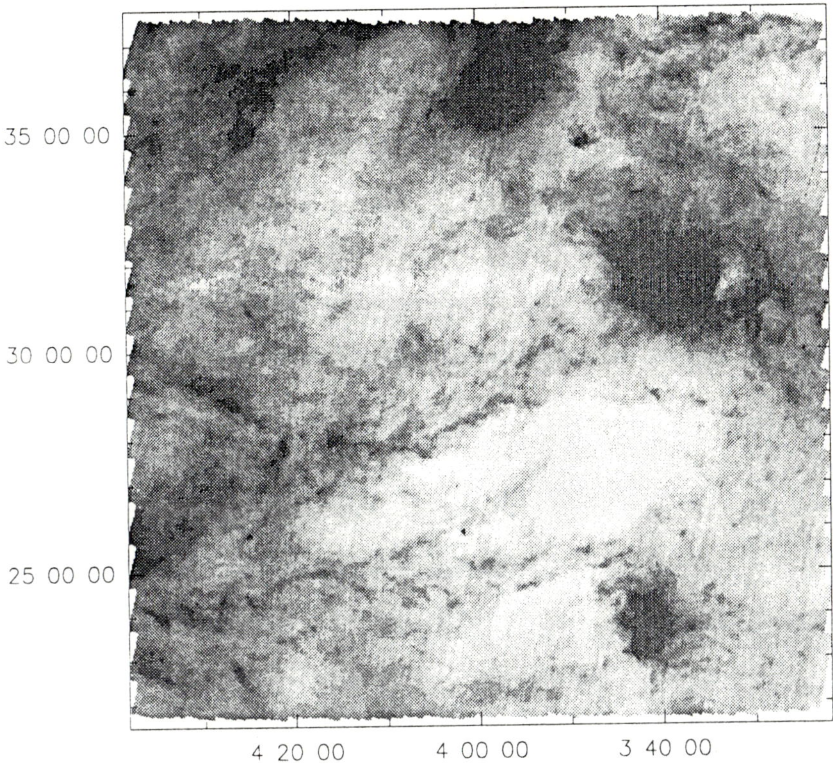

Fig. 2. Negative print of the IRAS 100 μm flux map of the region around L1495 in Taurus-Perseus (α,δ(1950) coordinates).

Studies in ^{12}CO of the more transparent interclump medium reveal structure that extends to still smaller scales, down to the current resolution limit of ~ 0.01 pc. Falgarone, Phillips, & Walker (1991) find that nested, hierarchical structures appear in regions of low average CO surface brightness as angular resolution sharpens. Most of the emission comes from unresolved optically thick structures, with a varying beam filling factor over the mapped area. Since cloud maps follow a self-similar area-perimeter relationship of the form $P \propto A^{D/2}$ with $D \sim 1.4$, over scales 100-0.1 pc, the authors further suggest that the emitting matter is organized in a fractal geometry, perhaps related to the turbulence believed to be present in molecular clouds (see below). Independent, indirect evidence for unresolved clumpy structure in clouds is the spatial extension of 158 μm emission from ionized carbon in photo-dissociation regions, which requires an enhanced UV penetration, as expected in the presence of clumps (cf. Stutzki *et al.* 1991 and references therein).

Kinematics. Another striking property of molecular clouds is that, on all but the smallest scales, spectral line widths greatly exceed those expected from thermal broadening alone. To study this effect quantitatively, one can estimate an *rms* line-of-sight velocity dispersion σ_o from the shape of the observed line profile; for a gaussian line, it is given by $\Delta V/2\sqrt{2\ln 2}$ where ΔV is the full width at half maximum (see Bertoldi & McKee 1992, appendix C, for more general cases). The non-thermal part of line broadening is then derived through

$$\sigma_{NT}^2 = \sigma_o^2 - kT_k/\mu_o m_H, \tag{1}$$

where μ_o is the molecular weight of the observed species.

Non-thermal dispersions σ_{NT} range from 2 to 10 kms^{-1} in GMCs, from 0.4 to 1 kms^{-1} within individual ^{13}CO clumps, and from $\sim$ 0.4 kms^{-1} in massive dense cores down to 0.2 kms^{-1} or less in the more numerous low-mass cores. With the exception of the latter, these values are larger than the isothermal sound speed in the molecular gas, $c_s = kT_k/\mu m_H = 0.19$ kms^{-1} (for $T_k = 10$K and a mean particle of molecular weight $\mu = 2.33$). Hence it appears that all structures in molecular clouds but low-mass dense cores harbor supersonic internal motions.

The nature of these supersonic motions is not yet clearly identified (e.g. Dickman 1985). Radial infall or expansion seem ruled out by velocity maps at various scales. Global rotation could explain some of the observed systematic velocity gradients, however the velocity distributions are usually too complex to fit this simple description, e.g. in the colliding ^{13}CO filaments near L1495 (Duvert *et al.* 1986). More plausible explanations are: random motions of unresolved substructures in a turbulent flow, oscillations coupled to non-linear hydromagnetic waves propagating through the cloud, or a combination of these. Falgarone & Phillips (1990) have shown that molecular lines exhibit self-similar shapes on all scales, with low-level wings deviating from the central gaussian profile, and they relate it to the intermittent behavior of turbulent flows. In the following, I will refer to all supersonic motions in clouds as "turbulence", taken in a broad sense.

Observations of Magnetic Fields in Molecular Clouds. Our knowledge about magnetic fields in clouds, though still limited, has improved dramatically over the last decade. A recent review of this subject can be found in, e.g. Heiles *et al.* (1991). Most results come essentially from two observing techniques: The first is measurement of the polarization of background starlight by magnetically aligned needle-like dust grains in the cloud. These grains tend to lie in a plane perpendicular to **B**, and absorb preferentially light that is polarized along their long axis. Therefore, the *transmitted* light has a net polarization parallel to the component of **B** in the plane of the sky ($B_\perp$), and large-scale maps of the orientation of $B_\perp$ can be constructed, like shown in Fig. 3 for the Taurus complex.

From these maps, clouds appear permeated by a large-scale magnetic field that is organized over scales of several parsecs. Hence, magnetic energy must be sufficiently strong so that the field does not get tangled by turbulent motions. Direct estimates of field strengths cannot be made with this technique, however,

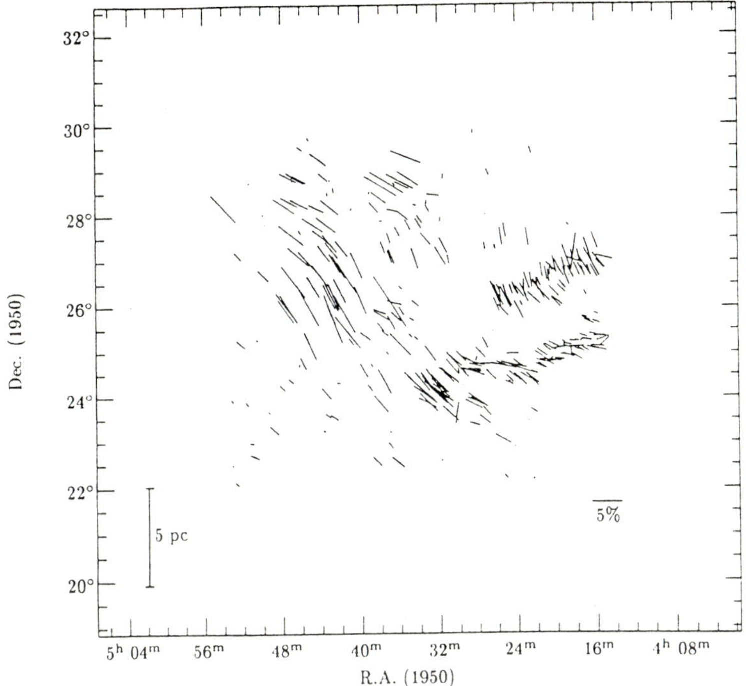

Fig. 3. Polarization of background starlight in the Taurus molecular cloud complex (from Goodman *et al.* 1990). The segments are parallel to the projected direction of the B-field and have a length proportional to the polarization percentage.

because the efficiency of the grain alignment mechanism is too poorly known. In addition, optical polarization can only trace the field in the outer, more transparent parts of clouds ($A_V \leq 1$ mag). In dense, opaque regions, one can in principle use the FIR and submm thermal *emission* of the same dust grains, which this time will be polarized perpendicular to $B_\perp$. When available, such observations agree with the large scale orientation of $B_\perp$ defined by optical data (e.g. Novak *et al.* 1989).

The second technique involves measurement of the Zeeman splitting of radio lines caused by coupling between the molecular moment and the magnetic field. The frequency split between right and left circularly polarized components is always very small, which makes it a challenging technique. For an unpaired electron spin moment (e.g. the OH line at 18cm and the H I line at 21cm) it is:

$$\Delta\nu = \frac{eB_\parallel}{2\pi m_e c} = 2.8\text{Hz}(B_\parallel/\mu\text{G}), \tag{2}$$

For a set of clouds at random orientations, the average total field strength will be twice the median of $B_\parallel$ (Heiles 1988). Recent OH measurements give $B \sim$

8-50 μG at the scale of molecular clouds, and higher values $B \sim 120\mu$G in OH absorption against massive dense cores associated with H II regions (Crutcher *et al.* 1987). This trend for increasing field strengths toward higher densities goes roughly as $B \propto n_{H_2}^{0.5}$, consistent with flux freezing in flattened structures (see Mouschovias 1991 for a review). The stronger fields in molecular clouds when compared to more diffuse atomic HI clouds, where $B \sim 3 - 10\mu$G, is also consistent with magnetic flux freezing during the condensation of GMCs from the interstellar gas (Kazès & Crutcher 1986).

Because the field is frozen in molecular gas at moderate densities, energetic turbulent motions in the cloud will cause magnetic perturbations and excite a spectrum of Alfvén waves which propagate along field lines at the Alfvén speed

$$v_A = \frac{B}{\sqrt{4\pi\rho}} = 1.3\text{km/s}\frac{(B/30\mu G)}{\sqrt{n_{H_2}/10^3\text{cm}^{-3}}}. \tag{3}$$

Several tentative correlations between line-of-sight velocity and $B_{\|}$ or spatial displacement have been noted in filamentary structures, which could indicate the presence of such MHD waves (Heiles 1988; Puget 1991). More observations are needed to confirm this phenomenon.

Because super-Alfvénic turbulence should dissipate very rapidly through magnetic shocks, one expects that σ_{NT} will not exceed v_A in molecular clouds. Hence, at scales where velocities are dominated by non-thermal motions, magnetic energy should always be significant compared to turbulent support.

2.3 Virial Equilibrium in Molecular Clouds

Basic Theoretical Framework. The fact that all scales in molecular clouds down to 0.1 pc do not show infall or expansion suggests that they are in global dynamical equilibrium with their surroundings. A condition expressing the global equilibrium of a physical system (i.e. no net change of inertia on average) is the well-known *Virial Equilibrium Theorem*:

$$\frac{1}{2}\frac{D^2 I}{Dt^2} = 2T + W + \mathcal{M} - \int_S (p + \frac{B^2}{8\pi})\mathbf{r}.\mathbf{dS} + \frac{1}{4\pi}\int_S (\mathbf{r}.\mathbf{B})\mathbf{r}.\mathbf{dS} \equiv 0 \tag{4}$$

which results from integrating the momentum equation over the volume of the system (e.g. Spitzer 1978). Here, I is the generalized moment of inertia in the volume, T is the *total* kinetic energy, including both thermal and "turbulent" fluid motions, W is the gravitational binding energy, $\mathcal{M}$ is the magnetic energy $B^2/8\pi$ integrated over the volume, and p is the total thermal and turbulent pressure. Virial equilibrium (i.e. $D^2I/Dt^2 = 0$) can be simply understood as a global balance between the negative gravity and surface pressure terms, which tend to compress the cloud, and the positive kinetic and magnetic energy terms, which tend to make it expand.

In order to keep the analysis simple and focus on the physics, we will consider only the *extremely idealized* case of a spherical, homogeneous, isothermal cloud with isotropic non-thermal motions, threaded by a mean field B, and surrounded by a much less dense ambient medium of pressure P_0 and uniform field B_0, i.e. we assume all field gradients occur inside the cloud (this is *not* correct, but the true field gradients depend on the equilibrium state and accurate solutions must be derived numerically). The volume terms of the virial theorem then become

$$2T = 3M(c_s^2 + \sigma_{NT}^2) = 3M\sigma^2 \tag{5}$$

$$W = -3GM^2/5R \tag{6}$$

$$\mathcal{M} = \frac{4\pi R^3}{3}\frac{B^2}{8\pi} \tag{7}$$

while the surface terms combine to give: $-4\pi R^3(P_0 + B_0^2/24\pi)$.

In the following paragraphs, I first discuss possible equilibria and their stability in terms of ratios between W, T, and $\mathcal{M}$. Then I will compare with observations of the various scales, using the values of R, M, σ_{NT}, c_s, and B derived in the previous sections.

***W* vs. *T*: The Jeans Criterion.** Let us for the moment neglect the dynamical role of magnetic fields and set $\mathcal{M}=0$, $B_0=0$. The virial equilibrium condition becomes

$$f(R) \equiv \frac{1}{2}\frac{D^2I}{Dt^2} = 3M\sigma^2 - \frac{3}{5}\frac{GM^2}{R} - 4\pi R^3 P_0 \equiv 0. \tag{8}$$

We can show that this equilibrium will be stable if and only if $f'(R) < 0$: Let us impose a small perturbation δR to the equilibrium state where $f(R) = 0$, and assume a smooth internal cloud density distribution. The inertia I is then proportional to MR^2 and first order expansion of the virial equilibrium equation gives

$$M\frac{D^2R^2}{Dt^2} \propto \frac{1}{2}\frac{D^2I}{Dt^2} = f'(R)\delta R \tag{9}$$

If $f'(R) < 0$, we see that a compression of the cloud, $\delta R < 0$, will produce a positive acceleration of R^2; therefore the cloud will reexpand (there is an outward force opposing contraction), and the equilibrium is stable. On the other hand, if $f'(R) > 0$, then the second derivative of R^2 will be in the same sense as the perturbation, and the contraction will accelerate: the equilibrium is unstable to gravitational collapse.

For a cloud of given mass M, and assuming a constant velocity dispersion σ (i.e. the cloud behaves like an "isothermal" sphere), we have

$$f'(R) = -12\pi R^2 P_0\left[1 + \frac{1}{3}\times\frac{W}{4\pi R^3 P_0}\right] = -12\pi R^2 P_0\left[1 - \frac{GM^2}{20\pi R^4 P_0}\right] \tag{10}$$

and the condition for stability is

$$\frac{M}{\pi R^2 m} \equiv N_{H_2} < N_J \tag{11}$$

where

$$N_J = \sqrt{\frac{20P_0}{\pi G}\frac{1}{m}} = 8\times 10^{21}\mathrm{cm}^{-2}\sqrt{\frac{P_0}{10^5\mathrm{K\ cm}^{-3}}}. \tag{12}$$

$N_{H_2} = N_J$ defines a "critical state" on the verge of gravitational instability. The corresponding maximum stable mass for a given ρ or σ, the famous "Jeans mass", is then

$$M_J = (N_J\, m)^3 \pi \left(\frac{3}{4\rho}\right)^2 = \left(\frac{15}{4G}\right)^{3/2} \frac{\sigma^4}{\sqrt{16\pi P_0/3}} \tag{13}$$

$$\sim 790 M_\odot (\sigma/\mathrm{km\ s^{-1}})^3\ (n_{H_2}/10^3 \mathrm{cm^{-3}})^{-1/2} \tag{14}$$

The second and third expressions make use of the fact that $\rho\sigma^2 = 4P_0$ in the critical state. This can be readily seen from the virial equilibrium equation, by dividing by $4\pi R^3 P_0$ and rearranging into

$$(\rho\sigma^2/P_0) = 1 + 3(N/N_J)^2. \tag{15}$$

Hence clouds closer to the Jeans critical state are denser.

Even if the values of P_0 and N_J are not known, the stability of a cloud can be estimated from the ratio of its kinetic and gravity terms (usually denoted by the *virial parameter* α) which has the advantage of involving only directly observable quantities:

$$\alpha \equiv \frac{2T}{|W|} = \frac{5\sigma^2 R}{GM} = 5\left(\frac{\sigma}{v_{ff}}\right)^2, \tag{16}$$

where $v_{ff} \equiv \sqrt{GM/R}$ is the free-fall speed. The virial equilibrium condition, divided by $|W|$, yields

$$\alpha = \frac{2T}{|W|} = 1 + 1/3(N_J/N)^2 \tag{17}$$

and we see that α decreases monotonically as the cloud is closer to instability. "Strongly self-gravitating" clouds, close to their critical state, will have $\alpha \sim 4/3$, or equivalently will show a velocity dispersion close to their free-fall velocity. On the other hand, stable clouds with $N_{H_2} \ll N_J$, where gravity is unimportant and where kinetic pressure serves mainly to balance P_0, will have large values of $\alpha \gg 1$ (they are "pressure confined"; cf. Bertoldi & McKee 1992). Non-magnetic clouds with $\alpha \ll 1$ are unstable and are not expected to be seen.

The above treatment is very approximate because a self-gravitating cloud in virial (hydrostatic) equilibrium will not be homogeneous but centrally condensed. An accurate solution is presented by Chièze (1987) in the general case of a polytropic pressure law ($P \propto \rho^\gamma$). For an isothermal cloud, M_J is reduced by a factor ~ 0.66 from the above value, $\rho_J\sigma^2 \sim 2.46 P_0$, and $\alpha_J \sim 2.0$ (see Bertoldi & McKee 1992).

***W* versus *M*: Magnetic Support.** The effect of magnetic forces on cloud equilibrium depends on how much they can oppose gravity. With our crude assumptions for the field geometry, the ratio of magnetic to gravitational energy is

$$\frac{\mathcal{M}}{|W|} = \frac{5}{6}\left(\frac{v_A}{v_{ff}}\right)^2 = \frac{5}{18G\pi^2}\left(\frac{\Phi}{M}\right)^2 \equiv \left(\frac{N_B}{N_{H_2}}\right)^2 \equiv \left(\frac{M_B}{M}\right)^{2/3} \quad (18)$$

where v_A is the Alfvén speed, $\Phi = \pi R^2 B$ is the magnetic flux through the cloud, and

$$N_B = \frac{B}{\pi}\sqrt{\frac{5}{18G}\frac{1}{m}} = 4.3 \times 10^{21}\mathrm{cm}^{-2}\left(\frac{B}{30\mu\mathrm{G}}\right) \quad (19)$$

$$M_B = \left(\frac{5}{18G}\right)^{3/2}\left(\frac{3}{4\pi\rho}\right)^2 B^3 \quad (20)$$

$$\sim 300 M_\odot (B/30\mu G)^3 (n_{H_2}/10^3\mathrm{cm}^{-3})^{-2}. \quad (21)$$

Since magnetic field is frozen in the gas at densities typical of molecular clouds (as long as there is no significant ambipolar diffusion), the flux-to-mass ratio Φ/M does not change as a given cloud contracts, and therefore $\mathcal{M}/|W|$ is an intrinsic characteristic of that cloud. This is also true for N_B/N_{H_2} and M_B/M. In addition, the critical magnetic mass M_B will be *the same* for a population of clouds of various masses and sizes, provided they all formed from a medium of similar B and ρ (e.g. the diffuse ISM) and conserved flux during their contraction.

With this parameterization of $\mathcal{M}$, virial equilibrium implies

$$3M\sigma^2 - \frac{3}{5}\frac{GM^2}{R}\left[1 - \left(\frac{M_B}{M}\right)^{2/3}\right] \equiv 4\pi R^3\left(P_0 + \frac{B_0^2}{24\pi}\right) = 4\pi R^3 P_0'. \quad (22)$$

Hence the net effect of magnetic fields is similar to multiplying the "gravity" term by a constant factor $\epsilon \equiv [1 - (M_B/M)^{2/3}]$, and we can deduce from the previous section the stability properties of magnetic clouds by replacing G with ϵG:

If $M \leq M_B$, magnetic pressure is stronger than gravitational attraction ($\epsilon W > 0$), hence $f'(R)$ will always be negative and the cloud will always be stable against gravitational collapse, even if turbulent support disappears. Such clouds of high flux-to-mass ratio are said to be *magnetically sub-critical.* They have $N_{H_2} < N_B$.

If $M > M_B$, the cloud is *magnetically super-critical,* i.e. it still feels an attractive gravity (though smaller than without magnetic forces), and any loss of turbulence will lead to contraction. Since $\epsilon W < 0$, there exists a critical state at a maximum value N_{cr} of the column density, which can be deduced from the non-magnetic case:

$$N_{cr} = N_J/\sqrt{\epsilon_{cr}} = N_J[1 - (N_B/N_{cr})^2]^{-1/2} \quad (23)$$

$$\text{or} \quad N_{cr} = \sqrt{N_J^2 + N_B^2} \quad (24)$$

and similarly, the maximum stable mass for a magnetic cloud is

$$M_{cr} = M_J \left[1 - (M_B/M_{cr})^{2/3}\right]^{-3/2} \tag{25}$$

$$= \left[M_J^{2/3} + M_B^{2/3}\right]^{3/2} \tag{26}$$

where N_J and M_J are the Jeans parameters defined earlier (but with P_0' instead of P_0). We also have

$$(\rho\sigma^2/P_0') = 1 + 3\epsilon(N/N_J)^2 \tag{27}$$

$$\alpha = \epsilon + 1/3(N_J/N)^2 \tag{28}$$

Therefore we find that the kinetic pressure $\rho\sigma^2$ equals $4P_0'$ in the critical state (cf. the non-magnetic case), and P_0' in a cloud that is just magnetically critical; however, the values of α for these limiting cases are not fixed and depend on an additional parameter which is (N_J/N_B). This parameter is directly related to the ratio of kinetic versus magnetic support through:

$$\frac{\mathcal{T}}{\mathcal{M}} = \frac{3\sigma^2}{v_A^2} \equiv m_A^2 = \frac{\rho\sigma^2}{6P_0'}\left(\frac{N_J}{N_B}\right)^2 \tag{29}$$

where m_A^2 is the Alfvén Mach number of the velocity dispersion in the cloud. Then, α can be written as

$$\alpha = \frac{2\mathcal{T}}{\mathcal{M}}\frac{\mathcal{M}}{|W|} = 2m_A^2\left(\frac{N_B}{N_{H_2}}\right)^2 = 2m_A^2\left(\frac{M_B}{M}\right)^{2/3} \tag{30}$$

and we see that magnetically subcritical clouds will have $\alpha \geq 2m_A^2$ while magnetically supercritical ones will have $\alpha < 2m_A^2$. In the critical state, the virial parameter α_{cr} will be:

$$\alpha_{cr} = \frac{4}{3}\epsilon_{cr} = \frac{4}{3}\left(\frac{N_J}{N_{cr}}\right)^2 = \frac{4}{3}\left[1 + \frac{2}{3m_A^2}\right]^{-1} \tag{31}$$

Thus, only critical magnetic clouds with the *same* Alfvénic Mach number will have the same *fixed proportions* between kinetic, gravitational, magnetic, and ambient pressures.

More accurate calculations taking into account gradients in **B**, central density concentration, and flattening along field lines, give the same overall behavior but show that M_B is reduced by about 0.4 from the value derived above, and that another small correction factor must be included in the definition of M_{cr} (Mouschovias & Spitzer 1976; Tomisaka *et al.* 1988). Then, the limit between magnetically sub- and supercritical clouds should occur around $\alpha_B \sim 3.9\ m_A^2$, while $\alpha_{cr} \sim 2.1/(1.12 + 2/m_A^2)$ and $\rho_{cr}\sigma^2 \sim 3P_0$ (see Bertoldi & McKee 1992 for a detailed discussion). A value of $\alpha < 2$ therefore requires the presence of a magnetic field ($m_A < \infty$).

2.4 Comparison with Observations

Observational investigations of virial equilibrium in molecular clouds involve many sources of error (see Heiles *et al.* 1991): Zeeman measurements are not yet available for all scales, and give only $B_{\|}$; uncertainties are present in converting observations of molecular tracers into H_2 column densities, in defining the characteristic source size for non-spherical clouds, and in estimating cloud distances; Finally, a turbulent support dominated by Alfvén waves would make σ both orientation and scale-dependent. It could also invert the Jeans criterion, making small scales more unstable than large ones if most of the wave power is at long wavelengths (Bonnazzola *et al.* 1987). Nevertheless, if done carefully and over a statistical sample, virial analysis remains an informative tool which can reveal a number of important trends. Here we will use average physical parameters for each scale, and adopt for B the typical value of $2B_{\|}$ observed in cloud structures of the corresponding density (Crutcher *et al.* 1987).

GMCs. Typical parameters for an "average" size GMC (e.g. L1641) are $M \sim 10^5$ $M_\odot$; $R \sim 26$ pc; $\sigma_{NT} \sim 2$kms^{-1}; $n_{\bar{H}_2} \sim 30$ cm^{-3}; and $B \sim 16\mu$G. Then, we infer $\alpha \sim 1$, and $m_A \sim 0.9$. Similar results are found for GMCs with $M \sim 10^4 - 10^6$ $M_\odot$. These values imply that GMCs are strongly self-gravitating and magnetically supercritical, but stable, and that they are supported by a turbulence that is close to Alfvénic (cf. McKee 1989). The ambient pressure consistent with virial equilibrium is $P_0 \sim \rho\sigma^2/3 \sim 2 \times 10^4$ K cm^{-3}, close to the total (thermal + turbulent + magnetic) pressure of the general ISM (see e.g. Falgarone 1992).

^{13}CO Clumps — Dark Clouds. Individual clumps identified in ^{13}CO maps of a given complex have similar values of $n_{\bar{H}_2} \sim 10^3$ cm^{-3} and $\sigma \sim 0.7$ kms^{-1}, over a much broader range of sizes and masses; their virial parameter α varies as $M^{-2/3}$, and ranges from 30 at $M \sim 1 - 10$ $M_\odot$ down to 1.2-1.5 at $M \sim 200 - 1000 M_\odot$ (see e.g. Loren 1989). Bertoldi & McKee (1992) have shown that the above properties are best explained if all clumps within a GMC share a similar value of $B \sim 30\mu$G, fixed by flux-freezing, and have $m_A \sim 1$ (cf. Eq. [30]). OH Zeeman measurements give an average $B \sim 30\mu$G in dark clouds (Crutcher *et al.* 1987), consistent with this result.

Since turbulence appears close to Alfvénic in ^{13}CO clumps, only the few most massive ones in a GMC, with $\alpha < 4$, are magnetically supercritical and strongly self-gravitating; the majority of ^{13}CO clumps, with $\alpha \geq 4$, are pressure confined and magnetically subcritical: they cannot spontaneously contract (and are presently unlikely to form stars). The ambient pressure consistent with virial equilibrium is $P_0 \sim 10^5$ K cm^{-3} for the whole clump population, which could be provided by the increased turbulent and magnetic pressure inside the self-gravitating GMC (Bertoldi & McKee 1992).

Dense Cores. Typical low-mass dense cores observed in NH_3, e.g. in Taurus, have $n_{H_2} \sim 2 \times 10^4$ cm^{-3}, $R \sim 0.07$pc, $M \sim 1.5 M_\odot$, $T_k \sim 11$K, and $\sigma_{NT} \sim 0.12$

kms^{-1}< $c_s \sim 0.2$ kms^{-1}(Benson & Myers 1989). Therefore these structures have more thermal than turbulent support (their turbulence is subsonic), and have $\alpha \sim 2$. Their amount of magnetic support cannot be firmly assessed because field strengths are not yet well determined on that scale. Still, σ_{NT} is much less than would be expected for Alfvénic turbulence if the cores were magnetically subcritical ($B > 50\mu$G); the value of α is also very close to that for non-magnetic spheres in critical equilibrium with gravity. This combination of facts suggests that low mass cores have lost magnetic support during contraction, probably through ambipolar diffusion (cf. §3), and are now very close to gravitational instability. Their estimated ambient pressure is then $P_0 \sim 10^5$ K cm^{-3}, similar to that of ^{13}CO clumps.

Higher mass dense cores have been observed in CS and NH_3; the latter species allows more accurate mass determinations, the former being optically thick and subject to complex radiative transfer effects (see Cernicharo 1991). The inferred physical properties cover a broad range of values (Benson & Myers 1989). Here we will adopt the average properties of NH_3 cores with $M > 20$ M$_\odot$ in L1641, measured with the same linear resolution as for Taurus cores (Harju *et al.* 1993): $M \sim 45$ M$_\odot$, $R \sim 0.2$ pc, $n_{H_2} \sim 2-7\times10^4$ cm^{-3}, $T_k \sim 14$K, and $\Delta V \sim 0.8$kms^{-1}. We see that, contrary to low-mass ones, massive cores have supersonic turbulence ($\sigma_{NT} = 0.33$ kms^{-1}> $c_s = 0.24$kms^{-1}) and have $\alpha \sim 0.9$, suggesting significant magnetic support against gravity. Relatively strong fields $B \sim 120\mu$G have been inferred toward massive dense cores associated with HII regions (Crutcher *et al.* 1987). If we adopt this value as a reasonable upper limit for B, then $(m_A)_{NT} \geq 0.5-1$ for this range of densities, compatible with Alfvénic turbulence, and massive dense cores are magnetically supercritical, very close to gravitational instability. The inferred ambient pressure is $P_0 \sim 3-9\times10^5$ K cm^{-3}.

Very Small Scale Structure. So far, parameters are available only for the small scale structure observed in CO toward edges of clouds (Falgarone *et al.* 1991). The CO(2-1), CO(3-2), and ^{13}CO (2-1) line ratios and line intensities are best reproduced for $T_k \sim 10$K and a column density $N_{H_2} \sim 0.18-1.1\times10^{21}$ cm^{-2}. With an observed size $R \sim 0.02$ pc and $\Delta V = 1.8$kms^{-1} ($\sigma \sim 0.75$ kms^{-1}), one infers $n_{H_2} \sim 3-20\times10^3$ cm^{-3}, and $\alpha \sim$ 450-2700. Therefore these structures are far from being gravitationally bound. Their turbulent pressure is in the range $4\times10^5 - 2\times10^6$ K cm^{-3}.

2.5 Origin of Cloud Structure

The hierarchical structure of molecular clouds appears more and more complex as observations improve, and neither its complete characteristics nor its origin are yet understood. The properties reviewed above show that several ingredients need to be included in any successful theory:

• *Self-Gravity*: Molecular clouds exhibit a wide range of structures; the most contrasted ones are gravitationally bound (GMCs as a whole; massive ^{13}CO clumps; dense cores) but others are unbound and pressure-confined (most ^{13}CO clumps; small-scale structure). Most, if not all, dense cores are found inside clumps that are already gravitationally bound. Hence self-gravity must play a role, directly or indirectly, in the formation of nested bound sub-structures. In particular, the large pressure in massive cores can be more easily understood if they formed deep in the potential well of the parent ^{13}CO clump.

• *Magnetic Fields*: At all scales where B can be reasonably estimated ($R >$ 0.2pc), and in *both* bound and unbound structures, the observed turbulent velocity is very close to the Alfvén speed. This equipartition indicates a strong coupling between magnetism and turbulence, maybe through hydromagnetic waves (cf. Arons & Max 1975), and suggests that the magnetic field strength might regulate turbulent support at most scales.

• *"Pressure" equilibrium*: We find that ambient pressures have similar values $2 \times 10^4 - 10^5$ K cm^{-3} in most structures over scales 0.1-50 pc (or equivalently, B is found to stay within a factor of 2 of $\sim 30\mu$G). The existence of a nearly constant characteristic pressure could be fundamental in that it readily explains the average scalings $\sigma \propto R^{0.5}$, and $n \propto R^{-1}$ found by Larson (1981) and subsequent authors for *bound, critical* structures over the same range of scales: since turbulent and magnetic energy are in equipartition, the pressure K entirely fixes the critical column density through $N_{cr} \propto N_J \propto N_B \sim (K/G)^{0.5}$ (cf. §2.3) and critical gravity requires $\sigma^2 R/GM \sim 1$, implying $\sigma \propto (KG)^{1/4} R^{0.5}$ and $n \propto (K/G)^{0.5} R^{-1}$. (Henriksen & Turner 1984; Chièze 1987; Myers & Goodman 1987; Henriksen 1991). Hence, Larson's velocity-size relation could be strongly dependent on both self-gravity and the effective "pressure" in molecular clouds. Support for this idea is that gravitationally *unbound* structures such as low-mass ^{13}CO clumps do *not* follow Larson's scalings (their values of σ and n vary little with size; cf. Loren 1989) although they share the same pervasive pressure and have supersonic turbulence. Similarly, strongly *bound* structures with a *higher* boundary pressure (i.e. the massive dense cores; cf. 2.4.3) also depart from the general scalings (Myers & Goodman 1987; IV*e*).

In summary, the origin of the hierarchical structure of molecular clouds probably involves a complex interplay between turbulence, magnetic fields, pressure gradients, and self-gravity. At present, MHD turbulence, in the form of a spectrum of waves or as a more stochastic field, appears as the most natural candidate for creating internal density enhancements inside clouds, leading eventually to the formation of self-gravitating dense cores. Detailed numerical simulations are required to improve our theoretical understanding of this problem. Current progress toward this goal is very promising (see e.g. Carlberg & Pudritz 1990; Pouquet *et al.* 1991; Puget 1991).

On the observational side, the velocity-size relation does not seem a good discriminant of the nature of supersonic motions in clouds. More useful indications on the turbulent velocity field, and in particular on the presence of Alfvén waves versus stochastic motions, can be obtained through the use of spatial correlation methods, ideally including cross correlation with magnetic fields (see e.g. Dickman 1985; Gill & Henriksen 1991), and through analysis of higher order moments of the line profiles over various scales (Falgarone & Phillips 1990).

Finally, clues about the fragmentation process and the fractal properties of turbulent molecular gas can also be gained by better characterizing its spatial and density structure, with less bias toward self-gravitating regions (e.g. Falgarone *et al.* 1991).

3 Star Formation in Molecular Clouds

That most galactic GMCs appear to be presently forming stars is fascinating as it suggests that star formation is an unavoidable outcome of the intrinsic structure of molecular clouds. The details of this physical link are not yet fully understood. The various problems involved can be summarized by the following basic questions:

1. Where and how do stars form?
2. Why is star formation so inefficient globally (the star-to-gas mass ratio, known as the "*star formation efficiency*" ε, is only a few percent on the scale of molecular complexes; see e.g. Evans & Lada 1991; Fukui & Mizuno 1991)?
3. What is the impact of star formation on cloud turbulence and overall structure? Could energy injection by stellar winds explain the observed longevity of clouds (statistics on stellar ages indicate a GMC mean lifetime of 30 Myr, much longer than their free-fall time of 5 Myr)?

Over the last decade, there has been important observational and theoretical progress on these issues, mostly following technical advance in the infrared and millimetric domain, and a unified scenario of star formation is gradually emerging. Extensive observations of newly formed stars, and a more complete (though still crude) picture of cloud structure, outlined in §2, have provided at least partial answers to the above questions. The next sections summarize these results (specific aspects relevant to protostellar collapse and pre-main sequence evolution are covered by S. Beckwith and C. Bertout, this volume).

3.1 Dense Cores as the Sites of Star Formation

A key discovery has been the recognition in the early 1980s that gravitational collapse leading to star formation occurs only very locally, inside "dense cores" already much more condensed than the bulk of molecular gas within GMCs (see §2). More recently, it has appeared that, while low-mass dense cores result in the formation of individual stars, massive dense cores often give rise to rich stellar clusters; these clusters apparently obey the same initial mass function (IMF) as that established for field stars, pointing to a local origin for this important property. Finally, systematic trends linking core and parent cloud properties have been identified. Several studies illustrating these results are described below.

Low-mass Cores and Individual Star Formation. Low-mass dense cores were discovered through observations, in the density-sensitive lines of NH_3, of small opaque patches within nearby dark molecular clouds, mostly the Taurus complex and Bok globules (Benson & Myers 1983). Several arguments indicate that they are the likely sites of low-mass star formation in these clouds: (1) they are supported mostly by thermal motions, are on the verge of gravitational instability, and have masses close to stellar values (cf. §2); (2) they show a strong tendency to be found near groups of optically visible young stars such as T Tauri stars; (3) the number of NH_3 cores in the Taurus-Auriga complex (10-20) is consistent with the number of stars expected to form there within the next core free fall time $t_{ff} \sim 2 \times 10^5$ yrs.

This conclusion is supported by the fact that almost half of low-mass NH_3 cores harbor an IRAS source within their boundaries, 57% of them without optical counterpart (Beichman *et al.* 1986). These *embedded* IRAS sources ("Class I" sources in the classification introduced by Lada, 1988) are located closer to the core center, and present colder dust emission, than optically visible T Tauri stars, indicating that they are probably very young, with an age $\leq 1 - 3 \times 10^5$ yr. Follow-up observations at near-infrared wavelengths confirmed the deeply embedded nature of these stellar objects (Myers *et al.* 1987).

T Tauri stars in Taurus-Auriga are grouped around individual dark clouds and have a mean separation of 0.3pc, greater than the typical diameter of a low-mass core, $\sim$ 0.1pc (Gomez *et al.* 1993). This loose clustering is consistent with each low-mass core forming only one stellar system (single or binary/triple star), but the modest separation between stars indicates that truly isolated star formation is a rare event, perhaps limited to low-mass Bok globules.

Massive Dense Cores and Cluster Star Formation. The loose mode of star formation from individual low-mass cores observed in low-mass nearby dark clouds may not be dominant in the Galaxy; In large GMCs (M $> 10^4\ M_\odot$), massive OB stars are frequently found in tight clusters. Such a "cluster" mode of star formation could be important for low-mass stars as well.

Testing this hypothesis requires a complete record of recent, low-mass star formation within large GMCs. Until recently, this could only be partially attempted using (1) deep optical and X-ray surveys, which trace the T Tauri population of age 0.5-20 Myr (see Bertout, this volume), but are limited to low A_V; (2) IRAS data, which give access to much younger ($\geq$ 0.1 Myr) and embedded objects, but have a low angular resolution making it difficult to pinpoint counterparts at other wavelengths, especially in regions heated by nearby massive stars (e.g. Wilking *et al.* 1989 for ρ Oph).

The advent of sensitive near infrared arrays has allowed for the first time to fully investigate this question by conducting extensive, high-resolution surveys for invisible young stars in several nearby GMCs, down to K magnitudes of 13-15 (Lada 1991). Particularly illustrative of the potential of this technique are two recent studies of the Orion complex, 450 pc away: one in the northern region (Orion B = L1630; Lada *et al.* 1991a,b) and one south of Orion A (L1641; Strom *et al.* 1993). These studies reveal that a significant fraction — and possibly the majority — of young stars in these complexes form in clusters, and that stellar clusters are associated with massive dense cores.

Clusters and dense gas in Orion B: Four stellar clusters were identified in a 2.2μm survey of a significant portion of Orion B by Lada *et al.* (1991b): a small group (20 members in a radius of 0.3pc) and three very rich ones, with more than a hundred members inside a radius of 0.6-0.9 pc. These clusters are all associated with previously known regions of active star formation, including the HII region NGC 2024. After correction for background and field stars, it is estimated that up to 96% of all the infrared sources associated with the entire Orion B cloud are contained in the four clusters. Hence, present star formation in this cloud is highly localized in dense cluster environments.

To better understand the origin of these clusters, Lada *et al.* (1991a) carried out an extensive survey for dense molecular gas in Orion B using the CS(2-1) line ($n > 3 \times 10^4$ cm^{-3}). Forty-two individual dense clumps of size > 0.2pc (mass > 8 $M_\odot$) were identified. The 5 most massive ones, with $M \geq 200$ $M_\odot$, together contain 50% of the total mass in dense gas over the region surveyed.

Strikingly, the newly found infrared stellar clusters are coincident or nearly coincident with 4 of these 5 most massive CS clumps, as illustrated in Fig. 4. The estimated star formation efficiency $\varepsilon = M_{stars}/(M_{stars} + M_{core})$, ranges from 7% up to 20-40% in the rich clusters. The latter values are reminiscent of the closer-by infrared cluster in the ρ Ophiuchi cloud core, which contains 600 $M_\odot$ of dense gas and has $\varepsilon \geq 20\%$ (Wilking & Lada 1983 and references therein). Therefore, it appears that star formation can proceed *very efficiently* at gas densities $> 3 \times 10^4$ cm^{-3}, so that massive dense clumps often give rise to rich stellar clusters. The overall low star formation efficiency in GMCs ($\varepsilon \sim 4\%$ in Orion B) is then explained as a result of the small fraction of gas mass in the form of dense cores (less than 19% in Orion B; Lada *et al.* 1991a).

Multi-color study of the embedded young population in L1641: The survey of representative regions of L1641 by Strom *et al.* (1993) is complementary from the previous study in that it involved simultaneous imaging in the J, H, and K photometric bands (1, 1.6, and 2.2 μm). Color-color diagrams and intrinsic luminosity functions could then be constructed, allowing a detailed study of the nature and evolutionary status of the young stellar population (presence of circumstellar disks, ages, IMF).

Several stellar density enhancements were identified, comprising seven small clusters ("aggregates"), and one rich cluster, all associated with other signposts of star formation. Hence, small clusters seem to be more frequent than rich ones in L1641, although both span a similar overall range in size and stellar members as those in Orion B. In addition, a *distributed* population, amounting to roughly 70% of all detected infrared sources, is found throughout L1641, while no such population was identified in Orion B. The colors of the infrared sources indicate that they are obscured low-mass T Tauri stars, and that many of them possess optically thick circumstellar disks (cf. Lada & Adams 1992). The fraction of

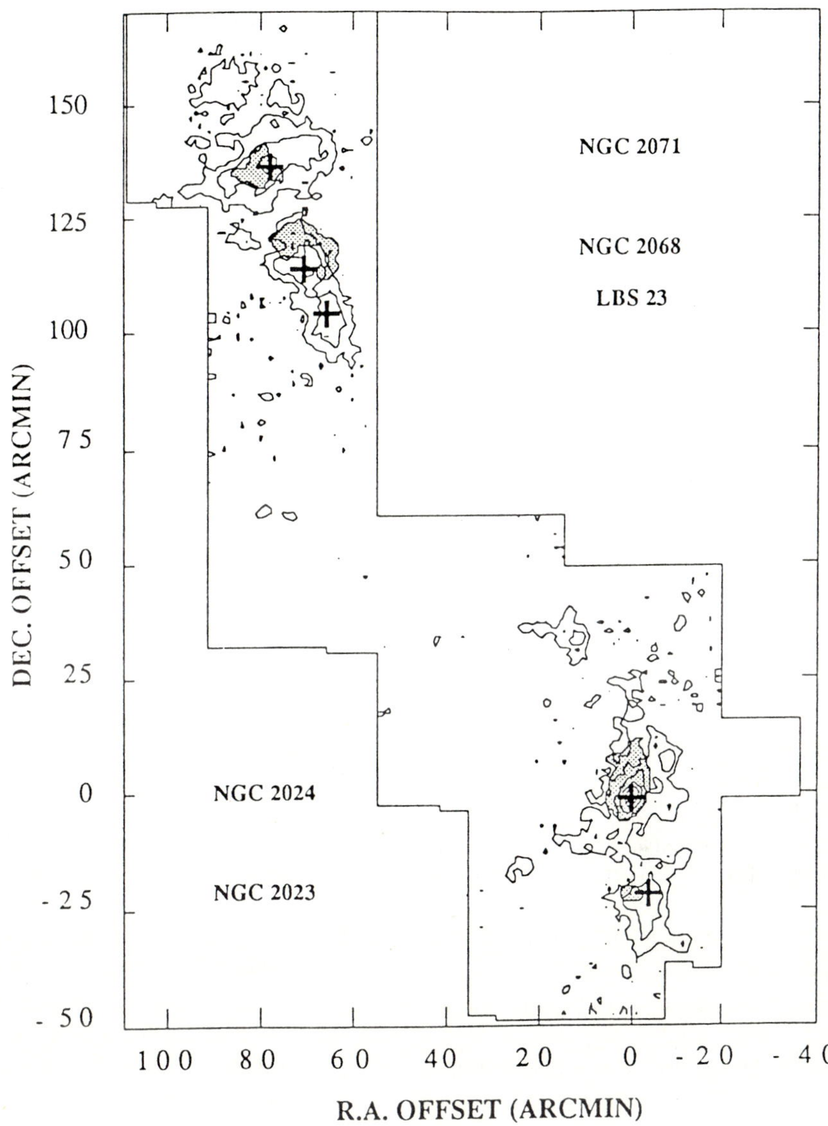

Fig. 4. The distribution of dense gas in the L1630 molecular cloud is presented as contours of CS(J=2-1) emission, while the location and extent of the 4 embedded infrared clusters are shown as shaded areas. The clusters are associated with 4 of the 5 most massive CS clumps (marked by crosses) from Lada (1992).

stars surrounded by disks decreases from $\sim$ 60% in the aggregates to 30-50% in the rich cluster and the distributed population, suggesting that the aggregate population is on average younger.

Another important use of IR surveys is that the observed luminosity function can constrain both the age of the young stellar population under study (low-mass stars younger than 10^7 yr decrease rapidly in luminosity with time) and the shape of the IMF (Lada 1991). Strom *et al.* conducted this analysis in the J-band, which is less affected by excess infrared disk emission in T Tauri stars (Hartigan *et al.* 1992). They estimate ages of respectively $t \leq 0.7$ Myr for the aggregates, $t \sim 3$ Myr for the rich cluster, and $t \sim 7$ Myr for the distributed population, consistent with the decreasing frequency of optically thick disks among the three groups. The J-band luminosity function for the rich cluster favors a standard IMF that decreases below 0.3 $M_{\odot}$, in agreement with other recent studies of field stars and young clusters (Scalo 1986; Lada *et al.* 1993).

It is not yet clear what fundamental factor(s) cause the difference between small aggregates and rich clusters. The age estimates obtained in L1641 suggest that the average star formation rate could be roughly the same in the two cases (25 stars in 0.7 Myr or 125 stars in 3 Myr); an attractive explanation would be that aggregates are cluster sites that are not as tightly bound together, so that older stars tend to migrate away and only the few youngest members show detectable clustering. The progressive dispersion of such "stellar nurseries" over 7 Myr could in fact explain all of the distributed stellar population in L1641. Unbound clusters in general occur after rapid dispersion of the surrounding gas by stellar winds (see e.g. Verschueren & David 1989 for a summary). However, all aggregates in L1641 harbor deeply embedded IRAS sources indicating that they are associated with dense cores more massive than those in Taurus (Strom *et al.* 1989), as confirmed in several cases by NH_3 observations (Harju *et al.* 1993). A firm test of the above hypothesis awaits more complete surveys for dense gas in L1641 allowing to compare the potential well and star formation efficiency in aggregates and clusters.

A second possibility is that the differing characteristics of aggregates and clusters reflect intrinsic differences in the physical properties of their parent massive dense clumps (e.g. magnetic field, rotation, velocity dispersion, evolutionary stage), while the distributed population of L1641 would have formed in situ, in isolated low-mass cores throughout the cloud. Sensitive surveys of representative regions of L1641 and Orion B for low-mass dense cores might shed some light on the origin of the distributed population seen in L1641, and on the reasons for its absence in Orion B.

Interrelations among clouds, cores, and stellar content. The differences noted above between Taurus, L1641, and Orion B can be put in a broader perspective, if one compares the properties of dense cores and young stars in the well-studied star forming molecular complexes within 500 pc of the Sun.

Myers (1991) (see also Wootten 1991) conducted such a comparison, and pointed out several significant trends : On average, the size, velocity dispersion, and mass of dense cores all increase systematically with increasing size, velocity dispersion, and mass of the parent molecular complex. This can define a global ranking for nearby molecular clouds, from Lupus to Taurus, to Ophiuchus, to Perseus, to Orion B, to Orion A (Trapezium cluster and Kleinman-Low nebula). L1641 would fall between Perseus and Orion B.

The number of young stars, the maximum mass among the stars formed, and the amount of stellar clustering, also increase according to the same ranking between molecular clouds. This is consistent with the fact, illustrated by the Orion B results above, that stellar clusters form with high efficiency in massive dense cores, which contain most of the dense gas mass; then, the total number of stars and, through the IMF, the maximum stellar mass achieved, both depend strongly on the maximum core mass in the cloud.

The properties of dense cores also appear to differ according to their location in the parent complex: massive cores are found close to major concentrations of gas corresponding to strongly self-gravitating, magnetically super-critical ^{13}CO clumps (cf. §2; Bertoldi & McKee 1992), while less massive cores are either satellites of the massive ones (see Lada *et al.* 1991b) or are found in the low-mass filaments extending away from the major clumps.

These trends suggest that the potential well in the parent molecular complex is crucial in determining the location and mass of dense cores, and thereby the global properties of star formation in the cloud. This effect can be understood qualitatively if one considers the various physical factors possibly controlling the formation of dense cores. A brief examination of these factors, especially the important process of ambipolar diffusion, is the topic of the next section.

3.2 The Path to Instability: Role of Ambipolar Diffusion

For typical conditions inside ^{13}CO clumps, i.e. $n_{H_2} \sim 10^3$ cm^{-3}, $B \sim 30\mu$G, and $\sigma \sim 0.9$ kms^{-1}, the magnetically critical mass is $M_B \sim 130$ M$_\odot$ and the Jeans mass is $M_J \sim 500$ M$_\odot$. Therefore, any spherical gas parcel with the mass of a dense core (≤ 100 M$_\odot$) is initially magnetically sub-critical — it has a large magnetic flux excess — so that it cannot contract isotropically under its own gravity, or become unstable. To collect into a self-gravitating dense core, the gas must become magnetically super-critical, i.e. its mass to flux ratio must increase. This will happen mainly through the process of ambipolar diffusion.

Ambipolar diffusion is an intrinsic, *unavoidable* process in magnetically supported, poorly ionized media (Mestel & Spitzer 1956): Because they do not directly feel the magnetic field pressure, neutrals have a tendency to drift with respect to charged particles (magnetic support against gravity is experienced only through ion-neutral collisions), and they progressively condense toward the center while field lines remain tied to the ionized component. The effective mass to flux ratio thus increases in the central parts of the slowly condensing core. In quasi-static equilibrium, the characteristic timescale for ambipolar diffusion is (Spitzer 1978; Mouschovias 1991):

$$t_{AD} \sim \frac{R}{v_{drift}} \sim \frac{n_i}{n_H} \frac{< u\sigma_s >}{2\pi G m_H} \sim 8 \times 10^{13}\text{yr} \times \xi \qquad (32)$$

where $< u\sigma_s >$ is the rate coefficient for ion-neutral collisions, n_i and n_H are the number density of positive ions and hydrogen atoms, which provide the dominant collisions, and ξ is the ionization fraction.

Ionization in molecular clouds is provided mostly by UV, X-rays and cosmic rays, in order of increasing degree of penetration. Photoionization dominates at $A_V < 4$ mag, and cosmic ray ionization in opaque regions with $A_V > 4$ mag (McKee 1989). Although the reaction chains are complex and quite uncertain, involving e.g. metal recombination onto grains, it appears that the ionization fraction by cosmic rays is very low, approximately given at moderate densities by: $\xi \sim 4.5 \times 10^{-6}(\zeta_{-17}/n)^{1/2}$, where ζ_{-17} is the cosmic ray ionization rate in units of 10^{-17} s^{-1} (McKee 1989; Nishi *et al.* 1991). The ambipolar diffusion timescale t_{AD} is then only about a factor of 10 longer than the free-fall time

$$t_{AD} \sim 8\text{Myr} \left(\frac{\zeta_{-17}}{n_{H_2}/10^3\text{cm}^{-3}} \right)^{1/2} \sim 10 t_{ff}. \tag{33}$$

The above properties of ambipolar diffusion have important consequences:

First, the strong dependence of t_{AD} on ionization fraction shows that 1ambipolar diffusion will be significant in the shielded interior of clouds with $A_V > 4$, but will be too slow to be important in diffuse regions penetrated by the external UV field (where ξ is up to a hundred times larger).

This would readily explain why dense cores and the ensuing star formation predominantly occur in massive, strongly self-gravitating ^{13}CO clumps, which are centrally condensed and opaque, while they are not as frequent in less massive, magnetically subcritical ^{13}CO clumps, which are mostly transparent to the interstellar UV flux (Fukui & Mizuno 1991). It also explains the other general trend noted in the previous section that larger GMCs harbor more massive dense cores: ambipolar diffusion will be more efficient inside large concentrations of mass because a larger fraction of the volume will be UV shielded.

Second, the dependence of t_{AD} on the inverse square root of density (Eq. 33) shows that any density enhancement inside a cloud will loose magnetic flux faster, and thus condense faster, than its surroundings, amplifying the density fluctuation. This instability could presumably lead to the formation of dense cores (Nakano 1984). The initial density enhancements could be created e.g. by self-gravity, by compressible MHD turbulence, or by Jeans-thermal instability along field lines, as turbulent support by Alfvén waves damps by ion-neutral friction (see Carlberg & Pudritz 1990; Mouschovias 1991 for a review).

Once ambipolar diffusion is under way in the condensing core, the Alfvén velocity in the medium drops, and thus the maximum turbulent velocity that can be maintained without excessive dissipation decreases. Therefore, the turbulent-Jeans mass, the thermal-Jeans mass, and the magnetic critical mass all decrease, and eventually the core will be on the verge of instability (e.g. Mouschovias 1976; 1991).

In low-mass cores, this leads to a critical state where thermal pressure dominates, and the density law asymptotically approaches that of an isothermal sphere (Lizano & Shu 1989). In massive dense cores, criticality will be reached after little flux loss, while turbulent support is still important, and one expects further fragmentation steps followed by local ambipolar diffusion, that will ultimately lead to the formation of a cluster. Another possible outcome is that a sudden increase of pressure (e.g. due to the propagation of a shock) or a lack of turbulent support will suddenly decrease M_J and drive the whole massive dense core gravitationally unstable. This mechanism has been invoked to explain the formation of clusters of massive stars (M > 10 M$_\odot$).

3.3 Self-Regulated Star Formation

While ambipolar diffusion is significant only in shielded regions, turbulence dissipation (through e.g. ion-neutral friction, hydrodynamical cascade, or cloud-cloud collisions) is an unavoidable process which affects the equilibrium of the *whole* cloud. Energy losses cause the cloud to progressively contract, and the value of M_J will decrease. Structures which cannot be maintained by magnetic fields alone (i.e. GMCs as a whole or massive ^{13}CO clumps) should ultimately cross the border of instability and turn entirely into stars, perhaps in a catastrophic event.

However, observations indicate that GMCs keep very low star formation efficiencies (a few %) over a typical timescale of 10-30 Myr, before they are disrupted by the action of massive OB stars (Blitz & Shu 1980). The implications of these facts depend on the scale considered, because the turbulence dissipation timescale and the possible sources for its replenishment are different.

Analytical models and numerical simulations of clumpy, Magnetized molecular clouds suggest that the turbulence dissipation timescale t_{turb} roughly scales as ηt_{ff} with η ranging from 3 to 10 (Elmegreen 1985; Pudritz 1990), i.e. it is close to the ambipolar diffusion timescale. At the GMC scale, the average density is low ($\bar{n}_{H_2} \sim 40$ cm^{-3}), and the expected t_{turb} is of order 20 Myr; this agrees relatively well with observed lifetimes, especially since Galactic differential shear may further help to maintain turbulence.

In ^{13}CO clumps, average densities are higher ($n_{H_2} > 300$ cm^{-3}) and the expected dissipation timescales are shorter: $t_{turb} \sim 5$ Myr. Low-mass clumps will not become unstable since they are magnetically sub-critical (and ambipolar diffusion there is slow enough to maintain the magnetic flux). In addition, turbulent support by resonant Alfvén waves can be fed into the clump from the outside (Puget 1991). On the other hand, massive clumps which are magnetically supercritical should be driven into global gravitational instability by turbulent dissipation in $\sim$ 5 Myr (refracted waves cannot become resonant because they damp too fast; Bertoldi & McKee 1993). Since these most massive clumps contain at least 10% of molecular gas in GMCs, the resulting star formation rate on a Galaxy-wide basis would be $\geq 10\% \times 2 \times 10^9 M_\odot / 5 \times 10^6$yr ~ 40 M$_\odot$ yr^{-1}, much higher than the few M$_\odot$ per year that are observed (e.g. McKee 1989).

Therefore, massive ^{13}CO clumps must have a process replenishing their turbulent energy to prevent contraction and excessive star formation. Most young stars go through energetic mass-loss phases (including bipolar outflows, FU Orionis eruptions, and T Tauri winds) where a sizable fraction of the stellar binding energy is released in supersonic motions. Such winds provide an *internal* source of turbulent energy which would naturally have the desired self-regulating effect (Norman & Silk 1980): As turbulent energy is injected into the clump, it will tend to "puff up". The higher UV penetration caused by cloud expansion will increase the ionization fraction ξ deeper in the envelope, and a smaller amount of the clump mass will be able to undergo efficient ambipolar diffusion, thereby decreasing the star formation rate. One envisions that a self-regulated steady-state equilibrium configuration could naturally arise, where kinetic energy from protostellar winds exactly compensate turbulent losses and maintains a constant rate of star formation in the clump.

Present estimates of mass-loss rates in young stars suggest that such a "photo-ionization regulated" equilibrium is indeed possible (McKee 1989). Only a small fraction $\sim 10 - 30\%$ of the gas mass is then undergoing significant ambipolar diffusion, at a rate fixed by its density and ionization. The resulting star formation rate is compatible with observations (§3.2).

This scenario of course depends crucially on the energetics of protostellar flows, on their duration and frequency of occurrence in stellar evolution, and on how efficiently they can transfer energy into the surrounding medium, either as "turbulent" motions or as MHD waves. Because they also have major implications on the star formation process itself, protostellar flows are an important phenomenon linking star formation and molecular cloud structure. They will be the topic of the last section of these lectures.

4 Molecular Outflows and Their Role in Star Formation

Since their discovery, more than a decade ago, molecular outflows from young stars have attracted a lot of attention, and have been the topic of numerous reviews (Lada 1985; Snell 1987; Fukui 1989; Bally & Lane 1991; Fukui *et al.* 1991; Edwards *et al.* 1993; Bachiller & Gómez-González 1992; Stahler 1993). Many important issues are raised by this phenomenon, including: the impact of outflows on molecular cloud equilibrium, their entrainment process, the nature of the ejection and collimation mechanisms, and their role in star formation. This section summarizes our present understanding of this phenomenon, starting with a few well established observational facts, then venturing into more speculative areas.

4.1 Basic Observational Properties

Molecular outflows from young stellar object (YSOs) are defined as *spatially-limited* regions of *high-velocity* molecular gas, detected as broad line wings in rotational emission lines of CO, HCN, CS, SiO, and HCO^+ (the latter three having possibly enhanced abundances in these regions). To ensure that the observed motions do reflect outflow, rather than enhanced turbulence or the presence of another cloud along the line of sight, at least two of the following three criteria must be fulfilled:

1. The maximum velocity in the line wing, V_{CO}, must exceed the escape velocity $\sqrt{GM/R_{CO}}$, where M is the total mass enclosed by the high-velocity region of size R_{CO}.
2. The high-velocity gas must be centered about the position of a young star (usually a deeply embedded infrared or radio source).
3. The wing emission must be relatively optically thin, with a ratio of ^{12}CO to ^{13}CO intensity greater than in the static cloud component (usually this ratio is ≥ 10 in high-velocity gas, e.g. Levreault 1988).

A striking property of molecular outflows is their frequent bipolar geometry (about 85% of known cases; Fukui 1989), with blueshifted and redshifted lobes symmetrically displaced on opposite sides of the central YSO. Since this common geometry is strong evidence for outflowing (rather than turbulent) motions, it is often used as an additional identification criterion in outflow searches.

A spectacular example of a highly-collimated bipolar outflow is the one found in the L1448 cloud, shown in Fig. 5 (Bachiller *et al.* 1990). However, many flows show less well separated lobes (Lada 1985). A few flows are monopolar, probably because the missing lobe is breaking out of the molecular cloud (Chernin & Masson 1991). Others show complex geometries that may arise from multiple sources or deflection by dense ambient gas (e.g. IRAS 16293, Cep A; Mizuno *et al.* 1990; Torrelles *et al.* 1993).

CO observations of a variety of outflows reveal a broad range of properties (e.g. Snell 1987; Fukui *et al.* 1991), with typical wing velocities $V_{CO} \sim 4$ to 60 kms^{-1} at the 0.1K sensitivity level, projected lobe sizes $R_{CO} \sim 0.04$ to 4 pc, and dynamical timescales $R_{CO}/V_{CO} \sim 3000$ to 5×10^5 yr. Inferred total masses in molecular gas range from $M_{CO} \sim 0.01$ to 200 $M_\odot$, often exceeding the estimated mass of the central YSO. Therefore it is currently believed that most of the flow is ambient material locally entrained by an underlying faster "primary wind". This is consistent with the low estimated temperature $\sim$ 10-50 K for the bulk of the flow (Snell *et al.* 1984). In addition, outflows are often associated with localized regions of warmer, shocked gas traced e.g. by Herbig-Haro (HH) objects, ionized jets, and H_2O masers (Bally and Lane 1991; Ray and Mundt 1993; Terebey *et al.* 1992) often moving at much higher speeds $\sim$ 100-500 kms^{-1}, and probably related to the primary wind.

4.2 Ubiquity of Outflows and Role in Cloud Support

Although they were totally unanticipated by theories of protostellar collapse and star formation, molecular outflows have proved a frequent phenomenon among YSOs. Up to 160 examples are known so far, around sources of bolometric luminosities ranging from L_{bol}= 0.5 $L_\odot$ to 3×10^5 $L_\odot$, of which about 100 lie within 1kpc of the Sun (Fukui *et al.* 1991). This suggest a very high frequency of occurrence among young stars.

Assuming an outflow duration similar to their average dynamical timescale $T_{CO} \sim 5 \times 10^4$ yr, one infers an outflow formation rate in the cylinder of radius 1 kpc centered on the Sun and perpendicular to the Galactic plane of $\sim 7 \times 10^{-4}$

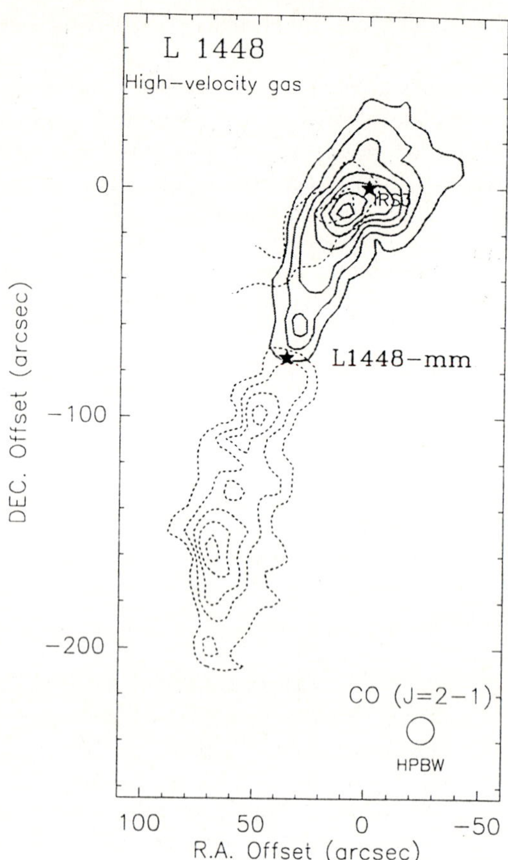

Fig. 5. CO(2-1) map of the molecular flow discovered with the IRAM-30m in L1448, showing contours of blueshifted (solid lines; V_{LSR}= -55 to 0 kms^{-1}) and redshifted (dashed lines; +10 to +65 kms^{-1}) integrated intensity (from Bachiller *et al.* 1990, 1991a). First contour and contour interval: 10K kms^{-1}. This spectacular outflow is driven by a cold object discovered at submm and radio wavelengths, which indicates that the outflow phase occurs earlier than suspected in the formation of a star (see §4.5).

kpc^{-2} yr^{-1}. This is similar to the formation rate of stars with $M \geq 1M_\odot$ (see Scalo 1986 for a comprehensive derivation of the Galactic star formation rate and initial mass function). Since many weaker outflows may have escaped detection, it is possible that *all young stars* go through a molecular outflow episode (Fukui 1989). The high detection rate of outflows in highly obscured infrared sources embedded in dense molecular cores (e.g. Myers *et al.* 1988; Morgan and Bally 1991) also supports this idea, and further indicates that this phenomenon traces a very early phase of stellar evolution (§4.5).

From their observed masses and velocities M_{CO} and V_{CO}, it is readily seen that the momentum and kinetic energy in molecular flows are extremely large, $\sim$ 1–100 $M_\odot$ kms^{-1} and $\sim 5 \times 10^{44} - 10^{47}$ ergs (Fukui 1989). For comparison, the turbulent energy of a typical ^{13}CO clump is 10^{46} ergs, and that of a whole GMC is a few 10^{48} ergs (cf. parameters in §2). Therefore outflows could play a significant role in feeding the turbulence in molecular clouds (Norman & Silk 1980), and possibly in establishing a regime of "self-regulated" star formation (McKee 1989; cf. §3.3).

This suggestion can be quantified in the following way: First, a conservative estimate of the turbulent energy injected by an outflow is obtained by assuming that only the momentum is conserved in the acceleration of ambient gas by the primary wind (the kinetic energy is mostly dissipated in radiative shocks). After the termination of the driving wind, the molecular flow coasts at constant momentum while continuing to sweep up the surrounding gas. Once the flow has decelerated to the cloud turbulent velocity $v_{rms} = \sqrt{3}\sigma$, the total flow energy remaining in the cloud in the form of turbulent motions is

$$E_{turb} \sim \frac{1}{2} P_{CO} v_{rms}. \tag{34}$$

The average rate of turbulent energy dissipation in a cloud of mass M_c is approximately $\frac{3}{2} M_c \sigma^2 / \eta t_{ff}$, where η is estimated to be $\sim 3-10$ (§3.3). The energy injection by outflows could balance the turbulent dissipation if the total number of stars formed per unit time, $\dot{N}_*$, were such that

$$\langle P_{CO} \rangle \dot{N}_* \geq \frac{M_c v_{rms}}{\eta t_{ff}}. \tag{35}$$

Here $\langle P_{CO} \rangle$ is an appropriate average of the wind luminosity over a mass distribution of stars with outflows, as P_{CO} depends strongly on the luminosity, and thereby the mass, of the exciting star; in terms of the normalized initial luminosity function $g(L_*) \equiv (1/N_*) dN(L_*)/dL_*$,

$$\langle P_{CO} \rangle = \int_{L_{min}}^{L_{max}} g(L_*) P_{CO}(L_*) p(L_*) dL_*, \tag{36}$$

where $p(L_*)$ is the fraction of stars of initial luminosity L_* that have outflows; we may well assume that $p(L_*) = 1$. The integral is not easily evaluated: CO flow observations trace only the line-of-sight velocity, they are affected by optical depth effects, and if the wind is still active, the flow momentum will not have yet reached its maximum value (Margulis *et al.* 1988). These effects tend to underestimate P_{CO}. On the other hand, surveys for outflows are biased toward detecting the strongest ones, which has the opposite effect (§4.5).

Previous studies have adopted $\langle P_{CO} \rangle \sim 20 - 80$ $M_\odot$ kms^{-1}(Fukui *et al.* 1986; Margulis *et al.* 1988; McKee 1989), and they have investigated mainly the possibility of supporting a whole GMC, where $\eta t_{ff}/v_{rms} \sim 5 \times 10^6$ yr/(kms^{-1}) (note that in a self-gravitating cloud, this ratio depends only on the mean column density; McKee 1989). The star formation rate required for equilibrium is then:

$$\dot{N}_* = 5 \times 10^{-4}\text{yr}^{-1} \left(\frac{M_c}{10^5 M_\odot}\right) \left(\frac{v_{rms}}{5\text{km s}^{-1}}\right) \left(\frac{40 M_\odot \text{kms}^{-1}}{\langle P_{CO}\rangle}\right) \left(\frac{25 \times 10^6 \text{yr}}{\eta t_{ff}}\right). \tag{37}$$

Note that this rate is independent of the duration of the outflow phase, which is observationally quite uncertain. We can now compare the derived equilibrium star formation rate with

1. the star formation rate in the Galaxy, SFR (in $M_\odot$ yr^{-1}), and the total Galactic molecular mass M_{H_2},
2. the number of outflows presently active in any GMC, N_{flow},
3. the number of young stars $N(t_*)$ of a given age t_* in a GMC.

In a steady state, one would expect that

$$\dot{N}_* \sim \frac{\text{SFR} \times M_c}{\langle m_* \rangle M_{H_2}} \sim \frac{N_{flow}}{T_{flow}} \sim \frac{N(t_*)}{t_*}. \tag{38}$$

The first comparison would assume that the star formation rate *per solar mass* of molecular material is relatively constant throughout the Galaxy (as argued e.g. by McKee 1989, Appendix A), while the other two depend on the adopted age for the given population (usually $T_{flow} \sim T_{CO} \sim 5 \times 10^4$ yr for outflows, and $t_* \sim$ 1-3 Myr for T Tauri stars). The observed Galactic SFR of 3 $M_\odot$ yr^{-1} in a total molecular mass $M_{H_2} \sim 2 \times 10^9$ $M_\odot$ (cf. McKee 1989) is consistent with the above star formation rate if $\langle m_* \rangle \sim 0.3$ $M_\odot$. The number of outflows $N_{flow} \sim 5 - 7$ found by recent surveys in $2 - 3 \times 10^4$ $M_\odot$ of molecular gas in the L1641 and Mon OB1 clouds also agrees with that expected from the equilibrium formation rate (Fukui *et al.* 1986; Margulis *et al.* 1988; Morgan *et al.* 1991). Hence one could conclude that molecular outflows are frequent enough and energetic enough to replenish the turbulence in GMCs and to explain their low star formation rate.

However, we have seen in §3.3 that the need for support is in fact more critical for the massive, star-forming ^{13}CO clumps, which are denser and dissipate turbulence faster than a GMC as a whole. Since the big clumps have typically $M_c \sim 10^3$ $M_\odot$, $\sigma \sim 1\text{kms}^{-1}$, and $t_{ff} \sim 10^6$ yr, the inferred equilibrium star formation rate per clump is $\dot{N}_* \sim 10^{-5}$ yr^{-1}. This predicts that such clumps would each harbor about 10-50 young stars after 1-5 Myr. This is consistent with the observed number of T Tauri stars in the dark clouds in Taurus-Auriga, $\sim 10 - 20$ (Gomez *et al.* 1993), and only slightly lower than the number of stars $\sim 30 - 200$ in the aggregates and rich clusters in L1641, Orion B, or the ρ Oph core (§3.1). The agreement is even better if $\langle P_{CO}\rangle$ is in fact weighted toward less energetic flows than those detected in CO surveys, or if clusters contain stars up to 10 Myr old. Thus the energy injected by outflows appears sufficient to replenish turbulent support in star forming clumps in the solar neighborhood (Bertoldi & McKee 1993) and may also explain their star formation rates.

4.3 Outflow Structure and Velocity Field

The previous section has shown that global arguments assessing the potential of cloud support by outflows suffer from uncertainties. To better understand the large-scale impact of molecular flows on the surrounding cloud, as well as to constrain their ejection and collimation mechanisms, it is crucial to determine both their internal structure and their velocity field. To extract this information from observations, models of CO line transfer in bipolar geometries have been recently developed (Cabrit & Bertout 1986,1990; Meyers-Rice & Lada 1991, hereafter ML91). Comparisons of synthetic and observed CO spectra have proved a very useful tool for interpreting outflow data. The main conclusions reached have been confirmed by subsequent detailed observational studies.

Overall geometry. Models have shown that, in addition to its intrinsic geometry, the appearance of the outflow is strongly dependent both on the viewing angle and on the maximum divergence θ_{max} between the velocity vectors and the flow axis (Liseau & Sandell 1986). For a conical, radial flow, four possible configurations of red and blue-shifted gas can be distinguished (adapted from Cabrit and Bertout 1986). Two important implications can be drawn:

(i) Velocity vectors must make a moderate, *non-zero* angle ($0 < \theta_{max} < 45^o$) to the flow axis in order to reproduce both the well-separated blue and red lobes that are observed in many well-collimated flows (e.g. L1448; Fig. 4), and the superimposed blue and red emission in the lobes of flows almost in the plane of the sky (e.g. B335; Cabrit *et al.* 1988). The same conclusion is reached for a parabolic geometry (ML91). The relatively rarity of Case 3 (Cabrit & Bertout 1986) flows indicates that θ_{max} is probably $\leq 15^o$ (the probability of observing this configuration is $\sin\theta_{max}$). Therefore, velocities in molecular flows are essentially forward-directed.

(ii) The statistical excess of "pole-on" flows reported by Lada (1985) is not consistent with the high degree of collimation seen in several outflows like L1448. If not due to selection effects or beam-smearing, this excess requires an additional population of intrinsically poorly collimated flows, which always appear as "pole-on". Such flows tend to be more frequent around luminous YSOs (Levreault 1988), suggesting that either the large-scale collimation process is less efficient, or the ejection itself is poorly collimated in these objects. High-resolution H_2 imaging of the outflow in Orion BN-KL reveals an explosive ejection event over a wide angle that would favor the latter interpretation (Allen & Burton 1993). An excess of pole-on flows could also reflect an evolutionary effect in the entrainment process of otherwise well-collimated winds (§4.5).

Internal Velocity and Density Structure. Bipolar flows usually exhibit very broad CO line profiles with sloping wings that are well described by a power law $T(v) \propto v^{\gamma}$ with $\gamma \sim -1.8 \pm 0.4$ (Rodríguez *et al.* 1982; Masson & Chernin 1992). This broad range of radial velocities implies that velocity vectors must vary in magnitude and/or direction within the flow. In particular, high-resolution observations of well-collimated flows that are inclined to the line of sight tend to show a complex structure with the following common features: (i) a hollow shell toward the flow edges at low velocities, (ii) an elongated lobe along the flow axis at high velocities, (iii) a quasi-linear increase of velocity with distance from the central source. Examples of this trend are L1551-IRS5 and Orion B (Moriarty-Schieven and Snell 1988; Richer *et al.* 1992). Figure 6 illustrates these features in Orion B.

One explanation first proposed in the context of L1551-IRS5 is that molecular flow lobes form a thin, hollow shell swept up by a moderately collimated wind, with the shell velocity increasing further from the source (Moriarty-Schieven & Snell 1988). In the following, I describe several theoretical problems encountered by this class of models. Then I will conclude that flows driven by highly-collimated jets which produce internal latitude velocity gradients avoid these drawbacks and better explain recent observations.

The simplest coherent dynamical model of a hollow swept-up flow is that proposed by Shu *et al.* (1991), where the shell is momentum-driven by a wind expanding *radially* into an ambient medium with density decreasing as $1/r^2$. The bipolarity is assumed to be caused by a separate dependence of the wind force and ambient density with the polar angle θ. A very nice, detailed study of this class of model is presented by Masson & Chernin (1992). In fact, the conclusions are valid for any power-law density profile $\rho(r,\theta) = \rho_*(\theta)r^\beta$ with $\beta > -3$. The radius of the swept-up shell is readily derived from momentum conservation (cf. Dyson 1984) and can be written as

$$r_s(\theta,t) = \phi_\beta \times \left[\frac{f_w(\theta)}{\rho_*(\theta)}\right]^{1/(\beta+4)} \times t^{2/(\beta+4)} \tag{39}$$

where $f_w(\theta) = d(\dot{M}_w v_w)/d\Omega$ is the wind force per solid angle, t the time elapsed, and ϕ_β a function of β only. If β is the same over all solid angles, the intrinsic shape of the shell is determined only by the variation of $[f_w(\theta)/\rho_*(\theta)]^{1/\beta+4}$ with θ, and therefore it remains *the same* independent of time (self-similar expansion).

The velocity of the shell is $v_s = 2r_s/t(\beta+4)$ and is radially directed. Hence, $\mathbf{v_S} \propto \mathbf{r_S}$ when $\beta =$ cst. Since one measures the line-of-sight component of the velocity vector, and the perpendicular component of the radius, the velocity-position diagram along the projected flow axis ("long-slit-spectrum") is going to reflect exactly the shape of the emitting shell. I have plotted in Fig. 7 the resulting velocity-position diagram for a shell with shape $r(\theta) \propto \cos^5(\theta)$ that reproduces the average collimation in molecular flows (cf. Masson and Chernin 1992), seen at various view angles.

As the figure demonstrates, two problems face this class of models: First, the relation between v_{rad} and r_{proj} is not strictly linear. The closed shell geometry produces a "loop" in the diagram, which is not observed. Next, the back side of the shell produces a second, lower-velocity ridge of emission which could remain hidden by cloud emission if $i \sim 75^o$, but should produce distinctive double-peaked profiles at inclinations $i \leq 45^o$. Such profiles have never been observed, except maybe in the complex outflow Mon R2 (ML91).

Finally, Masson and Chernin (1992) showed that the integrated CO line profile from the shell cannot reproduce observed profile shapes under realistic

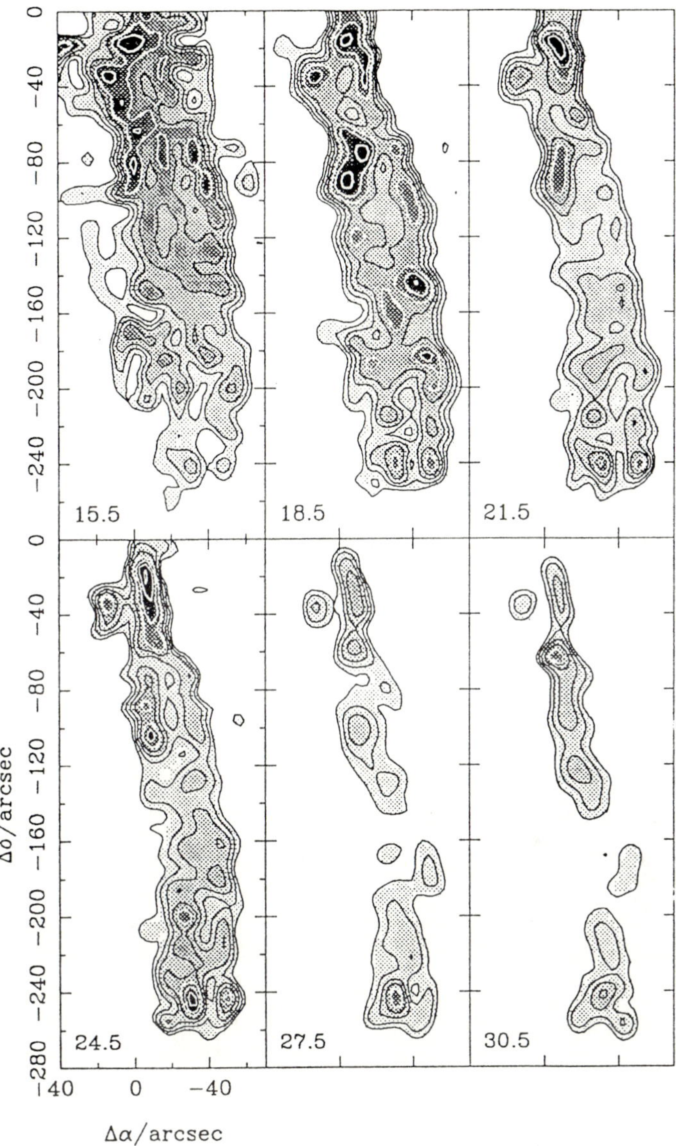

Fig. 6. IRAM-30m maps of CO(2-1) intensity in the red lobe of the Orion B outflow, integrated over successive 3 kms^{-1} intervals centered at the indicated LSR velocity. Contours levels increase with a scale factor of 1.5. Lowest contour is 16 K kms^{-1} in first map, and 4 kms^{-1} in others. Note the increasing collimation and the migration away from the source (located in (0,0)) with increasing velocity. From Richer *et al.* (1992).

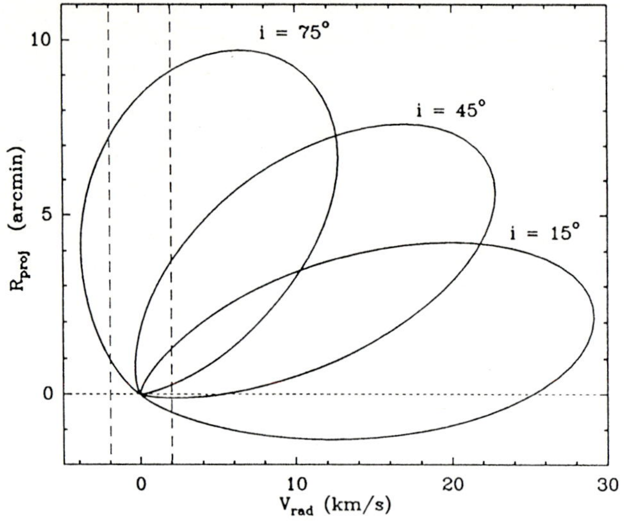

Fig. 7. Line-of-sight velocity v_{rad} as a function of projected distance along the flow axis r_{proj} for a shell of shape $r(\theta) \propto \cos^5(\theta)$ in self-similar expansion, viewed at various view angles i. Dashed lines denote the velocity domain hidden by ambient cloud emission. Note that the front and back sides of the shell produce emission at two distinct velocities.

assumptions, because it yields too much emission at large velocities. Again this result is quite general independent of β.

Therefore, general kinematic and dynamic models both show that a shell swept-up by a *radially directed* wind does not reproduce the properties of molecular flows (Cabrit & Bertout 1990; Masson & Chernin 1992). A natural alternative is that the flow is entrained by a *collimated jet-like component*, possibly linked to the optical jets observed in some YSOs (Ray & Mundt 1993).

While the optically emitting material of stellar jets has too low a momentum flux to drive CO outflows, strong support for this idea has come from the recent discovery of highly collimated and energetic *neutral* components reaching 100 kms^{-1} along the axis of several flows, and associated with diagnostics of entrainment (see below). Presumably, it is the complex interaction between this "neutral jet" and the ambient cloud which gives rise to the peculiar velocity-density structure of outflows. Simplified analytical models of this kind of interaction have been recently developed. They suggest that the surrounding cloud will be accelerated in a bowshock, forming a low-velocity expanding dense shell with high-velocity gas at its head (Masson & Chernin 1993). If the bowshock head contains a turbulent mixing layer, some of the gas may reexpand to refill the cavity, producing a wake of high-velocity, tenuous warm gas with an apparent acceleration (Raga & Cabrit 1993). The similarity with the observed features of outflows noted above is promising.

A jet in itself is not very efficient in sweeping mass, because a leading bowshock which rams into dense ambient gas is expected to be quite narrow. However, the very high-velocity jets show multiple discrete structures, attributed to either separate ejection pulses or to precession, which indicate that several bowshocks probably occur in the flow lifetime (see below). Their cumulative effect will progressively broaden the outflow and increase the swept-up mass. Less well-collimated outflow geometries might then correspond to later evolutionary stages of the jet interaction with the surrounding medium, as most of the ambient gas has been pushed aside or incorporated into a slow turbulent cocoon by many passing shock fronts (e.g. Raga *et al.* 1992; Stahler 1993). Ultimately, the polar regions will be mostly evacuated, leaving a cavity in the molecular cloud and, if the wind is still active, an exposed optical jet along the central axis. Detectable cavities are found e.g. in the L43 system (Matthieu *et al.* 1988), showing that outflows from low-mass stars can significantly disrupt their native dark cloud.

Observations of High-Velocity Jet Components. Perhaps the most exciting advance in the field of outflows from young stars has been the discovery, in both atomic (HI) and molecular species (CO, HCO^+, SiO), of faint "extremely-high-velocity" (EHV) components at velocities of 50-150 kms^{-1} several times higher than those in the "standard" slower molecular flow (Lizano *et al.* 1988; Mitchell *et al.* 1989; Koo 1989). They have been seen toward a dozen sources including both low-luminosity (10-90$L_\odot$) and high-luminosity ($10^3 - 10^4$ $L_\odot$) objects, suggesting that they may be a common phenomenon associated with molecular flows.

High-resolution observations using the VLA and large radiotelescopes have allowed us for the first time to probe the spatial distribution and velocity field of the EHV gas (see Bachiller and Gómez-González 1992 for a review). Figure 8 shows a velocity-position diagram of EHV CO components in the well-studied L1448 outflow (Bachiller *et al.* 1990). Such observations have shown that the EHV is highly collimated, warm ($T \sim 25 - 100$ K; Masson *et al.* 1990), and contains small clumps ("bullets") of mass $\sim 10^{-4} - 10^{-3}$ $M_\odot$ that emit at well-defined velocities and tend to be regularly spaced along the flow axis. The "standard" flow (SHV) is the most prominent close to the EHV clumps, indicating a strong local interaction between the two flow components. The EHV peaks are also closely associated with shocks traced by optical HH objects, enhanced SiO emission, and/or H_2 vibrational emission.

SiO emission in L1448 is detected *only* at high velocity, showing that the emitting gas has been processed through a shock that released SiO from grain mantles, enhancing its abundance by a factor $> 10^4$ over that in the undisturbed cloud (Bachiller *et al.* 1991b). H_2 images reveal arcs of emission along the axis of the flow, and bowshock structures with their head close to the EHV CO peaks (Terebey 1991; Bally *et al.* 1993). Another nice example of associated optical HH objects, H_2 emission, and EHV CO is the HH7-11 outflow (Garden *et al.* 1990; Bachiller & Cernicharo 1990). In fact, all flows having EHV CO emission show collimated H_2 emission provided the extinction is sufficiently low (Bally *et al.* 1993).

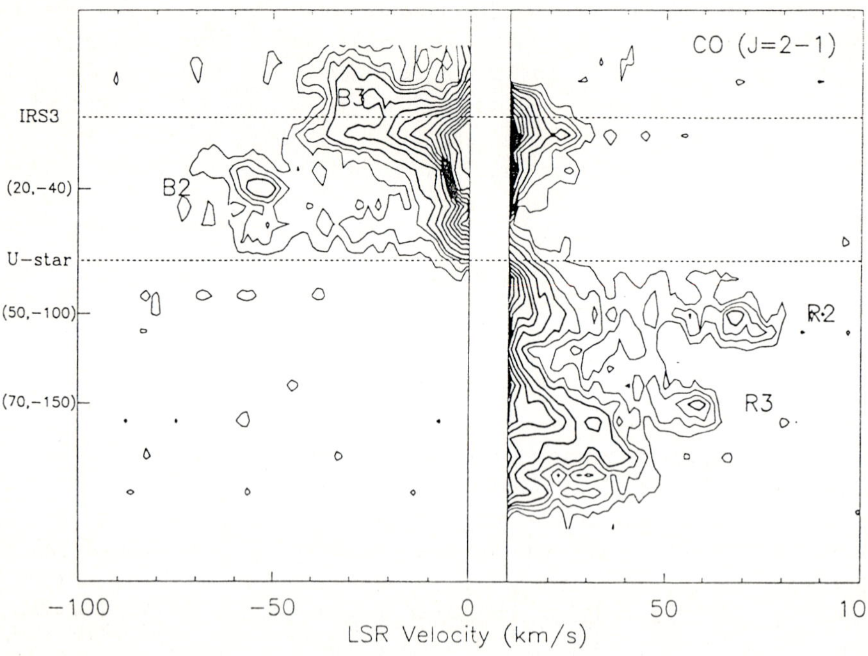

Fig. 8. Velocity-position diagram of CO(2-1) emission along the axis of the collimated outflow L1448 (light intensity contours: 0.2, 0.4, 0.6K; thick contours start at 0.8K, with steps of 0.4K). Cloud emission dominates at 0-10 kms^{-1}. The flow is driven by the U-star, later detected in the mm and cm range (§4.5). Note how the well-defined high-velocity clumps (B2, B3, R2, and R3) are regularly spaced and appear symmetrically in the blue and red lobes. The lower-velocity "standard" flow is enhanced at these positions. IRAM-30m data from Bachiller *et al.* (1990).

All of these facts agree quite well with the behavior expected for the simple entrainment models discussed above, and strongly suggest that EHV components trace a highly-collimated jet which is accelerating the slower "standard" molecular flow (e.g. Bachiller *et al.* 1990). The presence of several EHV peaks symmetrically placed along the flow axis is probably not due to previous inhomogeneities in the ambient cloud, and rather suggests that the ejection is *intrinsically time variable*, with either separate ejection pulses or precessing motions (S. Biro, student presentation, this volume). This is confirmed by interferometric SiO observations of the L1448 exciting source (Guilloteau *et al.* 1992), which reveal a laterally unresolved jet with two clumps again symmetrically displaced both in position and velocity on either side of the star; the apparent acceleration of the jet out to $\sim \pm 60$ kms^{-1} is achieved over 6" = 0.01 pc from the source, and the jet width is less than 700 AU, implying that the jet collimating mechanism must act very close to the ejection point.

4.4 Outflow Energetics and Ejection Mechanism

That outflows are so frequent suggest that they trace a central physical process in star formation, possibly related to the evacuation of excess angular momentum (Shu *et al.* 1987). To constrain the ejection mechanism, the energetics of protostellar winds must be determined, either from their time-integrated effect in the swept-up molecular flow, or from more direct tracers of the ejection activity like shocked gas or EHV components. I will first summarize results from the former approach, which was the only one available until a few years ago, and then turn to the latter, based on the recent results described above.

Global Energetics of Swept-Up Flows The force and mechanical luminosity in the swept-up molecular gas can be estimated from the flow mass M_{CO}, characteristic velocity V_{CO}, and dynamical time-scale $T_{CO} = R_{CO}/V_{CO}$ as $F_{CO} = MV_{CO}/T_{CO}$ and $L_{CO} = 1/2\ MV_{CO}^2/T_{CO}$. These latter two quantities are expected to be less dependent on time than momentum and energy, which should progressively increase.

F_{CO} and L_{CO} are both well correlated with the driving source bolometric luminosity L_{bol}, and the ratios F_{CO}/L_{bol} and L_{CO}/L_{bol} in principle set constraints on the ejection mechanism (Rodríguez *et al.* 1982; Lada 1985). However, ^{12}CO line opacity, projection effects, and velocity gradients introduce large uncertainties in CO flow parameters causing a scatter of several orders of magnitude in the observed ratios. Source confusion and beam dilution in distant flows further increase the scatter (Bally and Lane 1991). Tighter correlations are found e.g. by concentrating on well-resolved flows, and using theoretical flow models for estimating the inclination angle (Cabrit and Bertout 1992), giving:

$$cF_{\rm CO}/L_{\rm bol} \sim 2000(L_{\rm bol}/L_{\odot})^{-0.3} \quad \text{and} \quad L_{\rm CO}/L_{\rm bol} \sim 0.04(L_{\rm bol}/L_{\odot})^{-0.2}.$$

Since energy-conservation requires that the primary wind mechanical luminosity L_w exceeds L_{CO}, protostellar winds must be much more energetic than those encountered in main-sequence or red-giant stars, where $L_w < 10^{-3}\ L_{bol}$. YSOs with $L_{bol} \geq 10^4\ L_{\odot}$ have values of $cF_{CO} L_{bol}$ comparable to those in winds of Wolf-Rayet stars or in peculiar bipolar planetary nebulae like OH 231.8+4.2. Therefore a similar, possibly radiative, mechanism might play a role in these bright, deeply embedded objects (Bally 1986). However, a more efficient wind ejection is required in lower luminosity sources. Promising models involve MHD ejection from an accretion disk threaded by open magnetic field lines (e.g. Pelletier & Pudritz 1992), where L_w can be comparable to the accretion luminosity $\sim L_{bol}$; such winds have the additional advantage of being self-collimated, thus explaining the extreme collimation observed at high velocities and small scale, e.g. in the SiO jet of L1448 (§4.3.3).

Recently it was pointed out that the dynamical ages of outflows might underestimate their true ages by a significant factor (e.g. Parker *et al.* 1991). Then the above integrated energetic parameters would be too large by the ratio T_{flow}/T_{CO}. This may happen if the leading head of the flow propagates at a much smaller speed than the adopted characteristic velocity V_{CO}. A detailed model of jet propagation for the Orion B flow (Richer *et al.* 1992) suggests a mean velocity of 9 kms^{-1} over the flow lifetime, 4-5 times smaller than the velocity of the EHV feature, but comparable to the maximum velocity of the slower "standard" flow. Hence the above method, while very uncertain, might still give reasonable estimates of the lifetime and cumulated energetics of outflows provided one uses as V_{CO} the wing velocity of the standard swept-up component.

"Direct" Tracers of Mass Ejection. Indicators more closely related to the driving jet include radio centimetric emission, H_2 emission from shocked gas, and EHV molecular components, all with dynamical times at least ten times shorter than T_{CO}. Optical manifestations of the driving jet are rarely seen, owing to the high extinction toward most outflow sources.

(1) Centimetric free-free emission is frequently detected in outflow sources, and attributed to ionized gas in the wind; this is supported in a few cases by high-resolution VLA images revealing an elongation along the flow axis and proper motions at $\sim$ 300 kms^{-1} (e.g. Rodríguez *et al.* 1990). However, mass-loss rates derived assuming a spherical, constant velocity wind (Panagia & Felli 1985) or a shock-ionized region (Curiel *et al.* 1990) both fail by at least one order of magnitude to provide the momentum present in large-scale CO outflows (Snell and Bally 1986; Cabrit & Bertout 1992). The deficit is worse if one takes into account the collimated jet geometry, so that even an energy-conserving interaction would be insufficient to drive the flows (cf. Rodríguez *et al.* 1990). Therefore this ionized jet component appears dynamically unimportant in the ejection, with typical mass-loss rates $\dot{M}_{ion} \sim 10^{-8} - 10^{-7}$ $M_\odot$ yr^{-1} in moderate luminosity sources, and the driving jet must be *mostly neutral.*

(2) The H_2 emission in molecular flows traces radiative losses from post-shock regions associated with jet propagation. Therefore they reflect the impact of the whole jet, not only its ionized component. However, intrinsic jet energetics cannot be derived from H_2 intensities without a detailed shock model, like those developed e.g. for optical HH objects. The total H_2 luminosity at least provides a lower limit to the mechanical luminosity in present jet episodes. In L1448, $L_{H_2} \sim 0.2$ $L_\odot$, which is comparable to the average mechanical luminosity in the CO flow, $L_{CO} \sim 0.5$ $L_\odot$ (Bally *et al.* 1993). In less well-collimated and older flows, the ratio L_{H_2}/L_{CO} is lower $\sim 1/20$ (e.g. Bally 1986), as expected if the jet is running into more and more diffuse material. Therefore younger flows are a better laboratory for studying jet interaction with dense molecular gas.

(3) EHV components have the advantage of allowing direct mass and velocity estimates independent of shock models. Possible abundance variations introduce an uncertainty for molecular tracers, while the spatial resolution/sensitivity is the limiting factor for HI observations. Despite their very low total mass $\sim 10^{-3}-$

$10^{-2}M_\odot$, EHV molecular and atomic components contain 1-3 times the kinetic energy present in the slower "standard" flow. In other words, $P_{EHV}/T_{EHV} \geq P_{CO}/T_{CO}$, where $T = R_{CO}/V$ denotes the dynamical timescale for the EHV and the standard CO flow respectively. Therefore, the EHV gas could provide enough momentum to drive the large scale CO (Koo 1989). However, it is not yet firmly established which amount of the EHV component is part of the driving jet and which is made of shocked ambient gas entrained at the bow-shock head or along the jet. If we assume that all the EHV gas has been ejected from the central object, an estimate of the wind mass-loss rate can be derived. In L1448, interferometric SiO observations show 3×10^{-4} $M_\odot$ of gas in an ejection event of timescale $\sim$ 100 yr, giving an instantaneous mass-loss rate $\dot{M}_w \sim 3 \times 10^{-6}$ $M_\odot$ yr^{-1}(Guilloteau *et al.* 1992). Similar values are inferred from the more extended EHV, which has a crossing time scale $\sim$ 2000 yr, using either CO or HI (e.g. Bachiller *et al.* 1991c; Giovanardi *et al.* 1992). With EHV speeds of 50-200 kms^{-1}, the inferred wind mechanical luminosity range from 20% to 100% of the estimated bolometric source luminosity. This confirms the intrinsic high efficiency of the ejection mechanism in young stars, which points toward accretion as the ultimate energy source, and magnetic fields as the necessary accelerating and focusing agent (see Pudritz 1988 for a critical review of various mechanisms).

Another indication of the ejection process is the striking similarity between the properties of EHV "bullets" and the eruption events from FU Orionis stars, which are attributed to outbursts in accretion disks (see the review by Hartmann 1991). The typical CO bullet mass of $3 \times 10^{-4} - 10^{-3}$ $M_\odot$ is close to the mass expected for a typical FU Or outburst, with mass loss 10^{-5} $M_\odot$ yr^{-1} over $\sim$ 100 years. The time interval between successive bullets in L1448 is a few thousand years, comparable to current (but uncertain) estimates of the "duty cycle" between FU Ori eruptions. A similar timescale is inferred for multiple major bowshocks in optical jets, e.g. HH 111, where FU Or-like eruptions have been proposed as a possible explanation (e.g. Reipurth 1989). Molecular flows, optical jets and HH objects, and FU Ori outbursts might therefore be various manifestations averaged on decreasing timescales of the same powerful mechanism linked to the physics of accretion in YSOs. The crucial role of disks in the outflow process is substantiated by studies of the circumstellar environment of YSOs, as described in the next section.

4.5 Evolutionary Status of Outflow Sources

Molecular outflows are mostly found toward deeply embedded YSOs with infrared spectra that rise steeply from 2 to 60μm, indicative of a thick and cool dusty envelope (the so called "Class I" sources, cf. Lada 1988). The fraction of sources with outflows should then reflect the duration of this phase as compared with the Class I stage. Recent unbiased or IRAS-based surveys have revealed a high frequency of occurrence of molecular flows among luminous ($L_{bol} > 30$ $L_\odot$) Class I sources ($\sim$ 50% to 80%; e.g. Morgan and Bally 1991; Fukui *et al.* 1991). Therefore, most bright YSOs go through a molecular outflow episode, and the outflow stays visible for a large fraction of the embedded phase (consistent with the short evolutionary timescales expected for massive protostars).

The situation appears less clear-cut in low-luminosity sources ($L_{bol} \leq 30$ $L_{\odot}$): most known low-luminosity flows have been found in biased surveys of individual dark clouds or HH objects with a detection rate $\sim$ 50%, including many weak outflows (e.g. Myers *et al.* 1988; Parker *et al.* 1991). However, detection rates have been much lower in molecular complexes like Taurus-Auriga and ρ Oph, with only a few flows detected for several tens of low-luminosity Class I objects. Recent results suggest that *evolutionary effects*, together with sensitivity limitations, are probably the dominant cause of the low outflow detection rate in these complexes: it has appeared that well-developed outflows from low-luminosity YSOs in these complexes are excited by extremely cold sources previously undetected at $\lambda < 25$ μm, in contrast to typical Class I sources. A spectacular demonstration is the case of ρ Oph A (André *et al.* 1990). This fast, highly-collimated outflow, the only one discovered so far in the ρ Oph A cloud core, is not associated with any of the previously known Class I sources in the cluster. The driving object is identified as VLA1623, a weak VLA source with no IRAS counterpart and a submm spectrum characteristic of very cold dust (André *et al.* 1993). Another example is the fast bipolar outflow in L1448 (Fig. 5), which is driven by a previously unknown, obscured object discovered at cm and mm wavelengths (e.g. Bachiller *et al.* 1991). Several other outflow sources with similar properties are known (André *et al.* 1993). Although their cold spectra might be attributed to orientation effects, with a normal infrared source obscured behind a thick edge-on disk, their distinctive spectacular molecular outflows, with unusually high velocities (and their large circumstellar masses; see below) make this unlikely and show that they truly represent a distinct class of extremely embedded objects, presumably less evolved than other Class I sources. This in turn suggests that the outflow phase occurs earlier than anticipated, and that most known low-luminosity Class I sources detected at 2-25μm may have already evolved past the strong outflow stage; their weaker flows would then be hardly detectable inside large molecular complexes with intrinsic broad lines, explaining the low outflow detection rates when compared with smaller dark globules.

A survey of dust continuum emission at 1.3mm with the IRAM-30m of 25 nearby Class I sources with $L_{bol} \leq 100 L_{\odot}$ (Cabrit and André 1991) strongly supports this interpretation: Sources associated with highly collimated, well-developed outflows are found to have much stronger dust emission than sources with no, or only weak, detected flows (see Fig. 9). Since dust emission at 1.3mm is mostly optically thin, and comes primarily from cold dust at $T_d \sim 30$K, the flux density $S_{1.3}$ is essentially proportional to the total mass in the beam M, with $M \sim 0.05$ $M_{\odot}$ $(S_{1.3}/500\text{mJy}) \times (d/160\text{pc})^2$ for an opacity per mass column density $\kappa_{1.3mm} \sim 0.02$ g^{-1} cm^2. Hence the above result directly shows that strong outflow sources are surrounded by much more circumstellar mass than non-outflow sources within the radius $\sim$ 1000 AU encompassed by the telescope beam, with the border between the two groups around M $\sim$ 0.05-0.1 $M_{\odot}$. Probably, massive ($M \geq 0.1 M_{\odot}$) and centrally condensed circumstellar structures are intrinsically connected to the outflow driving engine in low-luminosity sources

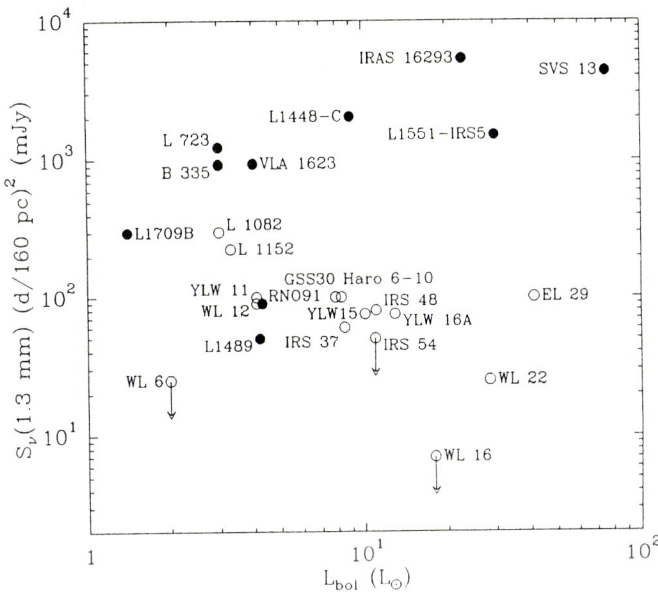

Fig. 9. 1.3mm flux in the IRAM-30m 12" beam, scaled to 160 pc, vs. the bolometric luminosity L_{bol} for 25 nearby Class I sources. Sources driving molecular outflows (filled circles) are much stronger than sources with no detectable outflows (open circles), implying a much larger circumstellar mass within 1000 AU. From Cabrit and André (1991).

(they might for example provide a mass reservoir for the primary wind and/or enforce various disk instabilities involved in the ejection process; cf. Hartmann 1991). Stars exciting optical HH objects or jets also have strong millimetric dust emission, comparable to that in molecular flow sources of similar luminosities (Reipurth *et al.* 1993), and this again favors a similar ejection mechanism in both kinds of objects.

In summary, the above results show that the outflow phase in low-luminosity YSOs starts at an extremely embedded and centrally condensed stage (one infers $n \sim 5 \times 10^8$ cm^{-3} within 1000 AU). According to theoretical models (e.g. Adams 1990) this may correspond to the first 10^4 years of protostellar collapse, in agreement with the short dynamical timescales of the associated flows. Since the typical circumstellar mass ~ 0.5 M$_\odot$ is close to stellar values, it is likely that the central star still has not assembled most of its final mass. Because of their very large extinction ($A_V \sim 1000$ mag), such young, cold objects (named "Class 0" sources) are very weak in the infrared and are most easily identified by their energetic outflows and by maps in the submm and cm wavelength range (André *et al.* 1993).

Surveys of star forming regions in the millimetric dust continuum can take us even further back in time by revealing possible *precursors* to the Class 0 stage; maps near VLA1623 show several cold emission peaks of similar mass $\sim$

$0.6M_\odot$, but amorphous in shape, and with no internal VLA source (Mezger *et al.* 1992; André *et al.* 1993). Their self-gravity is in close balance with their thermal pressure, and they could very soon start gravitational collapse ($t_{ff} \sim 10^4$ yr). They probably trace the last pre-stellar stages of NH_3 molecular cores, just before protostellar collapse sets in. Observations in the millimetric and submillimetric range therefore appear as a very powerful tool in the long lasting "quest for protostars" (i.e. objects with near free-fall motions in most of their mass). Given the rapid progress in this domain, in particular the advent of sensitive bolometer arrays, our understanding of the process of star formation and outflows from young stars will certainly dramatically increase in the coming years.

Acknowledgements. I am very much indebted to F. Bertoldi for numerous enlightening discussions about molecular clouds and for comments on this manuscript, and to E. Falgarone and F. Boulanger for sharing their views and reviews on the subject.

References

Adams, F.C.: The L-A_V diagram for protostars. ApJ **363** (1990) 578-588

Allen, D.A., Burton, M.G.: Explosive ejection of matter associated with star formation in the Orion nebula. Nature **363** (1993) 54-56

André, P., Ward-Thompson, D., Barsony, M.: Sub-millimeter continuum observations of ρ Oph A: The candidate protostar VLA 1623 and pre-stellar clumps. ApJ **406** (1993) 122-141

André, P., Martin-Pintado, J., Despois, D., Montmerle, T.: Discovery of a remarkable bipolar flow and exciting source in the ρ Ophiuchi cloud core. A&A **236** (1990) 180-192

Arons, J., Max, C.E.: Hydromagnetic waves in molecular clouds. ApJ **196** (1975) L77-L81

Bachiller, R., Cernicharo, J.: Extremely high-velocity emission from molecular jets in NGC 6334I and NGC 1333 (HH 7-11). A&A **239** (1990) 276-286

Bachiller, R., Gómez-González, J.: Bipolar molecular outflows. A&A Reviews **3** (1992) 257-287

Bachiller, R., André, P., Cabrit, S.: Detection of the exciting source of the spectacular molecular outflow L1448 at $\lambda\lambda$ 1-3mm. A&A **241** (1991a) L43-L46

Bachiller, R., Martín-Pintado, J., Fuente, A.: High-velocity SiO emission in the L1448 outflow. Evidence for dense shocked gas in the molecular bullets. A&A **243** (1991b) L21-L24

Bachiller, R., Martín-Pintado, J., Planesas, P.: High-velocity molecular jets and bullets from IRAS 03282+3035. A&A **251** (1991c) 639-648

Bachiller, R., Cernicharo, J., Martín-Pintado, J., Tafalla, M., Lazareff, B.: High-velocity molecular bullets in a fast bipolar outflow near L1448/IRS3. A&A **231** (1990) 174-186

Bally, J.: Massive bipolar flows around young stars. Irish Astr. J. **17** (1986) 270-279

Bally, J., Lane, A.P.: Molecular outflows: Observed properties. In C.J. Lada and N.D. Kylafis (eds.), The Physics of Star Formation and Early Stellar

Evolution, Kluwer Academic Publishers, Dordrecht, (1991) 471-495

Bally, J., Lada, E.A., Lane, A.P.: The L1448 molecular jet. ApJ **418** (1993) 322-327

Beichmann, C.A., Myers, P.C., Emerson, J.P. *et al.* : Candidate solar-type protostars in nearby molecular cloud cores. ApJ **307** (1986) 337-349

Benson, P. & Myers, P.C.: Dense cores in dark clouds. II. NH_3 observations and star formation. ApJ **266** (1983) 309-320

Benson, P. & Myers, P.C.: A survey for dense cores in dark clouds. ApJ Suppl. **71** (1989) 89-108

Bertoldi, F., McKee, C.F.: Pressure-confined clumps in magnetized molecular clouds. ApJ **395** (1992) 140-157

Bertoldi, F., McKee, C.F., (1993) in preparation

Blaauw, A.: OB associations and the fossil record of star formation. In C.J. Lada and N.D. Kylafis (eds.), The Physics of Star Formation and Early Stellar Evolution, Kluwer Academic Publishers, Dordrecht (1991) 125-154

Blake, G.A.: Chemistry in dense molecular clouds: theory and observational constraints. In R.L. Dickman, R.L. Snell and J.S. Young (eds.), Molecular clouds in the Milky Way and in external galaxies, Springer-Verlag, (1988) 132-150

Blitz, L.: Star forming giant molecular clouds. In C.J. Lada and N.D. Kylafis (eds.), The Physics of Star Formation and Early Stellar Evolution, Kluwer Academic Publishers, Dordrecht (1991) pp.3-33

Blitz, L., Shu, F.H.: The origin and lifetimes of giant molecular complexes. ApJ **238** (1980) 148-157

Bonnazzola, S., Falgarone, E., Heyvaerts, J., Puget, J.L.: Jeans collapse in a turbulent medium. A&A **172** (1987) 293-298

Cabrit, S., André, P.: An observational connection between circumstellar disk mass and molecular outflows. ApJ **379** (1991) L25-L28

Cabrit, S. Bertout, C.: CO line formation in bipolar flows. I. Accelerated outflows. ApJ **307** (1986) 313-323

Cabrit, S., Bertout, C.: CO line formation in bipolar flows. II. Decelerated outflow case and summary of results. ApJ **348** (1990) 530-541

Cabrit, S., Bertout, C.: CO line formation in bipolar flows. III. The energetics of molecular flows and ionized winds. A&A **261** (1992) 274-284

Cabrit, S., Goldsmith, P.F., Snell, R.L.: Identification of RNO 43 and B335 as two highly collimated bipolar flows oriented nearly in the plane of the sky. ApJ **334** (1988) 196-208

Carlberg, R.G., Pudritz, R.E.: Magnetic support and fragmentation of molecular clouds. MNRAS **247** (1990) 353-366

Cernicharo, J.:The physical conditions of low mass star forming regions. In C.J. Lada and N.D. Kylafis (eds.), The Physics of Star Formation and Early Stellar Evolution, Kluwer Academic Publishers, Dordrecht (1991) 287-328

Cernicharo, J., Bachiller, R., Duvert, G.: The Taurus-Auriga-Perseus complex of dark clouds. I. Density structure. A&A **149** (1985) 273-282

Chernin, L.M., Masson, C.R.: A nearly unipolar CO outflow from the HH46-47 system. ApJ **382** (1991) L93-L96

Chièze, J.P.: The fragmentation of molecular clouds. I. The mass-radius-velocity dispersion relations. A&A **171** (1987) 225-232

Combes, F.: Distribution of CO in the Milky Way. ARA&A **29** (1991) 195-237

Crutcher, R.M., Kazès, I., Troland, T.H.: Magnetic field strengths in molecular clouds. A&A **181** (1987) 119-126

Curiel, S., Raymond, J.C., Rodríguez, L.F., Cantó, J., Moran, M.: The exciting source of the bipolar outflow in L1448. ApJ **365** (1990) L85-L88

Dame, T.M., Ungerechts, H., Cohen, R.S., *et al.* : A composite CO survey of the entire Milky-Way. ApJ **322** (1987) 706-720

Dickman, R.L.: Turbulence in molecular clouds. In D.C. Black and M.S. Matthews (eds.), Protostars & Planets II, The University of Arizona Press, Tucson (1985) 150-174

Dutrey, A., Duvert, G., Castets, A. *et al.* : A multi-transition study of carbon monoxide in the Orion A molecular cloud. II. $C^{18}O$. A&A **270** (1993) 468-476

Duvert, G., Cernicharo, J., Baudry, A.: A molecular survey of three dark clouds in Taurus. A&A **164** (1986) 349-357

Dyson, J.E.: The interpretation of flows in molecular clouds. Ap. Space Science **106** (1984) 181-197

Edwards, S., Ray, T., Mundt, R.: Energetic mass outflows from young stars. In E.H. Levy and J. Lunine (eds.), Protostars & Planets III, University of Arizona Press, Space Science Series, (1993) 567-602

Elmegreen, B.G.: Energy dissipation in clumpy magnetic clouds. ApJ **299** (1985) 196-210

Elmegreen, B.G. : A pressure and metallicity dependence for molecular cloud correlations and the calibration of mass. ApJ **338** (1989) 178-196

Elmegreen, B.G.: The origin and evolution of giant molecular clouds. In C.J. Lada and N.D. Kylafis (eds.), The Physics of Star Formation and Early Stellar Evolution, Kluwer Academic Publishers, Dordrecht (1991) 35-59

Evans, N.J., Lada, E.A.: Star formation in three nearby cloud complexes. In E. Falgarone, F. Boulanger, and G. Duvert (eds.), Fragmentation of Molecular Clouds and Star Formation, Kluwer Academic Publishers, Dordrecht (1991) 293-315

Falgarone, E.: Interstellar Medium. In Th. Encrenaz and M. Kessler (eds.), Infrared astronomy with ISO, (1992), Nova Science Publishers, New York

Falgarone, E., Phillips, T.G.: A signature of the intermittency of interstellar turbulence: The wings of molecular line profiles. ApJ **359** (1990) 344-354

Falgarone, E., Phillips, T.G., Walker, C.K.: The edge of molecular clouds: Fractal boundaries and density structure. ApJ **378** (1991) 186-201

Falgarone, E., Puget, J.L.: Model of clumped molecular clouds. II. Physics and evolution of the hierarchical structure. A&A **162** (1986) 235-247

Fukui, Y., Sugitani, K., Takaba, H. *et al.* : Discovery of seven bipolar outflows by an unbiased survey. ApJ **311** (1986) L85-L88

Fukui, Y.: Molecular outflows: their implications on protostellar evolution. In B. Reipurth (ed.), Low mass star formation and pre-main sequence objects, ESO-Garching, (1989) 95-117

Fukui, Y., Mizuno, A.: A comparative study of star formation efficiencies in nearby molecular cloud complexes. In E. Falgarone, F. Boulanger, and G. Duvert (eds.), Fragmentation of Molecular Clouds and Star Formation, Kluwer Academic Publishers, Dordrecht (1991) 275-288

Fukui, Y., Iwata, T., Mizuno, A., Bally, J., Lane, A.P.: Molecular Outflows. In E.H. Levy and J. Lunine (eds.), Protostars & Planets III, University of Arizona Press, Space Science Series, (1993) 603-639

Garden, R.P., Russell, A.P.G., Burton, N.G.: Images of shock-excited molecular hydrogen in young stellar outflows. ApJ **354** (1990) 232-241

Genzel, R.: Physical conditions and heating/cooling processes in high mass star formation regions. In C.J. Lada and N.D. Kylafis (eds.), The Physics of Star Formation and Early Stellar Evolution, Kluwer Academic Publishers, Dordrecht (1991) , 155-219

Gill, A.G., Henriksen, R.N.: A first use of wavelet analysis for molecular clouds. ApJ **365** (1990) L27-L30

Giovanardi C., Lizano, S. Natta, A., Evans, N.J., Heiles, C.: Neutral winds from protostars. ApJ **397** (1992) 214-224

Goldsmith, P.F.: Temperatures and densities in interstellar molecular clouds. In R.L. Dickman, R.L. Snell, and J.S. Young (eds.), Molecular Clouds in the Milky Way, Springer-Verlag (1988) 1-25

Gomez, M., Hartmann, S., Kenyon, S.J., Hewett, R.: On the spatial distribution of pre-main sequence stars in Taurus. AJ **105** (1993) 1927-1937

Goodman A.A., Bastien, P., Myers, P.C., Ménard, F.: Optical polarization maps of star forming regions in Perseus, Taurus, and Ophiuchus. ApJ **359** (1990) 363-377

Guilloteau, S., Bachiller, R., Fuente, A., Lucas, R.: First observations of young bipolar outflows with the IRAM interferometer: 2" resolution SiO images of the molecular jet in L1448. A&A **256** (1992) L49-L52

Harju, J., Walmsley, C.M., Wouterloot, J.G.A.: Ammonia clumps in the Orion and Cepheus clouds. A&A Suppl. Ser. **98** (1993) 51-76

Hartigan, P., Kenyon, S.J., Hartmann, L. *et al.* : Optical excess emission in T Tauri stars. ApJ **382** (1992) 617-635

Hartmann, L.: Episodic phenomena in early stellar evolution. In C.J. Lada and N.D. Kylafis (eds.), The Physics of Star Formation and Early Stellar Evolution, Kluwer Academic Publishers, Dordrecht (1991) 623-647

Heiles, C.: L204: A gravitationally confined dark cloud in a strong magnetic environment. ApJ **324** (1988) 321-330

Heiles, C., Goodman, A.A., McKee, C.F., Zweibel, E.G.: Magnetic fields in dense regions. In E. Falgarone, F. Boulanger, and G. Duvert (eds.), Fragmentation of Molecular Clouds and Star Formation, Kluwer Academic Publishers, Dordrecht (1991) 43-60

Henriksen, R.N., Turner, B.E.: Star cloud turbulence. ApJ **287** (1984) 200-207

Henriksen, R.N.: On molecular cloud scaling laws and star formation. ApJ **377** (1991) 500-509

Irvine, W.M., Goldsmith, P., Hjalmarson, Å: Chemical abundances in molecular clouds. In D.J. Hollenbach and H.A. Thronson (eds.), Interstellar Processes,

Reidel, Dordrecht (1987) 561-609

Kazès, I., Crutcher, R.M.: Measurement of magnetic-field strength in molecular clouds: detection of OH-line Zeeman splitting. A&A **164** (1986) 328-336

Koo, B.C.: Extremely High-velocity molecular flows in young stellar objects. ApJ **337** (1989) 318-331

Lada, C.J.: Cold outflows, energetic winds, and enigmatic jets around young stellar objects. ARA&A **23** (1985) 267-317

Lada, C.J.: Infrared energy distributions and the nature of young stellar objects. In A.K. Dupree and M.T.V.T. Lago (eds.), Formation and Evolution of Low Mass Stars, Kluwer Academic Publishers, Dordrecht, (1988) 93-109

Lada, C.J.: The formation of low mass stars: Observations. In C.J. Lada and N.D. Kylafis (eds.), The Physics of Star Formation and Early Stellar Evolution, Kluwer Academic Publishers, Dordrecht (1991) 329-363

Lada, C.J., Adams, F.C.: Interpreting infrared color-color diagrams: circumstellar disks around low- and intermediate-mass young stellar objects. ApJ **393** (1992) 278-288

Lada, C.J., Young, E.T., Greene, T.P.: Infrared images of the young cluster NGC 2264. ApJ **408** (1993) 471-483

Lada, E.A.: Global star formation in the L1630 molecular cloud. ApJ **393** (1992) L25-L28

Lada, E.A., Bally, J., Stark, A.A.: An unbiased survey for dense cores in the L1630 molecular cloud. ApJ **368** (1991a) 432-444

Lada, E.A., DePoy, D.L., Evans, N.J., Gatley, I.: A 2.2 micron survey in the L1630 molecular cloud. ApJ **371** (1991b) 171-182

Larson, R.B. MNRAS **194** (1981) 809

Levreault, R.M.: A search for molecular outflows around pre-main-sequence objects. ApJ Suppl **67** 283-371

Liseau, R., Sandell, G.: The geometry of anisotropic CO outflows. ApJ **304** (1986) 459-465

Lizano, S., Shu, F.H.: Molecular cloud cores and bimodal star formation. ApJ **342** (1989) 834-854

Lizano, S., Heiles, C., Rodríguez, L.F. *et al.* : Neutral stellar winds that drive outflows in low-mass protostars. ApJ **328** (1988) 763-776

Loren, R.B.: The cobwebs of Ophiuchus. II. ^{13}CO filament kinematics. ApJ **338** (1989) 925-944

Margulis, M.S., Lada, C.J., Snell, R.L.: Molecular outflows in the Monoceros OB1 molecular cloud. ApJ **333** (1988) 316-331

Masson, C.R., Chernin, L.M.: Properties of swept-up molecular outflows. ApJ **387** (1992) L47-L50

Masson, C.R., Chernin, L.M.: Properties of jet-driven molecular outflows. ApJ **414** (1993) 230-241

Masson, C.R., Mundy, L.G., Keene, J.: The extremely high-velocity CO flow in HH 7-11. ApJ **357** (1990) L25-L28

Mathieu, R.D., Benson, P.J., Fuller, G.A., Myers, P.C., Schild, R.E.: L43: An example of interaction between molecular outflows and dense cores. ApJ **330** (1988) 385-398

McKee, C.F.: Photoionization-regulated star formation. ApJ **345** (1989) 782-801

Mestel, L., Spitzer, L., Jr. MNRAS **116** (1956) 503

Meyers-Rice, B.A., Lada, C.J.: The structure and kinematics of bipolar outflows: Observations and models of the Monoceros R2 outflow. ApJ **368** (1991) 445-462 (ML91)

Mezger, P.G., Sievers, A., Zylka, R. *et al.* : Dust emission from star-forming regions. III. The Rho Ophiuchus cloud - where are the low-mass protostars? A&A **265** (1992) 743-751

Mitchell, G.F., Curry, C., Maillard, J.P., Allen, M.: The gas environment of the young stellar object GL 2591 studied by infrared spectroscopy. ApJ **341** (1989) 1020-1034

Mizuno, A., Fukui, Y., Iwata, T., Nozawa, S., Takano, T.: A remarkable multilobe molecular outflow: ρ Ophiuchi East, associated with IRAS 16293-2422. ApJ **356** (1990) 184-194

Morgan, J.A., Bally, J.: Molecular outflows in the L1641 region of Orion. ApJ **372** (1991) 505-517

Morgan, J., Schloerb, F.P., Snell, R.L., Bally, J.: Molecular outflows associated with young stellar objects in the L1641 region of Orion. ApJ **376** (1991) 618-629

Moriarty-Schieven, G.H., Snell, R.L.: High-resolution observations of the L1551 molecular outflow. II. Structure and kinematics. ApJ **332** (1988) 364-378

Mouschovias, T.Ch.: 1976 Non-homologous contraction and equilibria of self-gravitating, magnetic interstellar clouds embedded in an intercloud medium: Star formation, ApJ **207** 141-158

Mouschovias, T.Ch. : Single-stage fragmentation and a modern theory of star formation. In C.J. Lada and N.D. Kylafis (eds.), The Physics of Star Formation and Early Stellar Evolution, Kluwer Academic Publishers, Dordrecht (1991) 449-468

Mouschovias, T.Ch., Spitzer, L.: Note on the collapse of magnetic interstellar clouds. ApJ **210** (1976) 326-327

Myers, P.C.: Clouds, cores, and stars in the nearest molecular complexes. In E. Falgarone, F. Boulanger, and G. Duvert (eds.), Fragmentation of Molecular Clouds and Star Formation, Kluwer Academic Publishers, Dordrecht (1991) 221-228

Myers, P.C., Goodman, A.A.: Magnetic molecular clouds: Indirect evidence for magnetic support and ambipolar diffusion. ApJ, **329** (1987) 392-405

Myers, P.C., Fuller, G.A., Mathieu, R.D., *et al.* : Near-infrared and optical observations of IRAS sources in and near dense cores. ApJ **319** (1987) 340-357

Myers, P.C., Heyer, M., Snell, R.L., and Goldsmith, P.F.: Dense cores in dark clouds. V. CO outflows. ApJ **324** (1988) 907-919

Myers, P.C., Fuller, G.A., Goodman, A.A., Benson. P.J.: Dense cores in dark clouds. VI. Shapes. ApJ **376** (1991) 561-572

Nakano, T.: Contraction of magnetic interstellar clouds. Fund. Cosmic Phys. **9** (1984) 139-232

Nishi, R, Nakano, T., Umebayashi, T.: Magnetic flux loss from interstellar clouds with various grain-size distributions. ApJ **368** (1991) 181-194

Norman, C., Silk, J.: Clumpy giant molecular clouds: a dynamic model self-consistently regulated by T Tauri formation. ApJ **238** (1980) 158-174

Novak, G., Gonatas, D.P., Hildebrand, R.H., Platt, S.R., Dragovan, M.: Polarization of far-infrared radiation from molecular clouds. ApJ **345** (1989) 802-810

Panagia, N., Felli, M.: The spectrum of the free-free radiation from extended envelopes. A&A **39** (1985) 1-5

Parker, N.D., Padman, R., Scott, P.F.: Outflows in dark clouds: their role in protostellar evolution. MNRAS **252** (1991) 442-461

Pelletier, G., Pudritz, R.E.: Hydromagnetic disk winds in young stellar objects and active galactic nuclei. ApJ **394** (1992) 117-138

Pérault, M., Falgarone, E., Puget, J.L.: ^{13}CO observations of cool giant molecular clouds. A&A **152** (1985) 371-386

Pouquet, A., Passot, T., Léorat, J.: Numerical simulations of turbulent compressible flows. In E. Falgarone, F. Boulanger, and G. Duvert (eds.), Fragmentation of Molecular Clouds and Star Formation, Kluwer Academic Publishers, Dordrecht (1991) 101-118

Pudritz, R.E.: The origin of bipolar outflows. In R.E. Pudritz and M. Fich (eds.), Galactic and extragalactic star formation, Kluwer Academic Publishers, Dordrecht, (1986) 135-158

Pudritz, R.E.: The stability of molecular clouds. ApJ **350** (1990) 195-208

Puget, J.L.: Magnetic fields and the dynamics of molecular clouds. In E. Falgarone, F. Boulanger, and G. Duvert (eds.), Fragmentation of Molecular Clouds and Star Formation, Kluwer Academic Publishers, Dordrecht (1991) 75-81

Raga, A., Cabrit, S.: Molecular outflows entrained by jet bowshocks. A&A **278** (1993) 267-278

Raga, A., Cantó, J., Calvet, N., Rodríguez, L.F., Torrelles, J.M.: A unified stellar jet/molecular outflow model. ApJ (1993) in press

Ray, T.P., Mundt, R.: Interpreting jets from young stellar objects. In D. Burgarella, M. Livio, C.P. O Dea (eds.), Astrophysical Jets, Space Telescope Science Institute Symposium Series, (1993), Cambridge University Press, 145-176

Reipurth, B.: Nature **340** (1989) 42

Reipurth, B., Pettersson, B.: Star formation in Bok globules and low-mass clouds. V. Hα emission stars near Sa101, CG13, and CG22. A&A **267** (1993) 439-446

Reipurth, B., Chini, R., Krügel, E., Kreysa, E., Sievers, A.: Cold dust around Herbig-Haro energy sources: A 1300 μm survey. A&A **273** (1993) 221-238

Richer, J.S., Hills, R.E., Padman, R.: A fast CO jet in Orion B. MNRAS **254** (1992) 525-538

Rodríguez, L.F., Carral, P., Ho, P.T.P., Moran, J.M.: Anisotropic mass outflow in regions of star formation. ApJ **260** (1982) 635-646

Rodríguez, L.F., Ho, P.T.P., Torrelles, J.M., Curiel, S., Cantó, J.: VLA observations of the Herbig-Haro 1-2 system. ApJ **352** (1990) 645-653

Scalo, J.M.: The fragmentation and hierarchical structure of the interstellar medium. In D.C. Black and M.S. Matthews (eds.), Protostars & Planets II, The University of Arizona Press, Tucson (1985) 201-296

Scalo, J.M.: The stellar initial mass function. Fund. Cosmic. Phys. **11** (1986) 1-278

Shu, F.H., Adams, F.C., Lizano, S.: Star formation in molecular clouds: Observation and theory. ARA&A **25** (1987) 23-81

Shu, F.H., Ruden, S.P., Lada, C.J., Lizano, S.: Star formation and the nature of bipolar flows. ApJ **370** (1991) L31-L34

Snell, R.L., Scoville, N.Z., Sanders, D.B., Erickson, N.R.: High-velocity molecular jets. ApJ **284** (1984) 176-193

Snell, R.L. Bally, J.: Compact radio sources associated with molecular outflows. ApJ **303** (1986) 683-701

Snell, R.L.: Bipolar outflows and stellar jets. In M. Peimbert and J. Jugaku (eds.), Star forming regions, Kluwer Academic Publishers, Dordrecht (1987) 213-237

Spitzer, L.: Physical Processes in the Interstellar Medium, Wiley, New York (1978)

Stahler, S.: The relation between optical jets and molecular outflows from young star. In D. Burgarella, M. Livio, C.P. O Dea (eds.), Astrophysical Jets, Space Telescope Science Institute Symposium Series, (1993), Cambridge University Press, 183-210

Strom, K.M., Strom, S.E., Merrill, K.M.: Infrared luminosity functions for the young stellar population associated with the L1641 molecular cloud. ApJ **412** (1993) 233-253

Strom, K.M., Margulis, M., Strom, S.E.: A study of the stellar population in the Lynds 1641 dark cloud: deep near infrared imaging. ApJ **345** (1989) L79-L82

Stutzki, J., Genzel, R., Graf, U. *et al.* : UV penetrated clumpy molecular cloud cores. In E. Falgarone, F. Boulanger, and G. Duvert (eds.), Fragmentation of Molecular Clouds and Star Formation, Kluwer Academic Publishers, Dordrecht (1991) 235-244

Terebey, S.: Molecular Hydrogen emission in L1448 associated with a highly collimated molecular outflow. Mem. S. A. It. **62** (1991) 823-828

Terebey, S., Vogel, S.N., Myers, P.C.: Probing the circumstellar environments of very young low mass stars using water masers. ApJ **390** (1992) 181-190

Tielens, A.G.G.M., Hollenbach, D.: Photodissociation regions. I. Basic model. ApJ **291** (1985) 722-746

Tomisaka, K., Ikeuchi, S., Nakamura, T.: Equilibria and evolution of magnetic, rotating, isothermal clouds. II. The extreme case: non-rotating clouds. ApJ **335** (1988) 239-262

Torrelles, J.M., Verdes-Montenegro, L., Ho, P.T.P., Rodríguez, L.F., Cantó, J.: From bipolar to quadrupolar: The collimation process of the Cepheus A outflow. ApJ **410** (1993) 202-217

Verschueren, W., David, M.: The effect of gas removal on the dynamical evolution of young stellar clusters. A&A **219** (1989) 105-120

Wilking, B.A., Lada, C.J.: The discovery of new embedded sources in the centrally condensed core of the rho Ophiuchi dark cloud: the formation of a bound cluster? ApJ **274** (1983) 698-716

Wilking, B.A., Lada, C.J., Young, E.T.: IRAS observations of the ρ Ophiuchi infrared cluster: spectral energy distribution and luminosity function. ApJ **340** (1989) 823-852

Wootten, A.: Dense core structure and fragmentation in the Rho Ophiuchi molecular cloud. In E. Falgarone, F. Boulanger, and G. Duvert (eds.), Fragmentation of Molecular Clouds and Star Formation, Kluwer Academic Publishers, Dordrecht (1991) 229-233

Yun, J.L., Clemens, D.P.: Star formation in small globules: Bart Bok was correct! ApJ **365** (1990) L73-L76

An Introduction to T Tauri Stars

Claude Bertout

Laboratoire d'Astrophysique
Observatoire de Grenoble
BP 53 X
F - 38041 Grenoble Cedex
France

Abstract

These lectures are meant as an introduction to some basic physical processes taking place in T Tauri stars. The first section is devoted to a brief description of the various optically visible objects encountered in molecular clouds, and also gives a short overview of early studies of young stellar objects. The second section shows how the spectroscopic criteria defining the class of T Tauri stars can be translated into relevant physical properties. The third section presents our current understanding of these stars, with emphasis on the role of their circumstellar disks. Because current knowledge of outflows is presented elsewhere in this book, no attempt is made here to discuss T Tauri winds in detail.

1 The Visible Stellar Content of Dark Clouds

1.1 Orion Population

The term Orion population was introduced by George Herbig in his classical 1962 paper to designate "*the mass of emission-Hα stars and irregular variables that one encounters in rich nebulous clusters below approximately M_{pg}=+4 (...) by analogy with the Orion Nebula*".

At 800 pc, the approximate distance of the Orion Nebula, the absolute photographic magnitude M_{pg} quoted by Herbig corresponds to m_{pg}=13.5; the Orion population is thus the faint stellar population of interstellar clouds, and comprises several classes of visible stellar objects: T Tauri stars, FU Orionis objects, flare stars, and Herbig stars.

1.2 T Tauri Stars

"*Eleven irregular variable stars have been observed whose physical characteristics seem much alike and yet are sufficiently different from other known classes of variables to warrant the recognition of a new type of variable stars whose prototype is T Tauri.*" Alfred H. Joy (1945)

Table 1. The original T Tauri stars

Star	Spectral type[1]	m_v variability[2]	L_{bol}[3]	$W_\lambda(H\alpha)$[4]
RW Aur A	K1:	9.6 - 13.6	4.3	84
UY Aur	K7 V	11.6 - 14.3	4.3	60:
R CrA	A5:e	9.9 - 14.4	79.4	-
S CrA	K6:	10.5 - 13.2	5.8	90
RU Lup	K	9.6 - 13.4	11.0	216
R Mon	-	11.0 - 13.8	1480	85:
T Tau	K0 IV,V	9.3 - 13.5	17.0	60:
RY Tau	K1 IV,V	9.3 - 13.0	13.0	-
UX Tau A	K2 V	10.6 - 13.7	1.9	4
UZ Tau e	M1,3 V:	11.7 - 15.0	1.7	82
XZ Tau	M3	10.4- 16.6	10.7	274

The youth of T Tauri stars (TTS) was recognized first by Ambartsumian (1947), mainly because of their association with short-lived, massive OB stars in interstellar dark clouds, but it took several years before this idea gained recognition in the US and in Europe. Joy cited the following selection criteria for T Tauri stars:

1. *irregular light variations of about three magnitudes,*
2. *spectral type F5-G5 with emission lines resembling the solar chromosphere,*
3. *low luminosity, and*
4. *association with dark or bright nebulosity.*

He added "*These stars differ from other known variables, especially in their low luminosity and the high intensity of bright H and K in their spectra*".

Classical selection criteria for T Tauri stars Selection criteria were refined by Herbig (1962) who noticed specific spectroscopic properties common to the T Tauri class.

1. *"The hydrogen lines and the H and K lines of CaII are in emission,*
2. *the fluorescent FeI emission lines λλ4068, 4132 are present (they have been found only in T Tauri stars),*
3. *the [SII] emission lines λλ4068, 4076 are usually, but not always, present. Probably [SII]λλ6717, 6731 and [OI] λλ6300, 6363 are also characteristic,*
4. *(...) the presence of strong LiI λ6707 absorption, in those stars in which an absorption spectrum can be seen at all, may constitute another primary criterion. The emission lines are usually superimposed upon a continuous spectrum which may range from a pure continuum through one with only vague depressions at the positions of the strongest late-type features, to an approximately normal absorption spectrum of late-type F, G, K or M."*

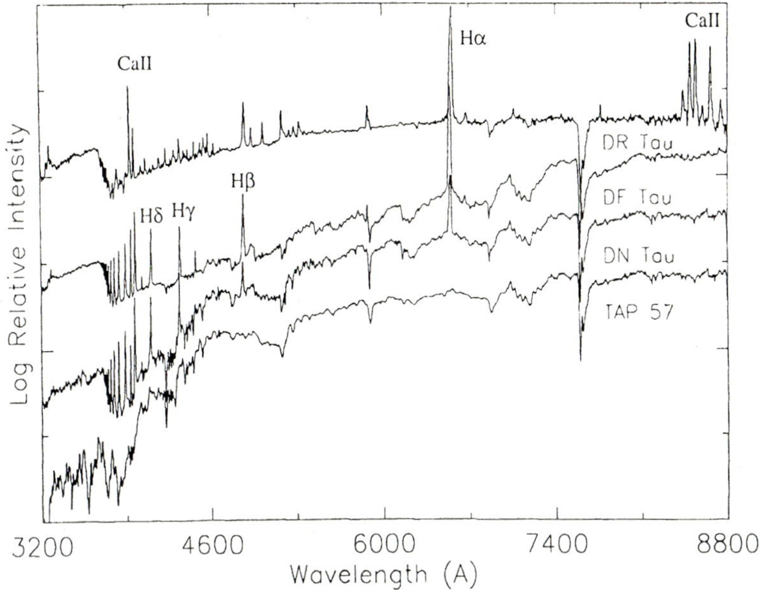

Fig. 1. Medium resolution spectrograms covering the spectral range 3200-8800Å of 4 late-K or early-M TTS, shown in order of increasing emission levels (kindly communicated by G. Basri.)

Classical vs. weak-emission line T Tauri stars Stars which obey Herbig's criteria are now called Classical T Tauri Stars (CTTS). They were discovered mainly during Hα surveys. In recent years, another population of visible solar-type young stars that had escaped detection so far was discovered because of their strong X-ray emission (Walter 1986, Feigelson *et al.* 1987, Walter *et al.* 1988). Because they display little activity in the optical range, and notably no line emission except for weak Hα emission, these objects are named weak-emission line T Tauri Stars (WTTS). According to current estimates, there is about as many WTTS as CTTS. Fig. 1 illustrates the various subgroups of TTS by displaying the 3200 to 8800Å spectral region of four stars. TAP 57 (045251 +3016) is a WTTS similar in many respects to a standard K7 dwarf and DN Tau is a moderate M0 CTTS with Balmer line and continuum emission. DF Tau is a *veiled* M0 CTTS, which means that continuum radiation is superimposed on its photospheric line spectrum, leading to anomalous line strengths. Finally, DR Tau is an extreme CTTS with probable K5 spectral type which displays strong CaII line emission both in the blue and in the infrared in addition to Balmer, FeII, TiII, and HeI line emission.

Photometric classification: nebular variables The class of T Tauri stars, as defined by Joy and Herbig, brings together nebular variables belonging to three categories: RW Aurigae variables, T Tauri variables, and T Ori variables (cf. Glasby 1974). Typical visual light curves of these variable stars are shown in Figs. 2 to 4.

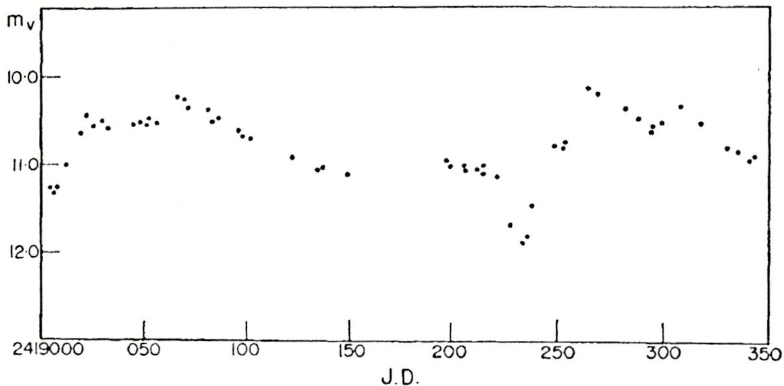

Fig. 2. RW Aurigae variables show rapid and totally irregular light variations.

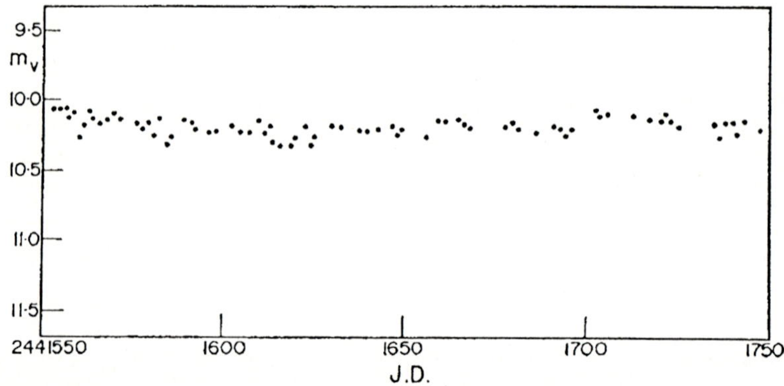

Fig. 3. T Tauri variables display slow, small amplitude sinusoidal variations with a general tendency toward stronger variations when the star is bright.

In his 1962 paper, Herbig notes "*Despite their ubiquity, the light variations of the T Tauri stars have shown no clearly distinctive group property, except that an occasional display of rapid activity has been reported for most well-studied objects*". In contrast, spectroscopic properties on which Herbig based his definition can be translated into physical properties specific to the class, as will be demonstrate in §2.

1.3 FU Orionis Stars

The three best known members of this class are: FU Ori, V 1057 Cyg, and V 1515 Cyg. FUOrs are characterized by a large increase in brightness over

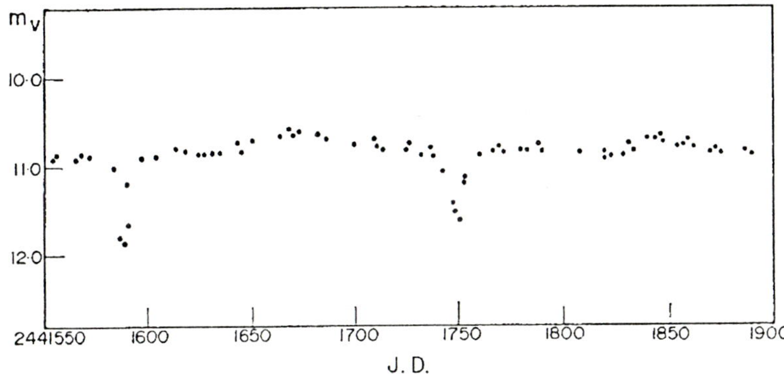

Fig. 4. T Orionis variables show small fluctuations around maximum light with sudden, irregular fading by several magnitudes.

several hundreds days. The light curves of these three objects are displayed in Fig. 5 together with the light curves of some extremely active CTTS. Comparison with the previous figures is instructive in showing the large range of variability of T Tauri stars, from small quasi-sinusoidal variations on a timescale of a few days to large scale variations on timescales of months to years.

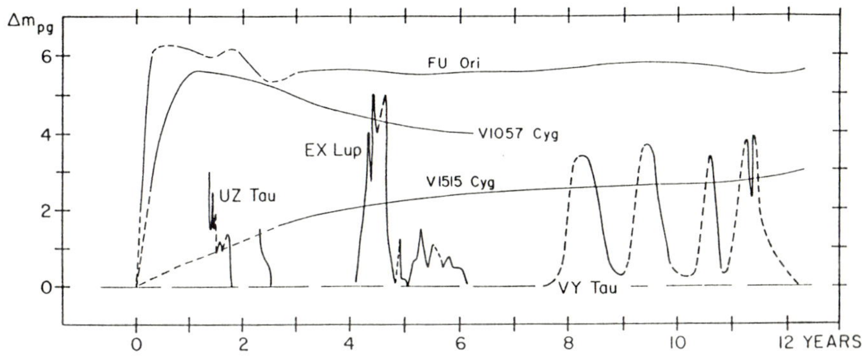

Fig. 5. Compared photographic light curves of FU Orionis objects and extreme T Tauri stars UZ Tau, EX Lup, and VY Tau (from Herbig 1977).

FUOr progenitors are believed to be T Tauri stars. This assumption relies on only one pre-outburst spectrogram of V1057 Cygni showing a spectrum characteristic of strong emission-line T Tauri stars. This historical spectrogram is reproduced in Fig. 6 together with post-outburst spectrograms which are of type F-G.

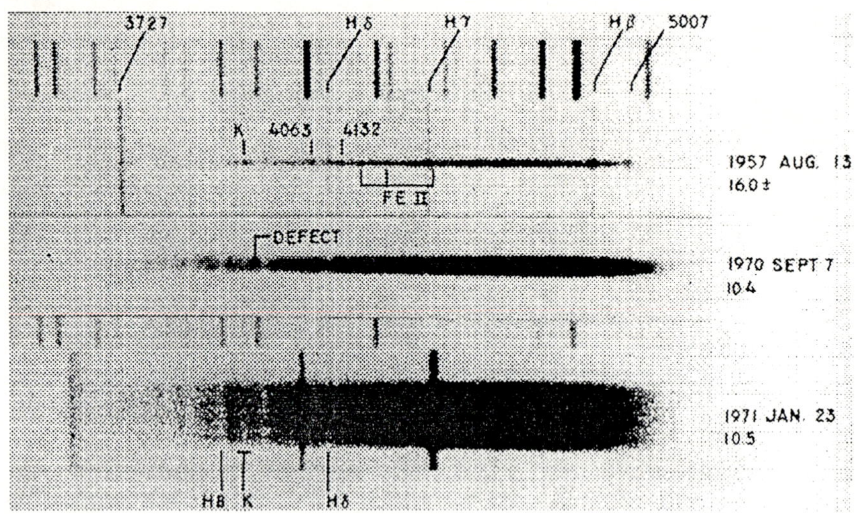

Fig. 6. Pre- and post-outburst spectrograms of V1057 Cyg showing the T Tauri nature of the star before outburst (from Herbig 1977).

1.4 Flare Stars

UV Ceti stars are Me dwarfs with masses roughly in the range 0.1 - $0.2M_\odot$, found in the vicinity of the Sun. They display sudden photometric variations called flares in analogy with solar flares. But while solar flares are localized events, flares on dMe stars are much more energetic. Haro (1968) discovered many flare stars in the Orion Nebula, which he called *flash* stars to distinguish them from the UV Ceti type stars. Criteria for classifying nebular variables as flare stars are:

1. *relative steadiness of the starlight during minima or "normal phase";*
2. *a sudden, unpredictable rise to maximum and a slower, but still rapid, decline toward minimum.*

Flare stars show little emission in their quiescent phases, but their line emission spectrum during flares resembles the more permanent emission-line spectrum of T Tauri stars. However, while large-scale velocity gradients are responsible for the widening of emission lines in T Tauri stars, pressure broadening can explain the emission line widths of flare stars. The observational boundaries between T Tauri stars and flare stars are fuzzy, and Haro estimated that 30% of

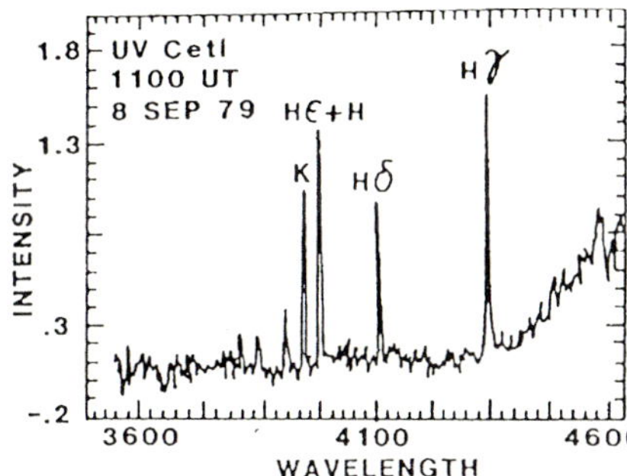

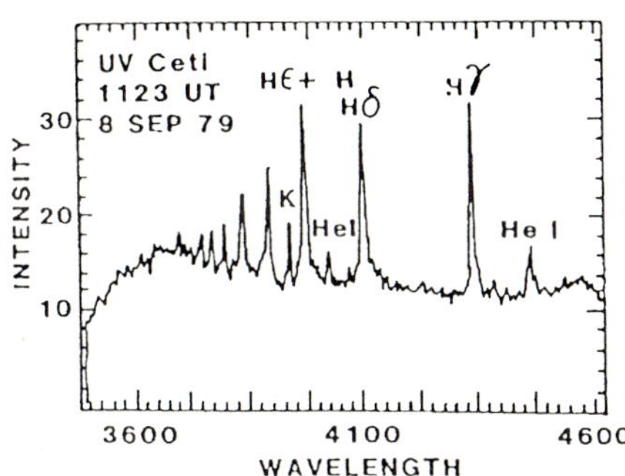

Fig. 7. The spectrum of UV Ceti at quiescence (left panel) and during a flare (right panel). Note the similarities with TTS spectra (Fig. 1)

the Orion flare stars were also T Tauri stars. Well-known T Tauri stars that were classified as flare stars on the basis on one flare only include YY Ori, NS Ori and SU Ori, DF Tau, DS Tau and DN Tau. It was proposed that T Tauri stars are permanently flaring flare-stars, or that flash stars are post T Tauri stars. None of these interpretations now appears likely (cf. §3).

1.5 Herbig Ae and Be Stars

This class was introduced by G. Herbig (1960), and is defined in the following way.

1. *The spectral type is A or earlier, with emission lines,*
2. *the star lies in an obscured region,*
3. *the star illuminates fairly bright nebulosity in its immediate vicinity.*

Herbig suspected these stars to continue the T Tauri group toward higher masses, and this was largely confirmed by later work, although some peculiar objects originally put in this class on the basis of the above criteria turned out later to be probable FU Orionis stars. This is notably the case for Z CMa, which will be discussed in some detail in §3. Extensive spectroscopic studies of Herbig Ae/Be objects were performed by Finkenzeller and Jankovics (1984), Finkenzeller and Mundt (1984) and Hamann and Persson (1992). Fig. 8 displays the blue spectrum of a moderately active star of the class, HD 250550.

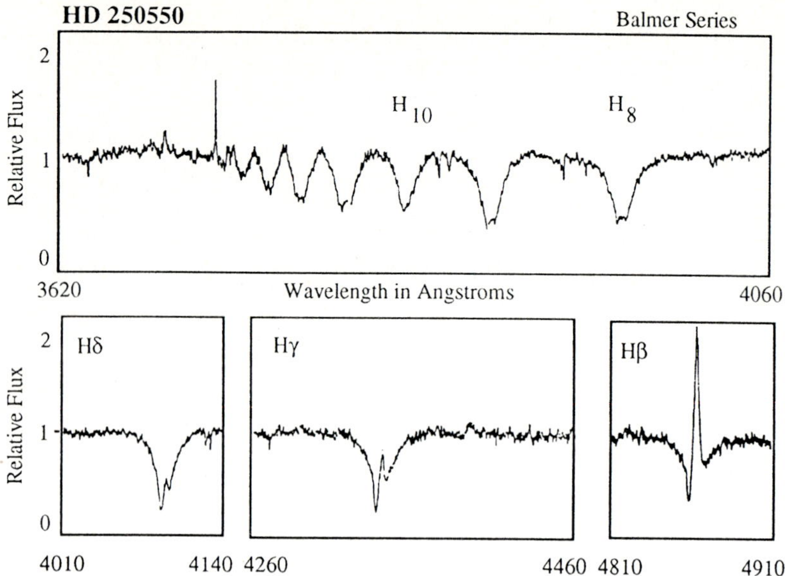

Fig. 8. The Balmer series of the modest emission line Herbig star HD 250550 (Finkenzeller and Jankovics 1984)

2 Interpretation Of Herbig's Spectroscopic Criteria For TTS

The interpretation of spectroscopic data is based on radiative transfer concepts and tools of various complexity levels. We now derive fundamental properties of T Tauri stars from Herbig's spectroscopic criteria without complicated modeling.

2.1 Formation of Emission Lines: a Simplified Approach

Emergent intensity We consider the equation of radiation transfer at frequency ν and in direction s

$$\frac{dI_\nu(s)}{ds} = \eta_\nu - \chi_\nu I_\nu(s) \tag{1}$$

where $I_\nu(s)$ is the radiation specific intensity in ergs cm^{-2} sec^{-1} Hz^{-1} sr^{-1}; η_ν is the emissivity of the gas in ergs cm^{-3} sec^{-1} Hz^{-1} sr^{-1}; χ_ν is the absorption coefficient in cm^2 per cm^3 = cm^{-1}. We define the source function by $S_\nu = \eta_\nu/\chi_\nu$ and the optical depth τ_ν by $d\tau_\nu = -\chi_\nu ds$. Integrating Eq. 1, we then get

$$I_\nu(\infty) = \int_0^\infty S_\nu(\tau_\nu)e^{-\tau_\nu} d\tau_\nu \tag{2}$$

where $I_\nu(\infty)$ is the emergent intensity at $s = \infty$, i.e., at $\tau_\nu = 0$.

Radiative flux in direction s at frequency ν The flux is found by integrating the emergent intensity over the emitting region. When the star is surrounded by an extended, spherically symmetric envelope, it is convenient to introduce the (p, z) coordinates, which are related to the usual coordinates r and θ by

$$r^2 = p^2 + z^2 \quad \text{and} \quad z = r\cos\theta \tag{3}$$

Lines of sight, defined by p = const., are often called impact parameters in analogy with atomic physics. Assuming that the observer is located at $z = \infty$, the flux emitted in his direction (in ergs cm^{-2} sec^{-1} Hz^{-1}) is

$$F_\nu = \int_0^{R_{max}} I_\nu(\infty) 2\pi p dp \tag{4}$$

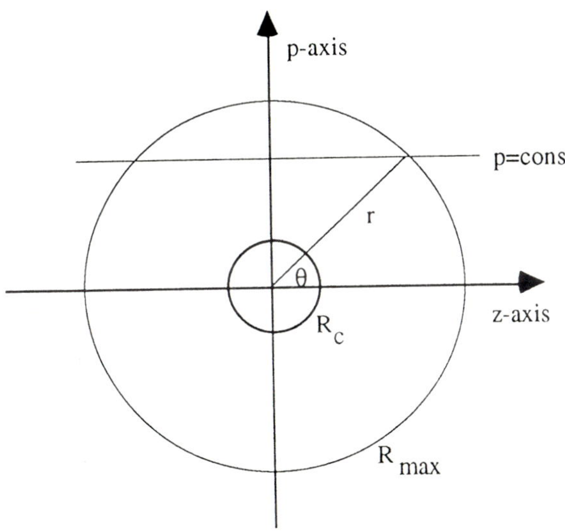

Fig. 9. Geometry of the emission region.

2.2 Application to T Tauri Stars: Hα and CaII H & K Line Emission

In TTS, the emission line spectrum is superimposed over the continuous and absorption spectrum emitted by the star, which we call the background flux. We now make the simplifying assumption that the formation of background radiation at a line frequency is independent of the formation of that line. More precisely, let us assume that the background radiation is emitted by an opaque stellar core and that the line is emitted in a region surrounding the stellar core. Whether a particular line appears in emission will then depend only on the relative fluxes in the line and in the background at the same frequency.

If the stellar surface emits a continuous spectrum I_ν^C [e.g., $I_\nu^C = B_\nu(T_{eff})$] and if we neglect limb darkening effects, we can evaluate the background flux at frequency ν by

$$F_\nu^C = \pi R_C^2 I_\nu^C \tag{5}$$

where $R_C = r(T_{eff})$ is the stellar core radius. The line absorption coefficient at frequency ν is given by $\chi_\nu = \phi_\nu \chi_l$, where χ_l is the line center absorption coefficient and ϕ_ν is the absorption line profile. We also assume for simplicity that both the line source function and absorption coefficient at line center stay constant over the emitting region. Integration of Eq. 2 in direction z yields

$$I_\nu(z = \infty, p) = S_\nu(1 - e^{-\Delta\tau_\nu(p)}) + \epsilon(p) I_\nu^C e^{-\Delta\tau_\nu(p)} \tag{6}$$

where

$$\Delta\tau_\nu(p) = \chi_\nu \Delta z(p) \tag{7}$$

is the optical depth along an impact parameter with extent Δz inside the line emitting medium and where $\epsilon = 1$ if $p < R_C$ and $\epsilon = 0$ otherwise.

Optically thick line emission In the case of the envelope having high optical depth, Eq. 6 reduces to

$$I_\nu = S_\nu \tag{8}$$

Integration of Eq. 4 then gives the line flux:

$$\frac{F_\nu}{F_\nu^C} = \frac{S_\nu}{I_\nu^C} \frac{R_{max}^2}{R_C^2} \tag{9}$$

Line emission will occur whenever $F_\nu / F^C > 1$. Eq. 9 shows that there are two ways for producing line emission, depending on the size of the emitting envelope.

1. **In extended envelopes,** $R_{max} \gg R_C$, and emission occurs whenever

$$\frac{S_\nu}{I_\nu^C} > \frac{R_C^2}{R_{max}^2} \tag{10}$$

2. **In an atmosphere of small extent** $R_{max} \approx R_C$, and the line appears in emission whenever

$$\frac{S_\nu}{I_\nu^C} > 1 \tag{11}$$

i.e., when the line excitation temperature (defined by $S_\nu = B_\nu(T_{exc})$) is larger than the star's effective temperature.

Optically thin line emission The emergent intensity is given by

$$I_\nu(z=\infty,p) = S_\nu \Delta\tau_\nu(p) + \epsilon(p) I_\nu^C [1 - \Delta\tau_\nu(p)] \tag{12}$$

Using the following expressions for Δz in Eq. 7:

$$\Delta z(p < R_C) = \sqrt{R_{max}^2 - p^2} - \sqrt{R_C^2 - p^2}$$
$$\Delta z(p > R_C) = 2\sqrt{R_{max}^2 - p^2} \tag{13}$$

we integrate Eq. 4 to get:

$$F_\nu = \frac{2}{3}\pi\chi_\nu\{I_\nu^C[(R_{max}^2 - R_C^2)^{1.5} - (R_{max}^3 - R_C^3)] + S_\nu[(R_{max}^2 - R_C^2)^{1.5} + (R_{max}^3 - R_C^3)]\} + \pi R_C^2 I_\nu^C \tag{14}$$

We consider the same limiting cases as above:

1. **Extended envelope**: Eq. 14 simplifies to

$$\frac{F_\nu}{F_\nu^C} = 1 + S_\nu \chi_\nu \frac{4}{3}\pi R_{max}^3 \frac{1}{I_\nu^C \pi R_C^2} \tag{15}$$

   That is, the line appears always in emission and the flux is proportional to the envelope volume and inversely proportional to the background flux emitted by the star $\pi R_C^2 I_\nu^C$

2. **Envelope of small extent**: we get

$$\frac{F_\nu}{F_\nu^C} = 1 + \chi_\nu[\frac{S_\nu}{I_\nu^C} - 1]2\pi(R_{max} - R_C)R_C^2 \frac{1}{\pi R_C^2} \tag{16}$$

   Now, the line can appear in absorption provided its excitation temperature is lower than the star's effective temperature, and the flux is proportional to $2\pi R_C^2(R_{max} - R_C)$, i.e., to the volume of the part of the envelope that faces the observer.

In Case (1), emission is often referred to as *geometric*, which means that it is merely a consequence of the large emission volume of the envelope. In the Sun, Case (2) holds e.g. for the chromospheric CaII H and K lines and one often speaks by analogy of *chromospheric* emission whenever the excitation temperature of the line is higher than the radiation temperature of the background continuum.

Hα and CaII line excitation We approximate the Hα and CaII H and K transitions by the bound transition taking place in an atom with two bound energy levels and a continuum. This model is a fair approximation for resonance lines such as H and K, and for subordinate lines when the resonance lines are in radiative detailed balance, a condition that we will assume for the Lyα and Lyβ lines of hydrogen. Processes taking place in this atom model include: (i) photo-excitation and collisional excitation from the lower to the upper state, (ii) photo-de-excitation and collisional de-excitation from the upper to the lower state, (iii) photo-ionization and collisional ionization from the bound states to the continuum, (iv) radiative and collisional recombination from the continuum to the bound states.

Collision-dominated lines: If collisional processes dominate line excitation, then the source function is close to LTE and the source function is linked to the *local electron temperature* in the line forming region (upper panel of Fig. 10).

Photoionization-dominated lines: If photo-ionization dominates line excitation, the source function depends instead on a *non-local radiation temperature* characteristic of the energy distribution in the continua of the bound levels and are, therefore, insensitive to the electron temperature in the line forming region as long as the free-bound opacity is low there (lower panel of Fig. 10).

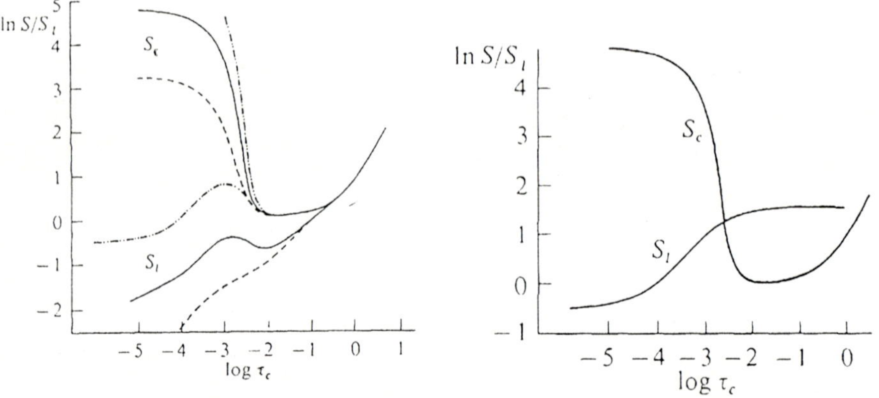

Fig. 10. Left Panel: Collisionally dominated line source for 3 different continuum source functions. Notice the coupling of the line to the continuum source. Right Panel: Photoionization dominated line source for one of the continuum source functions of the left panel.

The solar case As shown by Jefferies and Thomas (1959), the above distinction between two different excitation processes explains why the photo-ionization-dominated solar Hα line has no emission core while the collision-dominated CaII H and K line exhibit such cores, which are formed in the chromospheric temperature rise. *A chromospheric temperature rise is needed in order to produce the qualitatively different source functions of Hα and CaII.* If the electronic temperature was a monotonically increasing function of the optical depth in the atmosphere, the two classes of lines would have similar source functions and no emission core would be seen in the calcium lines.

The TTS Case N_e varies in the range 5×10^{10} to 10^{11} cm^{-3} for optical depths $10^{-8} < \tau_{5000Å} < 10^{-5}$ in the solar chromosphere. It can be shown that CaII H and K are collision-dominated at electron densities higher than about 10^9 cm^{-3} whenever T_e >3500K (Gebbie and Steinitz 1974). Since a typical T Tauri star has a K photospheric spectrum, corresponding to T_{eff} in the range 4000 - 5000K, and since electron densities of T Tauri chromospheres are probably comparable to solar chromospheric densities, it is likely that the source functions of CaII H and K are coupled to the electronic temperature rise and that *chromospheric emission* follows. The analogy with the Sun is not complete since the emission strength of H and K is often much stronger in T Tauri stars than in solar-type stars, which indicates that more extended regions have to be taken into account in addition to the chromosphere. In fact, the two contributions to the line flux (from chromosphere and moving envelope) can sometimes be separated in high resolution spectrograms of CTTS.

A detailed study of Hα excitation (Gebbie and Steinitz 1974) shows that there exists a critical N_e, roughly located in the 10^{11} to 10^{13}cm^{-3} range, above which Hα will be collision-dominated. Without going into detailed modeling of chromospheres, it is clear that Hα is photo-ionization-dominated in T Tauri stars if their chromospheres' densities are moderate. While we have few clues about the density of TTS chromospheres, we know that some magnetically active WTTS display Hα emission that appears to be formed in a chromosphere in a way similar to other active late-type stars (cf. §3). The flux level exhibited by these WTTS probably represents the maximum level that can be attributed to chromospheric emission in TTS. In particular, emission is likely to be *geometric* in CTTS, and the line flux depends on the extent and density of the emitting envelope.

2.3 FeI λ4063 and λ4132 Fluorescent Emission

In some T Tauri stars, two lines of multiplet 43 of FeI, namely λ4063 and λ4132 originating from the same sublevel yF are greatly enhanced with respect to other lines of the same multiplet. Herbig (1945) proposed that this anomalous emission was caused by the near coincidence of a third line originating from the same sublevel, λ3969, with the strong emission blend made up of CaII H at 3968.47Å and hydrogen Hϵ at 3970.07Å.

Willson (1974) showed that this near coincidence leads to a strong overpopulation of sublevel yF with respect to the other sublevels with J=2 and J=4, provided one of the pumping lines is suitably Doppler shifted onto the 3969.26Å FeI line. In a coordinate system moving with the FeI atoms, CaII H must appear red-shifted by about 60 kms^{-1}, or Hϵ must be blue-shifted by about the same amount.

Fluorescent FeI emission implies the presence of a velocity shift between the region of FeI line formation and that of CaII (and/or Hϵ) formation. If the

Balmer and CaII line forming regions are highly turbulent, both lines might contribute to the fluorescence process. In an organized velocity field, however, only a few combinations are possible depending on the type of flow in which the lines are formed. Assuming for now that all lines are formed so close to the star that sphericity can be ignored, we have the following possibilities.

1. In an accelerated outflow, CaII H is the exciting line.
2. In a decelerated outflow, Hϵ is the exciting line.
3. In an accelerated infall, both CaII H and Hϵ can excite FeI depending on the location of FeI emission; FeI atoms located in the infalling envelope are excited by CaII H, while FeI atoms located at the (static) stellar surface are excited by infalling Hϵ atoms.

An example: in the bright T Tauri star V1331 Cyg, the K-line is strong but the CaII H+Hϵ blend appears in absorption. The Balmer decrement is such that we expect little emission, if any, from Hϵ. Lack of emission in CaII H is however anomalous. We can conclude that H photons have been absorbed by overlying FeI atoms. The outflow around V1331 Cyg is thus accelerated, with FeI formed farther out in the envelope than CaII. While V1331Cyg is perhaps one of the few T Tauri stars for which such a clear-cut analysis is possible, we note that there has been no detailed investigation of fluorescent emission in recent times. Since these lines may represent the best opportunity to probe the inner parts of the T Tauri wind, a detailed study of a large sample of stars might prove rewarding.

2.4 Formation of Forbidden Lines

Forbidden lines are atomic transitions whose occurrence probability is zero at the first order of quantum-mechanical perturbation theory, but different from zero at higher orders. Thus, the spontaneous-emission probability of a forbidden transition is much lower than that of a permitted one. Consider an atom with only a ground state and an excited level, and assume for simplicity that the radiation field is so dilute that induced transitions can be neglected, the statistical equilibrium equation of the two levels can be written as

$$N_2 A_{21} + N_2 N_e \alpha(2,1) = N_1 N_e \alpha(1,2) \tag{17}$$

where the symbols have the following meaning

N_1: number of atoms in the ground state (per cm^3);
N_2: number of atoms in the excited state;
N_e: electronic number density;
A_{21}: Einstein coefficient;
$\alpha(i,j)$: rate of collisional transitions from level i to j.

The above equation can be rewritten as

$$A_{21} N_2 = \frac{N_1 N_e \alpha(1,2)}{1 + \frac{N_e \alpha(2,1)}{A_{21}}} \tag{18}$$

where the left hand side is proportional to the energy $E_{12} = h\nu A_{21} N_2$ emitted in the transition. Consider a permitted transition (denoted by p) for which $A_{21}(p) \gg 1$ and a forbidden line denoted by f for which $A_{21}(f) \ll A_{21}(p)$. We assume that the two lines have comparable wavelengths and collision strengths. Three emission regimes can then be distinguished depending on the magnitude of the second term of the denominator in Eq. 18.

1. $N_e \ll \frac{A_{21}(f)}{\alpha(1,2)}$ i.e., the energy $E_{12} = h\nu N_e N_1 \alpha(1,2)$ emitted in the transition does not depend on the Einstein coefficient anymore. Thus permitted and forbidden lines in a given spectral region will have comparable strength provided their collisional strengths are comparable.
2. $\frac{A_{21}(f)}{\alpha_f(1,2)} \ll N_e \ll \frac{A_{21}(p)}{\alpha_p(1,2)}$, that is, we have $E(p)/E(f) = N_e \alpha(2,1)/A_{21}(f) \gg 1$, and the permitted line now dominates over the forbidden one.
3. $N_e \gg \frac{A_{21}(p)}{\alpha(1,2)}$ then $E(p)/E(f) = A_{21}(p)/A_{21}(f) \gg 1$ and the permitted line again dominates over the forbidden one.

We thus expect forbidden emission to compete with allowed transitions when the electronic density is sufficiently low (Case 1 above). Otherwise, forbidden line emission will be suppressed with respect to allowed line emission, and forbidden lines will not be detected. For typical forbidden lines, this upper limit on N_e is in the range 10^4 - 10^7 cm^{-3}, and more detailed analysis of line ratios lead to precise estimates of the electron temperature and density in forbidden-line emission regions such as optical jets, Herbig-Haro objects, and the outer parts of T Tauri winds.

2.5 Lithium in TTS

Hunger (1956) noticed that the strong Li absorption in the spectra of both T Tauri and RY Tauri indicated overabundance of lithium by about two orders of magnitudes with respect to the solar atmosphere. Several extensive studies of lithium abundance in young stars were conducted recently (Strom *et al.* 1989, Basri, Martin, and Bertout 1991, Magazzù, Rebolo, and Pavlenko 1993). The procedure for deriving lithium abundance from the equivalent width of the LiI λ6707 resonance line goes as follows.

1. High resolution and high signal-to-noise spectrograms of the target star are secured and the lithium equivalent width is measured. In the case of T Tauri stars, there are difficulties due, e.g., to the inhomogeneity of their surfaces, which may lead to variable lithium absorption, or to veiling, a continuum radiation that affects the equivalent widths of TTS absorption lines (cf. Basri, Martin, and Bertout 1991 for a discussion of veiling). We assume in the following that we can find ways to deal with these problems.
2. The curves of growth (i.e., the variation of the line's equivalent width with the number of lithium atoms) is computed for different values of effective temperature, gravity, and metallicity, using an appropriate stellar atmosphere computer code, and theoretical curves relating the lithium equivalent width to its abundance are constructed.

3. The effective temperature, gravity and metallicity of the target star are found from an analysis of its atmosphere or from its colors in a standard photometric system.
4. The observed equivalent width of the lithium line then allows one to read the abundance from the theoretical curves.
5. The procedure is repeated for a metal with stronger lines in the spectrogram's wavelength range (e.g., CaI λ6717). The abundance of this metal is then used to calibrate the lithium abundance with respect to the Sun.

A major peculiarity of lithium abundance, first noted by Herbig (1965) is its strong decrease with increasing stellar age. Lithium is destroyed by mixing in the deep convection zones present during pre-main-sequence contraction, when temperature at the bottom of the convection zone is high enough to transform lithium into helium by the (p,a) reaction. The temperature needed to get significant depletion is about 2.1 - 2.5 10^6K and reaction rates are extremely sensitive to the temperature, varying as T^{22}. By the time it arrives on the main-sequence, a solar-type star will still have about one half of its original lithium, and lithium will be further depleted during main sequence evolution to a few percent of its initial value (cf. Profitt and Michaud 1989).

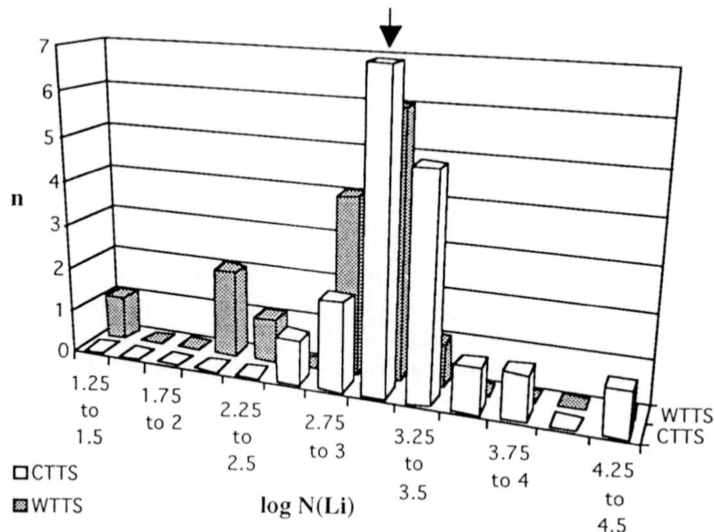

Fig. 11. Frequency distribution of lithium abundance N(Li) in CTTS and WTTS, on a scale where log N(H) is equal to 12 (after Magazzù, Rebolo and Pavlenko 1993). The vertical arrow indicates the interstellar value of lithium abundance.

As seen in Figure 11, which represents the best effort to date to determine lithium abundance in TTS, the average derived value is consistent with the primordial interstellar abundance, but the scatter is quite large, mainly because of uncertainties in assumed properties of TTS atmospheres. Basri, Martin, and Bertout (1992) have shown that these uncertainties conspire as to increase the apparent abundance in CTTS, which probably explains the high values found in some stars. The atmospheric parameters of the less exotic WTTS are presumably better known, and indeed the upper derived values are consistent with interstellar abundance. However, some WTTS seem to suffer more depletion than expected on the basis of their location in the HR diagram, which casts doubts on either age determination of young stars or theoretical models of pre-main sequence lithium depletion. In spite of these problems, the high lithium abundance observed generally in TTS is a clear indication of their youth. Because we do not understand the properties of TTS atmospheres well enough, it is however impossible to use lithium to calibrate their ages.

Summary Herbig's spectroscopic selection criteria for TTS lead to the following definition: a T Tauri star is a pre-main-sequence late-type star with a gaseous envelope where matter moves with velocities up to several hundreds kms^{-1}. This moving envelope includes both a chromosphere and an extended region spanning a large range of densities. Although translating the observed criteria into physical concepts may be somewhat involved, as the discussion of lithium shows, the main properties summarized in this definition are hardly controversial. However, this definition leaves open many possibilities for envelope structure and does not say much about the physics of the underlying object.

3 Nature Of T Tauri Activity

Early models tried to relate observational properties of young stars to magnetic activity, in analogy to the Sun. These models did not succeed in explaining the most exotic properties of CTTS but appear more appropriate for understanding WTTS. §3.1 summarizes briefly current ideas about magnetic activity in T Tauri stars, and §3.2 is devoted to a review of the current accretion disk model for CTTS.

3.1 Magnetic Activity in T Tauri Stars

Rotation and convection in interiors of solar-type stars combine to enhance the magnetic field through dynamo processes, creating a non-radiative energy flux leading to various manifestations:

creation of stellar spots above magnetically active regions (the magnetic field inhibits convective energy transport in photospheric layer, which results in the formation of a cool spot at the stellar surface).

heating of chromospheric layers above the stellar photosphere (UV and X-ray emission).

These phenomena are well documented for several classes of magnetically active stars such as RS CVn stars, but also dKe and dMe main sequence stars. Direct measurement of magnetic fields is possible only on a few pre-main sequence stars today, mainly because they are too faint for high-resolution spectroscopic studies, but comparative studies of young stars and magnetically active stars do suggest that magnetic fields of young stars are determined primarily by their rotation rates (10-30 kms^{-1}) and are quite comparable to fields seen in other active stars (cf. Bouvier 1990).

Rotational modulations in TTS light-curves There is a current worldwide effort to monitor the light curves of many TTS over periods of a few weeks, with the aim of determining the rotation period distribution, as well as properties of the spots responsible for the modulation, in a large sample of TTS (Fig. 12). Current results are the following (cf. Bouvier *et al.* 1993)

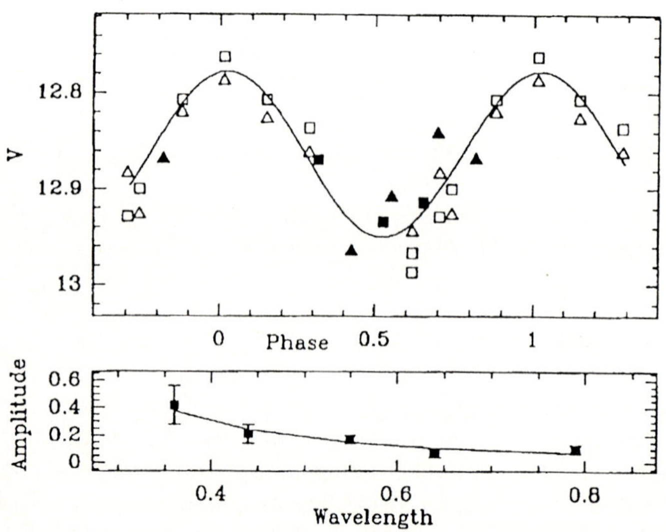

Fig. 12. An example of TTS light curve (here DE Tau) as determined from synoptic observations. The various symbols in the V light curve (upper panel) represent data from different observatories. The lower panel shows a variation of amplitude with wavelength. which is typical for cold spots (from Bouvier *et al.* 1993).

1. Dark spots are found on many (perhaps all) WTTS *and* CTTS. It appears likely that the quasi-sinusoidal variations that are commonly observed in T Tauri variables (§1.2, Fig. 3) are in fact due to cool spots.
2. Spot properties are comparable to those of spots found on other late-type active stars *with comparable rotation rates* (e.g., RS CVn stars).
3. The rotation periods of CTTS are, on the average, longer than the rotation periods of WTTS. This very exciting new result is illustrated in Fig. 13.

There also appears to be a correlation between the lithium equivalent width and the rotation period. Whether this correlation indicates a spin-up of the star as it nears the main-sequence is currently a matter of debate.

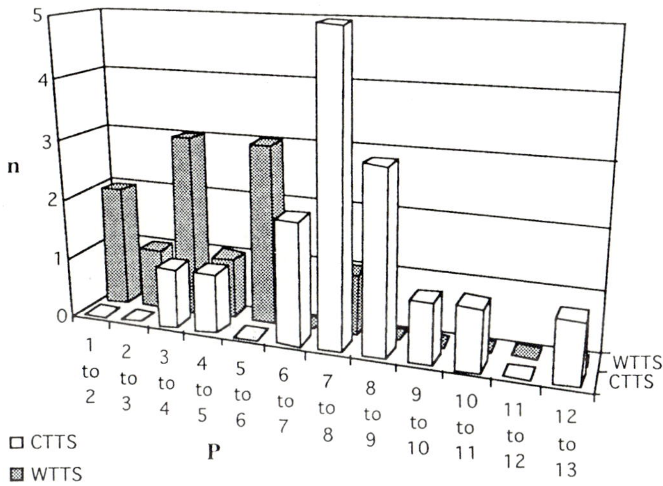

Fig. 13. Frequency distribution of the rotation period P (in days) for CTTS & WTTS (after Bouvier *et al.* 1993)

Other magnetic activity diagnostics In magnetically active late-type stars, coronal X-ray emission and Hα emission are correlated with the rotation rate. Bouvier (1990) found a similar anti-correlation between X-ray flux and rotation period, as illustrated in Fig. 14. The Hα flux, however, does not appear to be correlated with rotation in TTS. Except for some WTTS which fall on the same correlation as other active late-type stars, most TTS exhibit hydrogen emission that is far above the level expected from solar-type magnetic dynamo activity (cf. Fig. 15). Similar behavior is found for other strong emission lines such as CaII H and K or MgII.

Summary Several observed properties of WTTS and CTTS are probably caused by solar-type magnetic dynamo processes:

- X-ray emission level;
- Spot coverage and day-to-day light variability;
- CaII and Hα emission *in WTTS only.*

Note the key role of rotation in all aspects of dynamo-driven activity. From similarities between rotation rates of RS CVn stars and TTS, one expect a magnetic field strength of about 2 KGauss covering 30% of the stellar surface.

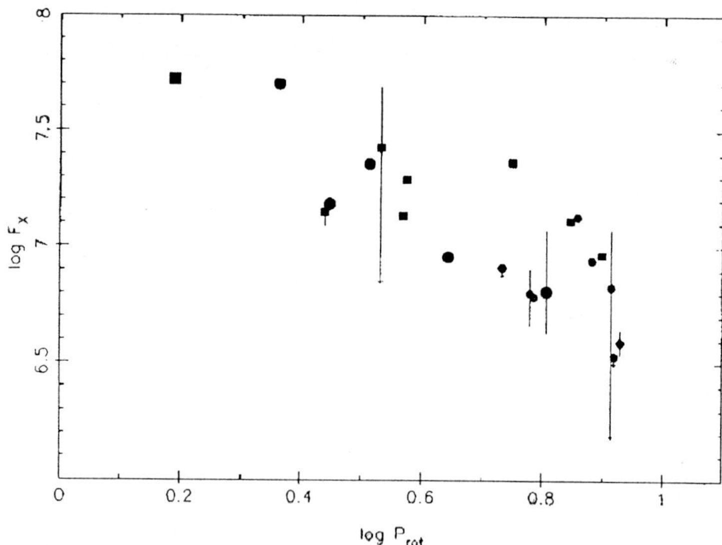

Fig. 14. X-ray flux vs. rotation period for TTS with directly measured periods. Squares denote WTTS and circles CTTS. The vertical bars indicate the degree of known variability (from Bouvier 1990).

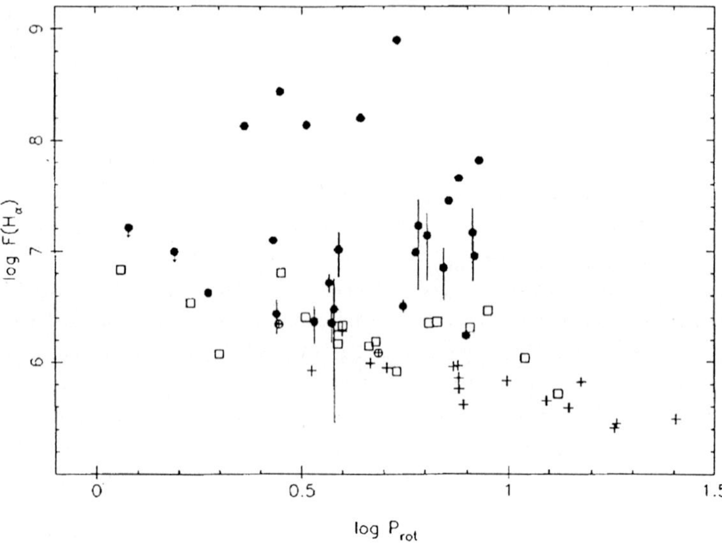

Fig. 15. Hα line flux vs. rotational period for TTS (dots), late-type dwarfs, dKe-dMe stars (crosses), and RS CVn systems (squares). The known variability of individual TTS is indicated by bars. Vertical arrows represent upper flux limits (from Bouvier 1990).

This order of magnitude is confirmed by the first direct measurement of magnetic field on a bright WTTS (Basri 1992).

However, solar-type dynamo activity cannot account for the most exotic properties of CTTS:

- Strong emission lines Hα, CaII, MgII;
- UV and IR continuum excesses;
- Large-scale irregular variability.

The failure of the magnetic model to explain key properties of TTS and the mounting evidence since 1986 for circumstellar disks around some of these objects leads to the following current picture:

- WTTS are diskless young stars whose properties reflect magnetic dynamo activity consistent with their rotation rates.
- CTTS also display magnetic activity but their more exotic properties are due to the presence of a circumstellar disk and to its interaction with the star.

3.2 Indirect Evidence for Circumstellar Disks

The evidence for the existence of circumstellar disks around young stars is fairly indirect. First, we should note that some 2-D computations of protostellar collapse do predict the formation of an equatorial disk at the same time as the central star, although the angular momentum distribution in the system remains a problem (see, e.g., the discussions in Tscharnuter 1985 and Bodenheimer 1991). However, these computations are unable at this point to reproduce the strong, collimated mass-ejection characteristic of the protostellar phase, so clearly there are important processes that are not yet understood in the physics of the collapse. In spite of the considerable efforts that go into performing these complex hydrodynamic computations, it is therefore obvious that their results cannot be considered compulsory for the presence, or absence, of disks in young stellar objects. We thus rely on observational evidence, but it will become clear in §3.3 that direct observations of solar-system sized disks at the distance of the closest star-forming regions are still extremely difficult today. There are, however, two main lines of *indirect* observational evidence for disks that we discuss in turn below.

Forbidden line profiles As illustrated in Fig. 16, forbidden lines are often blue-shifted in TTS, with little or no emission that is red-shifted (Appenzeller, Jankovics, and Östreicher 1985, Edwards *et al.* 1987). We have seen in §2 that forbidden line emission is able to compete with emission from allowed atomic transitions only in regions of relatively low electron density, e.g., $N_e \leq \frac{A_{21}}{\alpha(2,1)} = 1.3 \times 10^4 \mathrm{cm}^{-3}$ for [SII]λ6731. That is, forbidden emission probes the outer parts of the ionized TTS wind. The observed lack of red-shifted emission therefore shows that the part of the wind which goes away from us must be heavily obscured ($A_v > 6$mag), that is, there is a *disk-like screen* between us and the receding part of the wind. This result is based only on the assumption that

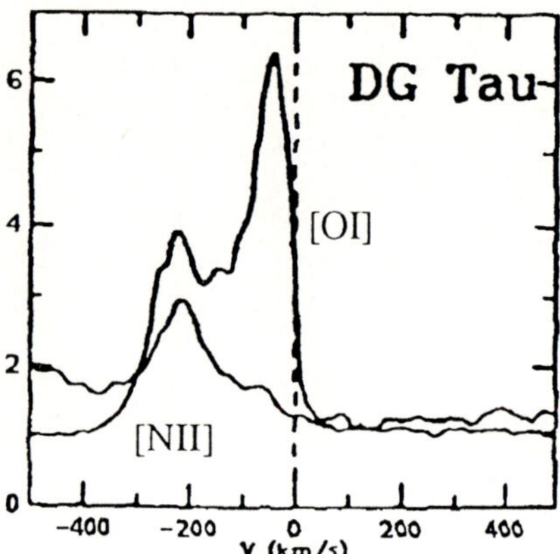

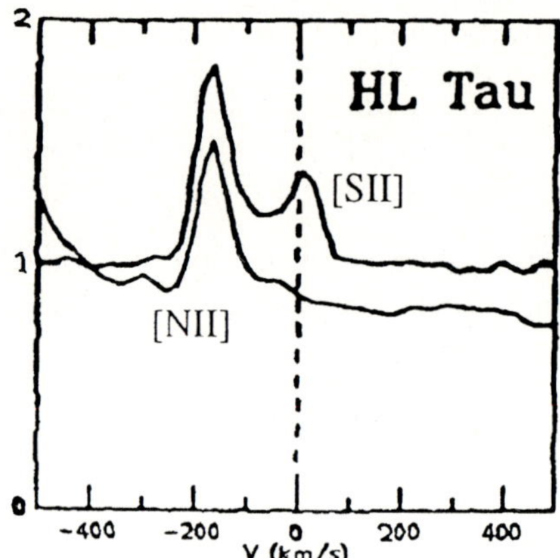

Fig. 16. Left Panel: The [OI]λ6300 and [NII]λ6584 forbidden lines in DG Tauri. The [OI] line displays broad, blue-shifted, double-peaked profiles. The [NII] profile does not show the low-velocity component. Right Panel: The [SII]λ6371 and [NII]λ6584 forbidden lines in HL Tauri. The [SII] profile is double-peaked and blue-shifted, and the low-velocity component is not seen in [NII] (from Edwards *et al.* 1987)

the forbidden lines are broadened by Doppler shifts due to the wind, a natural hypothesis since the line widths are several hundreds of kms^{-1}. In particular, this result is independent of any assumed wind geometry.

An estimate of the occulting disk size can be found from the observed [SII] line flux, which gives the emission measure

$$\mathrm{EM} = \mathrm{N_e^2 V} \approx 10^{54}\mathrm{cm}^{-3} \tag{19}$$

Assuming $\mathrm{N_e}$ to be equal to the critical value given above, and assuming a spherically symmetric emitting volume, one obtains a typical radius of the emission region $R \approx$ 10-100 AU The additional assumption that the gas has reached its terminal velocity in the forbidden line emission region gives $\dot{\mathrm{M}}_{wind} \approx 10^{-9} - 10^{-7}\mathrm{M}_{\odot}/\mathrm{yr}$ i.e., values comparable to the mass-loss rates derived from other diagnostics, notably Hα.

While lack of self-absorption makes it relatively simple to make detailed computations of the forbidden line formation in various velocity fields, we shall not go into this here. The interested reader is referred to the works of Appenzeller *et al.* (1984) for an analytical approach to this question (see also Fig. 17) and to the paper of Edwards *et al.* (1987) for a more detailed numerical computation. One result of these calculations is that T Tauri winds must be anisotropic. However, the exact nature of the velocity field in the forbidden line region remains controversial. In any case, the above order of magnitude estimates for the disk size and mass-loss rate do not depend on the details of the velocity field and provide a robust, albeit indirect, evidence for the presence of an optically thick, solar-system sized flattened structure surrounding many TTS.

Spectral energy distributions of TTS There is a large body of observational and theoretical work supporting the hypothesis that the observed spectral energy distributions of TTS is a combination of continuum emission from several sources: the star, a circumstellar disk, and an interaction region between disk and star (Lynden-Bell and Pringle 1974 (hereafter LBP), Rucinski 1985, Bertout 1987, Kenyon and Hartmann 1987, Bertout, Basri, and Bouvier 1988, Basri and Bertout 1989,, Bouvier and Bertout 1992 and others). Instead of describing all these results, which have been the topic of several reviews over the last years (e.g., Hartmann and Kenyon 1988; Bertout 1989, Appenzeller and Mundt 1989, Bertout, Basri, and Cabrit 1992), I merely discuss here some of the basic physical processes taking place in accretion disks and identify areas of current work.

Basic equations of steady accretion disks The idealized model of a T Tauri system consists, as sketched in Fig. 19, of (i) a central, late-type star surrounded by (ii) a geometrically thin, dusty accretion disk that interacts with the star via (iii) a boundary layer. The accretion disk model developed originally by LBP assumes that local processes induce a viscous coupling between neighboring disk annuli, thereby transporting angular momentum through the disk. Note that there could be other, more global ways of redistributing angular momentum, e.g., through density waves. Although potentially important, these non-local processes are not considered in the current disk model.

The basic angular momentum transport mechanism considered by LBP is kinematic viscosity. In a disk where the gas is rotating differentially, any chaotic motions in the gas will give rise to viscous forces (shear viscosity) although the required viscosity is much larger than normal molecular viscosity and is probably caused by turbulent motions within the disk (see the discussion in Frank, King, and Raine 1992). Gas particles moving along two neighboring streamlines at R and R+dR with angular velocities respectively $\Omega(\mathrm{R})$ and $\Omega(\mathrm{R+dR})$ have different amounts of angular momentum, and chaotic motions lead to angular momentum transport in the sense that a viscous torque N(R) is exerted on the outer streamline by the inner streamline. The rate of working done by the net torque on the ring between R and R+dR is

$$\mathbf{F}.\mathbf{v} = \Omega[\mathrm{N(R+dR)} - \mathrm{N(R)}] = \Omega\frac{\mathrm{dN}}{\mathrm{dR}}\mathrm{dR} = [\frac{\mathrm{d(N}\Omega)}{\mathrm{dR}} - \mathrm{N}\frac{\mathrm{d}\Omega}{\mathrm{dR}}]\mathrm{dR} \qquad (20)$$

The first right-hand side term of the above equation represents global transport of rotational energy by the torque and depends only on the assumed boundary conditions. The second RHS term represent the local rate of transforming mechanical energy into heat, which must be radiated away by the up and down faces of the annulus, with area $4\pi\mathrm{RdR}$ if one assumes that the radial radiative flux is zero. The energy dissipation rate per unit disk surface due to viscous torque is therefore

$$\mathrm{D(R)} = \frac{\mathrm{N(R)}}{4\pi\mathrm{R}}\frac{\mathrm{d}\Omega}{\mathrm{dR}} \qquad (21)$$

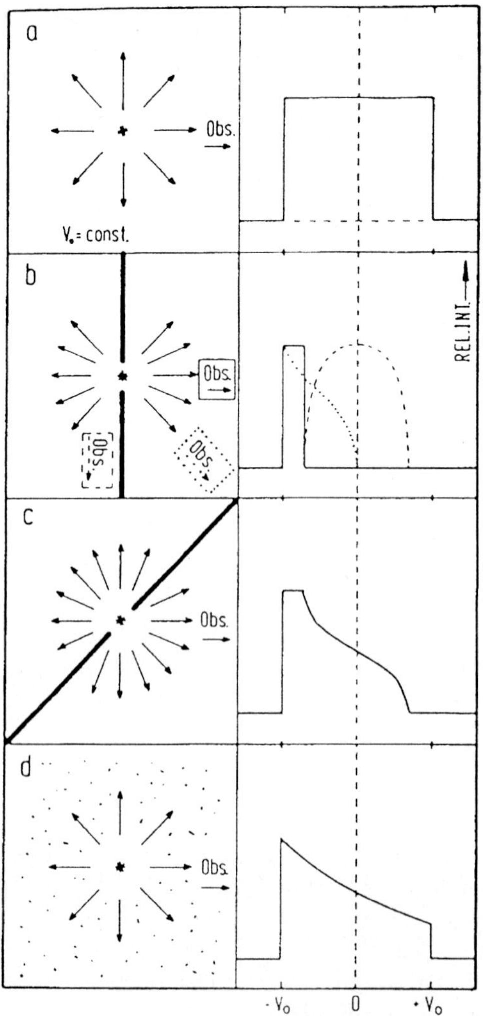

Fig. 17. Schematic wind models (left-hand side) and resulting optically thin line profiles, normalized to the same maximum intensity level. Panel a: constant velocity, spherically symmetric wind model produces a characteristic flat-topped line profile. Panel b: the vertical bar represents an optically thick, geometrically thin disk. The flow is assumed to be axially symmetric with the symmetry axis perpendicular to the disk's plane, and occurs in a cone with opening angle $\pi/4$. The three profiles shown correspond to different viewing angles, as indicated schematically. Panel c: same as Panel b but for an isotropic flow and only one viewing angle. Panel d: same as Panel a with the addition of a uniform dust distribution resulting in velocity-dependent extinction of the emitted line radiation. This figure appeared originally in Appenzeller, Jankovics, and Östreicher (1985), who give details of the analytical profile computations.

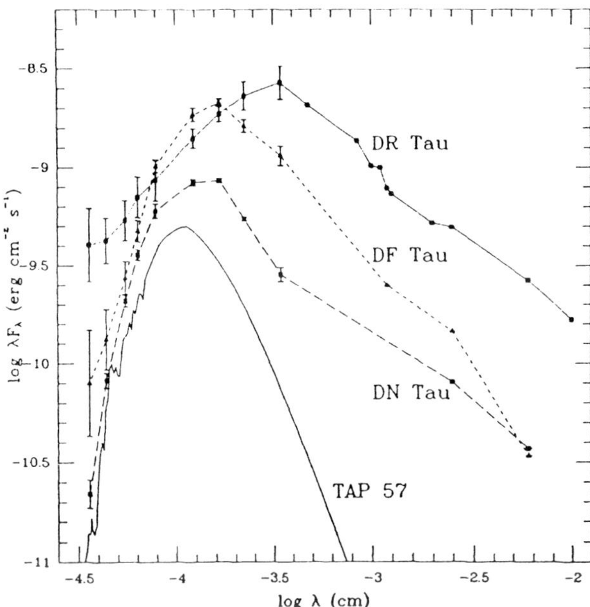

Fig. 18. Observed spectral energy distributions from 3600 Å to 100 μm of the stars whose spectra are shown in Fig. 1. The energy distribution of the WTTS TAP 57, shown as a solid line, has been displaced downward by 0.3 dex. The filled symbols are simultaneous (for DN Tau and DF Tau) or averaged (for DR Tau) photometric data. When available, observed variability is indicated by error bars. When compared with WTTS such as TAP 57, CTTS display prominent ultraviolet and infrared excesses. Excess continuum flux and optical emission-line activity are often correlated (from Bertout 1989).

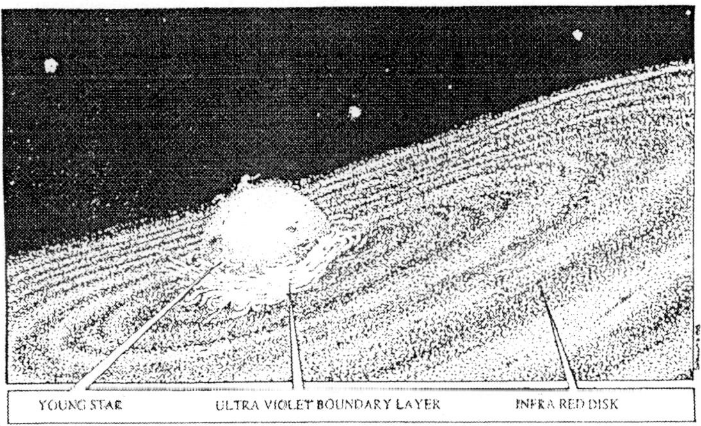

Fig. 19. An artist's conception of a T Tauri system

Assuming a steady-state, angular momentum conservation requires

$$\frac{dN}{dR} = \frac{d\dot{L}}{dR} = \dot{M}\frac{d(R^2\Omega)}{dR} \tag{22}$$

where represents the constant mass-accretion rate that accompanies angular momentum transport. Thus,

$$N(R) = \dot{M}R^2\Omega + C \tag{23}$$

where C is a constant determined by the inner boundary condition. At large R, C becomes negligible. If one then assumes that the disk is not self-gravitating, then its rotation law is quasi-Keplerian and

$$\Omega = (\frac{GM_s}{R^3})^{\frac{1}{2}} \tag{24}$$

The hypothesis of a quasi-Keplerian disk also implies that the disk must be geometrically thin. One then finds the following expression for the energy dissipation rate:

$$D(R) = \frac{3GM_s\dot{M}}{8\pi R^3} \tag{25}$$

No assumptions regarding the physical properties of the kinematic viscosity have been made in the above analysis, nor is the analysis restricted to Keplerian disks. If another rotation law was valid as would be the case, e.g., if the viscous torques were of magnetic rather than kinematic origin, then D(R) would not necessarily be proportional to R^{-3}. The disk temperature structure, which determines the emitted spectrum, can be computed from the energy dissipation rate by making an assumption about the transfer of radiation. As already apparent in the derivation of Eq. 21, LBP hypothesized that viscous energy released in a given disk annulus is radiated away through both faces of that annulus. Assuming additionally that the disk is optically thick, one then finds the disk spectrum $\lambda F_\lambda \propto \lambda^{-4/3}$.

Heating by stellar photons is also important for the disk temperature. Friedjung (1985), Adams and Shu (1986), Kenyon and Hartmann (1987), and Ruden and Pollack (1991) computed according to various approximations the resulting disk temperature at the photospheric level. At large distance R from the star, one finds that the local rate of heating of a flat disk due to reprocessing of photons originating from a star with radius R_s and effective temperature T_s is

$$F(R) = \frac{2\sigma T_s^4 R_s^3}{3\pi R^3} \tag{26}$$

At large distances from the central star, the above assumptions then lead to the following equation for the disk effective temperature $T_D(r)$:

$$\sigma T_D^4(r) = D(R) + F(R) = \frac{3GM_s\dot{M}}{8\pi R^3} + \frac{2\sigma T_s^4 R_s^3}{3\pi R^3} \tag{27}$$

where the RHS's first term represents the viscous energy dissipation rate and the second term takes into account the reprocessing of stellar photons. Note that both terms are proportional to R^{-3} in flat Keplerian disks, i.e., the overall emitted spectrum has the same spectral slope as that resulting from purely viscous heating.

Thus, there are two ways in which the central star influences disk properties:

- Its mass and radius determine the potential well seen by the disk matter, i.e., the viscous energy dissipation rate, and hence the disk temperature.
- The local disk temperature also depends upon the effective stellar temperature and radius, which determine, together with geometrical factors, the local rate of heating by the central star.

The last equation shows that the emitted spectrum of the optically thick accretion disk does not depend explicitly on the assumed viscosity. In order to determine the disk density, and hence its optical thickness, one must however assume something about the viscosity since its magnitude determines the angular momentum flow. LBP derive the disk density under the assumption that the kinematic viscosity is constant within the disk. Shakura and Sunyaev (1973) derive another analytical solution in which the kinematic viscosity is assumed to be proportional to the local scale height times the local sound speed, with the proportionality constant α being restricted to values smaller or equal to unity (this is the so-called α-prescription). Underlying this ad-hoc formulation are the ideas that turbulent eddies cannot be larger than the disk height and that any supersonic turbulence should rapidly become subsonic because of the formation of internal shocks in the disk. Both LBP and Shakura & Sunyaev formulations are ad-hoc prescriptions that simply parameterize our ignorance of turbulence physics and that give qualitatively comparable results for the run of density with radius. Within this framework, the assumption of optical thickness is verified for typical TTS disks parameters (cf. Bertout, Basri, and Bouvier 1988).

The disk/star interaction region Eq. 27 describes the temperature far away from the central star. When computing the inner disk structure, one must assume something about the inner boundary of the disk, and more specifically about the way angular momentum is transferred from disk to star. LBP imposed the condition that the star exerts no torque on the inner edge of the disk, which also implies the existence of a boundary layer between the slowly rotating star (typically 20 kms^{-1}at the equator) and the inner edge of the Keplerian disk, where matter is circling the star at about 250 kms^{-1}. Fig. 20 shows the expected angular velocity Ω in the boundary layer. There, r_b is the boundary layer radius in units of the stellar radius R_s, r_m is the radius where the derivative of Ω changes sign, i.e., where Ω reaches its maximum Ω_m, $\Omega_s (= \Omega_*)$ is the star's equatorial velocity, and Ω_{Ks} is the value of the Keplerian velocity at the stellar radius.

The LBP inner boundary condition assumes that the extent r_b of the boundary layer is infinitely small and that the angular velocity at the inner disk edge is comparable to the Keplerian velocity at the star, i.e., $\Omega_b = \Omega_m = \Omega_{Ks}$. The constant C in Eq. 23 can then be computed from the condition $NR_s = 0$. The

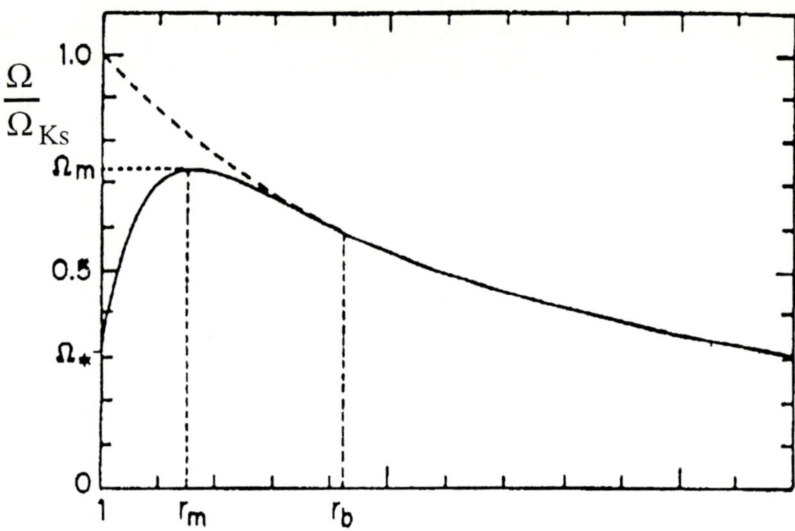

Fig. 20. Schematic illustration of the variations of Ω as a function of radial distance in the boundary layer. The various symbols are defined in the text (from Regev 1991).

general expression for the viscous torque is then

$$N(R) = \dot{M}(GM_sR_s)^{\frac{1}{2}}[1 - (\frac{R_s}{R})^{\frac{1}{2}}] \tag{28}$$

which leads to the well-known expression for the viscous energy dissipation rate:

$$D(R) = \frac{3GM_s\dot{M}}{8\pi R^3}[1 - (\frac{R_s}{R})^{\frac{1}{2}}] \tag{29}$$

Energetic properties of the disk/boundary layer system directly follow from this standard inner boundary condition. Integrating Eq. 29 from $r_b = R_s$ to ∞ one finds the disk luminosity

$$L_D = \frac{GM_s\dot{M}}{2R_s} = \frac{L_{acc}}{2} \tag{30}$$

Since only one half of the total accretion luminosity L_{acc} is dissipated in the disk, one thus concludes that the second half is advected from the disk into the boundary layer, corresponding to the Keplerian kinetic energy rate at the inner disk radius. Because of the many difficulties involved in modeling the star/disk interaction region, the ultimate fate of the incoming energetic matter is not known with certainty at this point, and several approaches are explored in current work. The conventional approach, pioneered by LBP, consists in assuming that the available energy is radiated away locally.

LBP boundary layer properties No attempt is made in the LBP approach to find a consistent solution of the hydrodynamic equations in the star/disk interaction region. Instead, one assumes that the incoming energy flux is dissipated radiatively in a small region at the star's equator. The radiative energy produced in the boundary layer is assumed to flow towards the optically thick disk where it diffuses isotropically to the disk surface; the size of the emitting region at the disk's surface is therefore comparable to the disk's scale height in this model. While temperatures in the disk vary from 10K far from the star to about 3000K near the star, the boundary layer's effective temperature range from 7000 to 12000K (LBP, Bertout 1987; Kenyon and Hartmann 1987). The boundary layer thus radiates in the ultraviolet and visible part of the spectrum a luminosity equal to half the accretion luminosity. In current models of T Tauri stars, this luminosity is considered to be the source of the UV excess and optical veiling observed in many CTTS.

It should be emphasized here that the LBP inner boundary condition maximizes the boundary layer's luminosity, which can be reduced in several ways. First, if the star rotates as some angular velocity Ω_s, then the boundary layer luminosity is reduced to

$$L_{bl} = \frac{GM_s\dot{M}}{2R_s} - \frac{\dot{M}R_s^2\Omega_s^2}{2} = \frac{L_{acc}}{2}[1 - \frac{\Omega_s^2}{\Omega_{Ks}}] \tag{31}$$

Second, part of the accretion power is used up to spin up the star via the shear at the stellar surface, where the derivative of Ω is non-zero (cf. Regev 1991). Eq. 31 then becomes

$$L_{bl} = \frac{L_{acc}}{2}[1 - \frac{\Omega_s}{\Omega_{Ks}}]^2 \tag{32}$$

Third, the rotational velocity in the disk may depart from its Keplerian value in a finite inner disk region. This possibility was considered in some detail by Duschl and Tscharnuter (1991), who demonstrated that the fraction of accretion luminosity dissipated in the non-Keplerian inner disk is indeed strongly dependent upon the assumed size of this region. Models of CTTS spectral energy distributions however rule out the possibility of an extended non-Keplerian inner disk in T Tauri disks (Bertout *et al.* , in preparation). Finally, some of the accretion energy could be released in non-radiative form, e.g., for driving a wind (cf. Pringle 1989).

Matched boundary layer solutions While the LBP approach presumably gives a useful albeit rough description of the radiative properties of the star/disk interaction region, it does not offer a consistent solution of the hydrodynamic equations in that region. Bertout and Regev (1992) recently proposed a 1D analytical solution of the hydrodynamics equations in the boundary layer, using the mathematical approach of matched asymptotic expansion (cf. Regev 1983) to overcome the difficulties resulting from the smallness of the boundary layer compared to the other parts of the star/disk system. They find that no steady-state solutions of the hydrodynamic equations are possible unless most of the accreting mass is allowed to escape from the boundary layer before reaching the central star. Allowing for such mass outflow, Bertout and Regev find that the radiative energy flux resulting from the dissipation of kinetic energy in the boundary layer is in fact directed radially toward the star's interior instead of outwards. This occurs because the characteristic thickness of the boundary layer is much smaller than its vertical extension Therefore, the boundary layer itself remains cool (its temperature is comparable to the disk temperature close to the star, say 3000 K), and the emitted flux comes from reprocessed radiation in the stellar atmosphere instead of the boundary layer proper. These results depend sensitively on the following assumptions:

- The disk and boundary layer are in dynamical steady-state.
- The radial radiative flux in the entire disk, and hence at the entrance of the boundary layer, is zero (this result from the LBP disk description).
- The viscosity is the same in the disk and the boundary layer and is assumed to result from turbulence and the size of the turbulent elements is given by the characteristic scale of the medium. Due to the small radial extent of the boundary layer, this implies that the viscosity is much lower there than in the disk, where the local disk height gives the characteristic eddy size.

While this 1D analytical approach is useful in demonstrating that boundary layer emission is probably not as straightforward as envisioned by LBP and others, it is obvious that the dynamical problem is a complex one that should be solved numerically (cf. Kley 1991).

Accretion columns The boundary layer and surrounding stellar atmo-

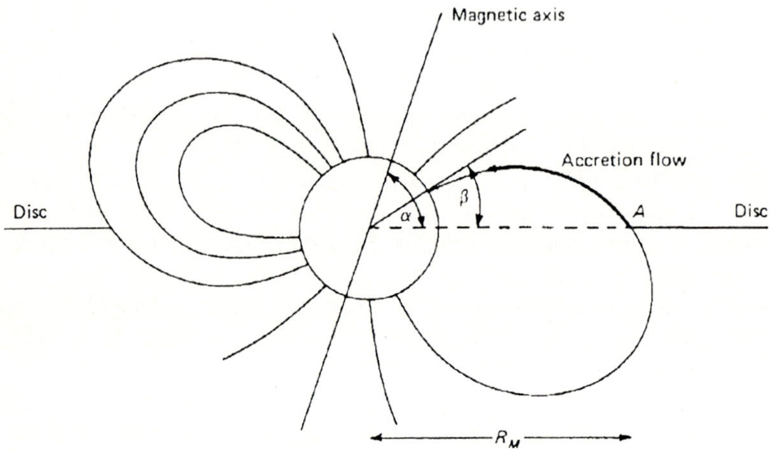

Fig. 21. Schematic view of accretion along the field lines of a strong dipolar magnetic field that disrupts the inner disk over a radial extent R_M (from Frank, King, and Raine 1992).

spheric layers are choice line emission regions. However, line profiles indicating *infall* velocities of several hundreds $kms^{-1}$are seen in many CTTS (and notably in the YY Orionis stars; cf. Walker 1972). These profiles cannot arise directly from the boundary layers of classical disk models, since accretion velocities are restricted to being subsonic there. However, strong magnetic loops at the surface of the star may interrupt the flow of the disk within a few stellar radii of the surface, and generate manifestations presumably similar to what we expect from boundary layers (Bertout, Basri, and Bouvier 1988; Königl 1991). A modulation of the accretion diagnostics (emission line intensity or veiling) with the stellar rotation period would establish that the emitting region is located at the interface between star and disk. There have already been a few reports of "hot spot" modulations which could be interpreted as accretion columns onto magnetic stellar regions, where accreting material could reach near free-fall velocities. However, recent work by Bouvier *et al.* (1993) suggests that the presence of an infrared companion in the vicinity of a star covered with cold magnetic spots can lead to light curves that mimic hot spots, so that some previously reported detections of hot spots should be viewed with caution.

Boundary layer or accretion column? Since it now appears that most of the small amplitude day-to-day variability of CTTS is caused by magnetic spots (Bouvier *et al.* 1993) which cover up to 30% of the stellar surface and may have a magnetic field strength of up to 2KGauss at the surface, it is clear that the field must have some influence on the disk in the vicinity of the star. However, the current picture of disk disruption through magnetic loops (e.g., Königl 1991) depends sensitively on the assumed dipolar field geometry, whereas it is likely that the surface magnetic field of a CTTS is in fact quite patchy. Also, current determinations of mass accretion and loss rates tell us that a sizable fraction of the matter accreted through the disk never reaches the star, or must be expelled again if it does reach the stellar surface. Therefore, if CTTS winds are driven magnetically, as appears likely, the field lines must be open over a large solid angle. These two arguments make it difficult to be readily convinced by a model calling for complete inner disk disruption due to a stellar dipole field. If, however, many smaller magnetic loops are present in the vicinity of the stellar equator, one can envision a complex situation in which non-axisymmetric accretion occurs in regions of low magnetic activity (these regions would be what we now call the boundary layer) while at the same time accretion along stellar field lines occurs in regions where closed loops are present. It is unclear at this point whether the wind is driven by the disk or the stellar magnetic field, but the complex accretion picture proposed here is compatible with both possibilities since it confines the accretion to regions close to the stellar equator.

Clearly, high-resolution spectroscopic studies will be decisive in testing the various possibilities for accretion discussed above. In this respect, renewed studies of the YY Orionis line profiles found in many TTS should bring crucial information. Whether these profiles arise in returning wind material that failed to reach escape velocity or in accreting disk matter along magnetic field lines remains a basic question that should be answered observationally.

Spectrum computation One should be aware that comparisons of observed and computed spectral energy distributions done so far assume an LBP boundary layer which radiates away half of the accretion luminosity in a region of radial extent comparable to the disk scale height near the star. This simplistic approach makes it easy to compute the emitted spectrum, while the more involved accretion scenarios discussed above do not allow, at least in their present state, for computations of the spectrum emitted in the star/disk interaction region.

The LBP inner boundary condition leads to a singular, isothermal boundary layer. If one then assumes that it is optically thick, its spectrum is a single blackbody with temperature T_{bl} given by

$$T_{bl}^4 = \frac{L_{bl}}{4\pi\sigma R_s r_b} \tag{33}$$

Observations to be discussed below indicate that the boundary layer must be optically thin at least in some spectral regions. Its temperature, found by solving

$$\int_0^{\infty} \pi B_\lambda(T_{bl})(1 - e^{-\tau_\lambda})d\lambda = \frac{L_{bl}}{4\pi\sigma R_s r_b} \tag{34}$$

is then higher than the value computed by Eq. 33. Notice that the viscosity does not enter in Eq. 33 but that its value is explicitly needed to compute the optical depth in Eq. 34. Current models of boundary layer emission use standard LTE stellar atmosphere models to compute T_{bl} iteratively (cf. Basri and Bertout 1989). Once the properties of the boundary layer are determined from the above equations, computation of the overall emitted spectrum by the system (star, disk, and boundary layer) can be performed (cf. Bertout, Basri, and Bouvier 1988 for details).

Comparisons with observations While the observed infrared spectral energy distributions of some CTTS are consistent with those of the LBP disk model, most are actually flatter; Rydgren and Zak (1987) showed that the average slope of CTTS infrared spectra beyond 5μm or so is proportional to $\lambda^{-3/4}$ rather than to $\lambda^{-4/3}$. Several suggestions were made to explain the fact that the observed infrared spectra are flatter than the LBP disk model predicts:

- Adams, Lada, and Shu (1988) assumed that the temperature distribution in T Tauri disks is flatter than in LBP disks. Following this suggestion, it has become common practice to parameterize the temperature law index and to use different values of this parameter to model infrared distributions of young stellar objects, particularly those with flat infrared spectra (e.g., Adams, Emerson, and Fuller 1990, Beckwith *et al.* 1990). Two objections can be made to this procedure, however. Hartmann and Kenyon (1988) showed that in order to get the required temperature gradient in the LBP disk theory one had to assume either massive disks unlike those observed, or unsteady accretion with strong accretion rates in the outer parts of the disk, leading to unrealistically small evolutionary timescales there. The second problem is that using the temperature law as a free parameter is equivalent to implicitly assuming that the energy dissipation mechanism in the disk differs from star to star, i.e., that the underlying physics of T Tauri disks is not the same for all disks.

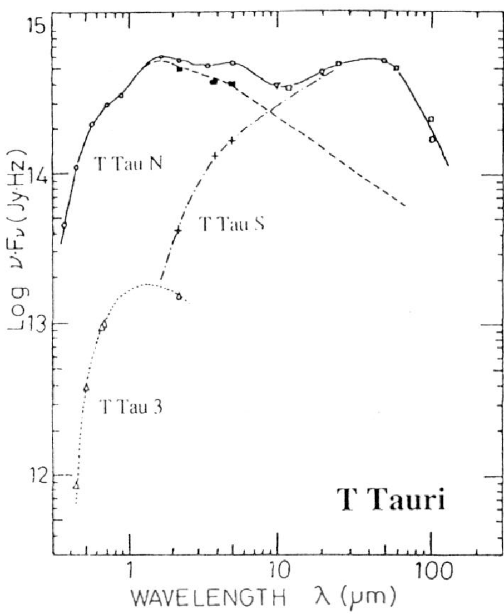

Fig. 22. Decomposition of T Tau's SED suggested by Maihara and Kataza (1991) on the basis of their speckle data. The optical star T Tau N is surrounded by an accretion disk; the infrared companion T Tau S has a Type I SED, and T Tau 3 is a possible third component of the system, the presence of which needs confirmation.

- High angular resolution observations have revealed that the integrated SED of some TTS is a composite spectrum of several objects. Cold companions contributing heavily to the infrared spectrum have been found in the vicinity, e.g., of T Tau itself (Maihara and Kataza 1991, cf. Fig. 22), Haro 6-10 (Leinert and Haas 1989 and XZ Tau (Leinert, and Zinnecker 1990). How many other SEDs are composite is of course uncertain but the large number of binary TTS (Reipurth and Zinnecker 1993) suggests that it may be a common occurrence.
- The canonical $\lambda^{-4/3}$ spectrum that results from reprocessing is a direct consequence of the unrealistic assumption that the disk is flat and infinitely thin. Any process leading to non-flat disks will thus result in changing the spectrum re-emitted by the disk. For example, if one takes into account the vertical structure of the disk, which "flares" at large distance from the star because of hydrostatic equilibrium, one increases the far-infrared flux since more stellar photons are absorbed and re-emitted from the outer parts of the disk (Kenyon and Hartmann 1987). Radiative transfer effects in disk atmospheres can also lead to deviations from the canonical law depending on the viewing angle of the disk (Malbet and Bertout 1991). It seems unlikely, however that any of these effects can give rise to really "flat" infrared spectra.
- That about 80% of all CTTS are members of binary systems (Reipurth and Zinnecker 1993) in fact suggests still another possibility that is now being

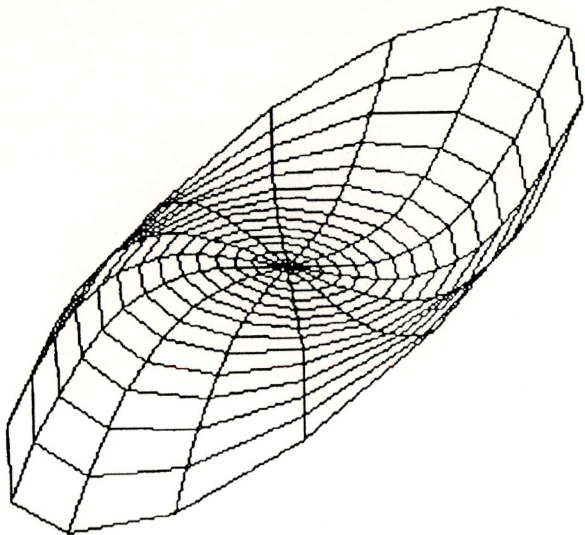

Fig. 23. The geometry of a tidally warped disk. While the warp magnitude is exaggerated for the purpose of illustration, the functional shape is the correct one in the framework of a first-order perturbation theory (from Terquem and Bertout 1993).

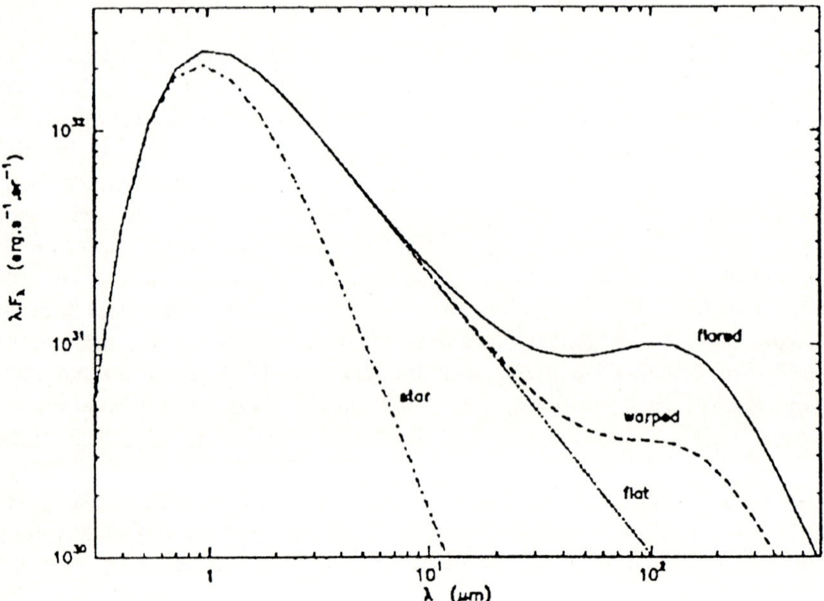

Fig. 24. Comparison of the spectrum emitted by warped (dashed line), flared (solid line), and flat (dotted line) accretion disks with otherwise comparable properties (from Terquem and Bertout 1993).

investigated by Terquem and Bertout (1993 and this volume), namely that their circumstellar disks might be appreciably warped because of tidal forces (cf. Fig. 23). As illustrated in Fig. 24, which shows the spectrum emitted by a warped disk and by a flaring disk with similar radial deformation but no angular dependence of the deformation, this geometry leads to anomalous infrared excesses just as flaring disks do. Again, however, one concludes that "flat" spectra are unlikely to be produced in this way.

One should point out that the last two possibilities discussed above produce sizable deviations from the canonical law only in systems where the accretion luminosity is smaller than the reprocessed luminosity, i.e., for $\dot{M}_{acc} \leq 10^{-8} M_{\odot}/yr$. For mass accretion rates in excess of this value, the spectrum starts to be dominated by accretion and the canonical accretion disk spectrum is eventually recovered. In order to produce flat infrared spectra, one must therefore assume either that some other component contributes to the spectrum (companion and/or circumstellar envelope) or that the effective temperature distribution in the disk is not given by the LBP model.

We conclude that the far-infrared spectral energy distributions of many T Tauri stars do not fully support the hypothesis that T Tauri disks are classical accretion disks, although there are a few stars, such as DF Tau, which fit the classic model perfectly. The strongest piece of evidence for the presence of *accretion* disks around CTTS in fact comes from the blue and ultraviolet spectral ranges. As it turns out, the observations of Balmer continuum emission jumps in CTTS can easily be explained if the boundary layer is optically thin in the Paschen continuum (Basri and Bertout 1989). Even more important, the amount of energy available depends solely on the accretion rate and not on the star's resources, which explains why the photospheric spectrum can be so veiled in some cases, while the disk model explains naturally the observed correlation between the respective amounts of infrared and ultraviolet excesses.

An example of the observations and fit to them for DE Tau appears in Fig. 25. Derived parameters for the model presented are $\dot{M}_{acc} = 6.10^{-8} M_{\odot}/yr$, $R_s = 3.2 R_{\odot}$, $T_{bl} = 7600$ K, $L_s = 1.3 L_{\odot}$, and $L_{acc} = 0.2 L_s$. Line emission from the Balmer lines with high quantum number appears roughly consistent with optically thick line emission from the boundary layer, whose emitting area is a few percent of the stellar surface. There is obviously a more extended region of emission which contributes to the flux in the lowest members of the Balmer series, since these are predicted to have very little emission contrast in this simple boundary layer model. Mass-accretion rates ranging from a few times 10^{-9} to a few times 10^{-7} $M_{\odot}/yr$ can be determined from such models of individual stars (Basri and Bertout 1989, Bouvier and Bertout 1992).

Another relevant accretion diagnostic is the amount of spectral veiling in the optical spectral lines. Veiling is defined by

$$r(\lambda) = f_{obs}(\lambda)/f_s(\lambda) - 1 \qquad (35)$$

where f_{obs} is the observed flux and f_s the photospheric flux, which can be estimated by comparing the veiled (observed) spectrum with an appropriate spectral

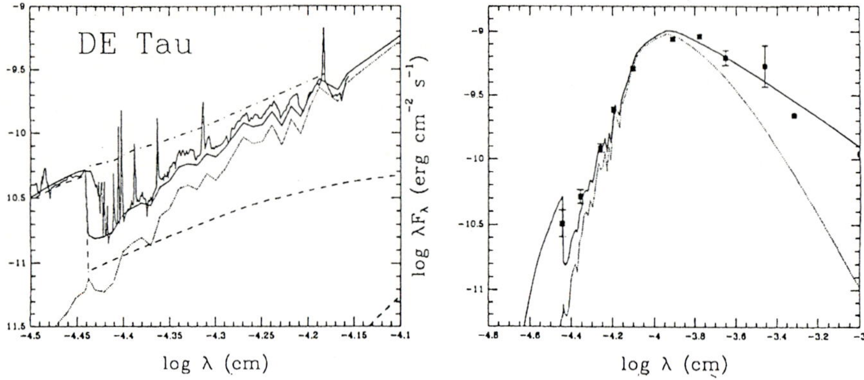

Fig. 25. Observations and model fitting for DE Tau (see also Fig. 12). The solid spectrum in the left panel is Lick Observatory UV Schmidt spectroscopy, and the squares with error bars in the right panel are simultaneous ESO photometric data. The solid smooth line in both panels is the final composite spectrum. The upper dashed line in the left panel is the boundary layer contribution, the lower one is the disk contribution. The dash-dotted line is the expected locus of Balmer line peaks, and the dotted line is the stellar photospheric flux. The left panel is for the optical range, and the right panel shows only the stellar and overall model spectra over the full spectral range (from Basri and Bertout 1989).

standard. One finds that the same absorption lines are present and in the same ratios in both the CTTS and the spectral standard, but all the line depths are reduced in the CTTS. One method to derive the veiling is then to add a flat continuum to the standard spectrum until it matches the CTTS spectrum. Basri and Batalha (1991) and Hartigan et al (1990) showed that veiling has the relation to infrared excess expected if both are due to accretion and also that the Hα emission flux is closely related to veiling. Cohen, Emerson, and Beichman (1989), and Cabrit *et al.* (1990) showed that the emission lines in general are correlated with infrared excess, which strongly suggests a close connection between accretion and mass loss (Fig. 26). While the origin of mass loss in TTS is still a matter of much work and controversy, it appears that stellar magnetic fields probably play a major role in driving the wind (e.g., Bertout, Basri, and Cabrit 1992).

The main advantages of the accretion disk idea are:

- to explain a variety of apparently unrelated phenomena (UV excess, IR excess, visual extinction, veiling, Balmer jump, forbidden line profiles) by the presence of a single physical entity;

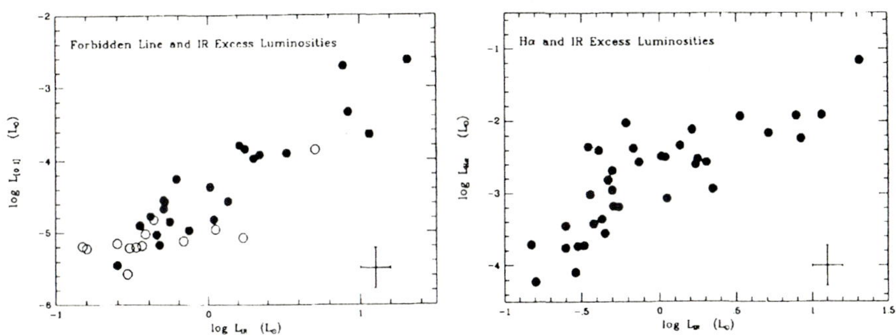

Fig. 26. Left Panel: correlation between the de-reddened [OI] luminosity and the IR excess luminosity. Open circles denote stars with upper limits of [OI] emission. Right Panel: relationship between the de-reddened Hα emission and the IR excess luminosity. In both plots, the two quantities are well correlated over two orders of magnitudes (from Cabrit *et al.* 1990).

- to provide T Tauri stars with a large reservoir of potential energy for covering their radiative expenditures and for driving their winds.

This apparent success should not hide the fact that basic problems about the disk physics remain to be solved. The nature of angular momentum transport within the disk and the physics of the disk/star interaction region are the most important questions to elucidate, since they are likely to determine the wind-driving mechanism and the ultimate shape of the emitted spectrum. But even in the framework of the α-theory for disk viscosity, there are still many aspects of disks to be investigated and many improvements to be made to the models. This is true in particular for studies of the radiative transfer in disk atmospheres, which are important for predicting how disks look like. I briefly review current work along this line in the following section.

3.3 Observations of Disks

There is an ongoing effort at Grenoble Observatory to develop accurate radiative transfer models for accretion disks and to observe disk candidates with the highest possible resolving power in all relevant wavelength domains. I describe the first results of these investigations in the following.

Synthetic disk images In order to develop accurate radiative transfer models of disks, one must first prescribe the vertical structure of the disk. Once the vertical optical depth and temperature stratification are known at all disk radii, it is then possible to integrate the transfer equation and to derive the emitted continuous spectrum at a given wavelength as a function of spatial position.

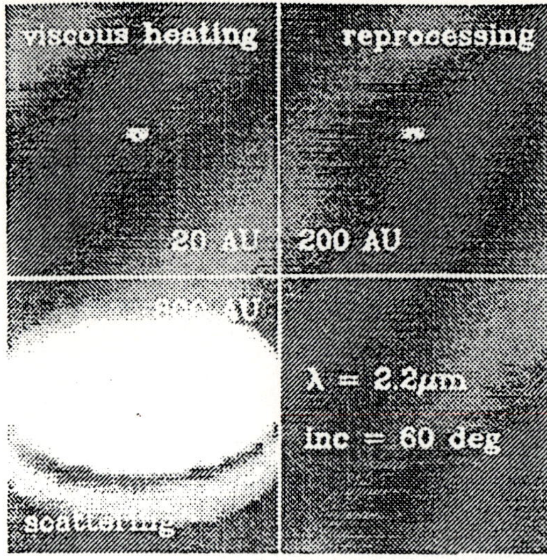

Fig. 27. Expected appearance of a T Tauri accretion disk at 2.2 μm depending on three different emission mechanisms. Upper left panel: emission due to viscous heating. Upper right panel: emission due to reprocessing of stellar radiation. Lower left panel: emission due to single scattering of stellar light in the upper disk layers. Note the different linear scales in the three panels.

Malbet and Bertout (1991), Malbet (1993), and Malbet *et al.* (1993) investigated these various aspects for T Tauri disks and constructed synthetic images of T Tauri disks using the basic assumption that the radiative flux in the radial direction is negligible compared to the vertical radiative flux. This assumption is a standard one in the framework of classical accretion disks and is verified in optically thick disks. The vertical disk structure is then computed assuming hydrostatic equilibrium in the vertical direction. The following emission processes are taken into account in the computation:

- thermal emission due to viscous energy dissipation in the disk;
- thermal emission resulting from reprocessing of stellar photons;
- single scattering of photons in the upper layers of the disk's atmosphere.

An example of synthetic map is shown in Fig. 27. There, the disk is assumed to have a radius of 400 AU and an accretion rate of 10^{-7} $M_\odot$/yr, the disk viewing angle is 60°, and the wavelength is 2.2μm. On large scales, the major emission process is scattering; thermal emission originates only in the warm inner parts of the disk. Such maps clearly demonstrate the difficulty in directly observing thermal emission from disks in the near infrared; a resolution on the order of one milliarcsecond would indeed be required in order to image the disk's central part. While is appears easier to resolve the scattered light emission, which extends up to the outer disk radius, problems arise because the intensity of scattered light is much lower than that of thermal emission, thus requiring a detector's dynamical range at the limit of current technology. In order to study the structure of a T Tauri accretion disk, one would ideally require an interferometer operating in the mid-infrared, where the disk thermal emission is more extended than in the near-infrared, and working with baselines up to 1 km.

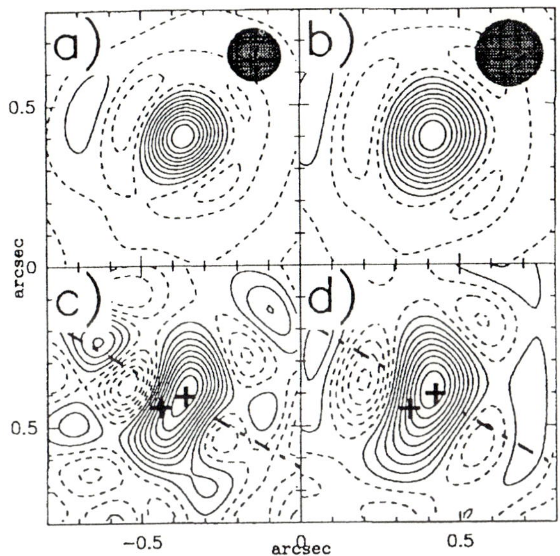

Fig. 28. Reconstructed images of Z CMa. Panel a: image at 3.87μm (Filter L'). Panel b: image at 4.75μm (Filter M). Panel c: geometry of the disk-like emission at L' after subtraction of stellar contributions. Panel d: same as c but for Filter M. The dash-dotted line indicates the direction of the outflow and the two crosses mark the positions of the binary components (from Malbet *et al.* 1993).

Of course, such an instrument is not available today; ESO's Very Large Telescope Interferometer, which will represent an important first step to reach the goal delineated above, will however be available before the end of this century. Current technology allows us to get a 0.1" resolution in the near-infrared using adaptive optics techniques on 4-meter class telescopes. In the millimetric range, current interferometers deliver a resolution of 2-3".

A disk around Z CMa? As an example of results that can be achieved today, I now present recent results obtained on the bright FU Orionis object Z CMa (see Hartmann 1991 for a review of current ideas about the nature of FUOrs). Z CMa displays an impressive outflow traced by numerous Herbig-Haro objects to a distance of more than 2 pc on each side of the star (Poetzel, Mundt, and Ray 1989). The light curve of Z CMa over the last decade shows sizable outbursts which are presumably caused by some kind of instability developing in the FUOr accretion disk, although the exact processes involved are not clearly identified yet. The Hα profile is also strongly variable, and indicates a maximum wind velocity of more than a thousand kms^{-1}. Speckle observations have shown Z CMa to be a binary with one optical component (the FU Orionis object) and a luminous infrared companion at a distance of about 100 AU from the optical object. Fig. 28 displays two images obtained at the near-infrared wavelengths L' and M using adaptive optics at the ESO 3.6m telescope (Malbet *et al.* 1993). In order to extract the low-level extended emission seen in this figure, the contribution of the two components of the Z CMa binary was subtracted from the original images. What is left is a 400 AU disk-like emission region extending in a direction perpendicular to the outflow that can be interpreted as scattered light from the FU Ori disk (cf. Malbet *et al.* 1993 for details and Malbet, this volume). Although suggestive, this result should be taken with a healthy dose of scepticism as long as it has not been confirmed by repeated observations since adaptive optics technology is still at an early stage.

4 Summary

We described the basic optical properties of various classes of young stellar objects, with emphasis on classical T Tauri stars. We then introduced the elementary radiative transfer concepts needed to give a physical interpretation of Herbig's criteria for defining the T Tauri class. We concluded from this study that optical spectroscopic properties can be translated into an initial physical description of the CTTS. We then used observational properties in other wavelength domains to gain more physical insight into the nature of the T Tauri phenomenon, regarding in particular solar-type magnetic activity. We discussed the various lines of evidence for the presence of accretion disks around these young stars and reviewed current disk and boundary layer models, focusing on their emission properties. Finally, we discussed some current high resolution work aiming at detecting disks.

In order to be reasonably rounded, this introduction should have included at least two more sections which time and space limitations did not allow me to present here. One section on observations and interpretation of high-resolution line profiles in the optical and near-infrared ranges, including a discussion of mass loss rate determinations and plausible wind mechanisms, and another one on the evolutionary status of CTTS and WTTS, including a discussion of theoretical rotation evolution. The interested reader is referred to the reviews by Hartmann (1986), Panagia (1991), Natta and Giavanardi (1991), and Bertout, Basri, and Cabrit (1992) for various viewpoints on T Tauri winds, and to Bouvier (1991) and Stauffer (1991) for discussions of pre-main sequence rotation evolution.

Acknowledgments While in Berlin, I enjoyed the hospitality of the TU Astrophysics Department, for which I thank Prof. Dr. E. Sedlmayr. It is a pleasure to acknowledge Oded Regev and Jérôme Bouvier for enlightening discussions and for a careful reading of the manuscript.

References

Beckwith, S.V., Sargent, A.I., Chini, R.S., Güsten, R.: 1990, AJ, 99, 924

Bertout, C.: 1987, in *Circumstellar Matter*, eds. I. Appenzeller and C. Jordan, Dordrecht, Reidel, p.23

Bertout, C.: 1989, ARA&A, 27, 351

Bertout, C., Basri, G., Bouvier, J.: 1988, ApJ, 330, 350

Bertout, C., Basri, G., Cabrit, S.: 1992, in *The Sun in Time*, eds. M.S. Giampapa and G.R. Sonnet, Univ. of Arizona, Tucson

Bertout, C., Regev, O.: 1992, ApJ, 399, L163

Bodenheimer, P.: 1991, in *Angular Momentum Evolution of Young Stars*, eds. S. Catalano and J.R. Stauffer, Dordrecht, Kluwer Academic Publ., p.1

Bouvier, J.: 1990, AJ, 99, 946

Bouvier, J.: 1991, in *Angular Momentum Evolution of Young Stars*, eds. S. Catalano and J.R. Stauffer, Dordrecht, Kluwer Academic Publ., p.41

Bouvier, J., Bertout, C.: 1992, A&A, 263, 113

Bouvier, J., Cabrit, S., Fernandez, M., Martin, E.L., Matthews, J.M.: 1993, A&A, 272 176

Cabrit, S., Edwards, S., Strom, S.E., Strom, K.M.: 1990, ApJ, 354, 687

Cohen, M., Emerson, J.P., Beichman, C.A.: 1989, ApJ, 339, 455

Cohen, M., Kuhi, L.V.: 1979, ApJS, 41, 743

Duschl, W., Tscharnuter, W.M.: 1991, A&A, 241, 153

Edwards, S., Cabrit, S., Strom, S.E., Heyer, I., Strom, K.M., Anderson, E.: 1987, ApJ, 321, 473

Feigelson, E.D., Giampapa, M.S., Vrba, F.J.: 1992, in *The Sun in Time*, eds. M.S. Giampapa and G.R. Sonnet, Univ. of Arizona, Tucson

Feigelson, E.D., Jackson, J.M., Mathieu, R.D., Myers, P.C., Walter, F.D.: 1987, AJ, 94, 1251

Finkenzeller, U., Jankovics, I.: 1984, A&AS, 57, 285

Finkenzeller, U., Mundt, R.: 1984, A&AS, 55, 109

Frank, J., King, A.R., Raine, D.J.: 1992, in *Accretion Power in Astrophysics*, 2nd edition, Cambridge University Press

Friedjung, M.: 1985, A&A, 146, 366

Gebbie, K.B., Steinitz, R.: 1974, ApJ, 188, 399

Glasby, J.S.: 1974, in *The Nebular Variables*, Oxford, Pergamon

Haas, M., Leinert, C., Zinnecker, H.: 1990, A&A, 230, L1

Hamann, F., Persson, S.E.: 1992, ApJ, 339, 1078

Haro, G.: 1968, in *Nebulae and Interstellar Matter (Stars and Stellar Systems, Vol. 7)*, eds. B.M. Middlehurst and L.H. Aller, Chicago, Univ. Chicago Press, p.141

Hartigan, P., Hartmann, L., Kenyon, S.J., Strom, S.E., Skrutskie, M.E.: 1990, ApJ, 354, L25

Hartmann, L.: 1986, Fundam. Cosmic Phys. 11, 279

Hartmann, L.: 1991, in *The Physics of Star Formation and Early Stellar Evolution*, eds. C.J. Lada and N.D. Kylafis, Kluwer Academic Publ., Dordrecht

Hartmann, L., Kenyon, S.J.: 1988, in *Formation and Evolution of Low-Mass Stars* ed. A.K. Dupree, Dordrecht, Reidel, p.163

Herbig, G.H.: 1945, PASP, 57, 166
Herbig, G.H.: 1960, ApJS, 4, 337
Herbig, G.H.: 1962, Adv. A&A, 1, 47
Herbig, G.H.: 1965, ApJ, 141, 588
Herbig, G.H.: 1977, ApJ, 217, 693
Herbig, G.H., Bell, K.R.: 1988, Lick Obs. Bull. No. 1111
Hunger, K.: 1956, Zs. f. Astrophys., 39, 36
Jefferies, J.T., Thomas, R,.: 1959, ApJ, 129, 401
Joy, A.H.: 1945, ApJ, 102, 168
Kenyon, S.J., Hartmann, L.: 1987, ApJ, 323, 714
Kley, W.: 1991 in *Structure and Emission Properties of Accretion Disks*, eds. C. Bertout, J.-P. Lasota, S. Collin, J. Tran Tranh Van, Editions Frontières, Gif-sur-Yvette, p.323
Königl, A.: 1991, ApJ, 370, L39
Leinert, C., Haas, M.: 1989, ApJ, 342, L39
Lynden-Bell, D., Pringle, J.E.: 1974, MNRAS, 168, 603
Magazzú, A., Rebolo, R., Pavlenko, Ya.V.: 1993, ApJ in press
Maihara, T., Kataza, H.: 1991, A&A, 249, 392
Malbet, F.: 1993, PhD thesis, Paris VII University
Malbet, F., Bertout, C.: 1991, ApJ, 383, 814
Malbet, F., Bouvier, J., Monin, J.-L.: 1993, preprint
Malbet, F., Rigaut, F., Bertout, C., Lèna, P.: 1993, A&A, 271, L9
Natta, A., Giovanardi, C.: 1991, in *The Physics of Star Formation and Early Stellar Evolution*, eds. C.J. Lada and N.D. Kylafis, Kluwer Academic Publ., Dordrecht, p.595
Panagia, N.: 1991, in *The Physics of Star Formation and Early Stellar Evolution*, eds. C.J. Lada and N.D. Kylafis, Kluwer Academic Publ., Dordrecht, p.565
Poetzel, R., Mundt, R., Ray, T.P.: 1989, A&A, 224, L23
Pringle, J.E.: 1989, MNRAS, 236, 107
Profitt, C.R., Michaud, G.: 1989, ApJ, 346, 976
Regev, O.: 1983, A&A, 123, 146
Regev, O.: 1991 in *Structure and Emission Properties of Accretion Disks*, eds. C. Bertout, J.-P. Lasota, S. Collin and J. Tran Tranh Van, Editions Frontières, Gif-sur-Yvette, p.311
Reipurth, B., Zinnecker, H.: 1993, A&A, 278, 81
Rucinski, S.M.: 1985, AJ, 90, 2321
Ruden, S.P., Pollack, J.B.: 1991, ApJ, 375, 740
Rydgren, A .E., Zak, D.S.: 1987, PASP, 99, 141
Shakura, N.I., Sunyaev, R.A.: 1973, A&A, 24, 337
Stauffer, J.: 1991, in *Angular Momentum Evolution of Young Stars*, eds. S. Catalano and J.R. Stauffer, Dordrecht, Kluwer Academic Publ., p.117
Strom, K.M., Wilkin, F.P., Strom, S.E., Seaman, R.L.: 1989, AJ, 98, 1444
Terquem, C., Bertout, C.: 1993, A&A, in press
Tscharnuter, W.M.: 1985, in *Birth and Infancy of Stars*, eds. R. Lucas, A. Omont and R. Stora, Amsterdam, Elsevier, p.601
Walker, M. F.: 1972, ApJ, 175, 89

Walter, F.M.: 1986, ApJ, 306, 573
Walter, F.M., Brown, A., Mathieu, R.D. Myers, P.C., Vrba, F.J.: 1988, AJ, 96, 297
Willson, L.A.: 1974, ApJ, 191, 143

Massive Stars and Their Interactions with Their Environment

J. E. Dyson

Department of Physics and Astronomy, University of Manchester,
Manchester M13 9PL, England

1 Introduction

This set of lectures deals mainly with the consequences of the birth of massive stars on the circumstellar environment. We therefore adhere to Leitherer's (1991) definition of such a star, namely one with an initial main sequence mass greater than around 8–10 $M_\odot$. Such stars possess powerful Lyman continuum radiation fields and hypersonic winds which together disrupt surrounding material. These stars and their descendants are important sources of mass, momentum and energy for the interstellar medium at large.

The evolution of massive stars can be strongly affected by mass loss and as a result, their tracks in the H-R diagram can be extremely complex. Maeder (1984; 1990) has, for example, shown that initially very high mass stars, ($M_i \gtrsim 60 M_\odot$), evolve from O stars to a final supernova event via stages in which they are successively blue supergiants, luminous blue variables and Wolf-Rayet stars. Lower mass stars replace the luminous blue variable phase by one where they become red supergiants. Fast and slow mass loss can occur at different stages of evolution, and the possibilities for hydrodynamic interactions are legion (e.g. Dyson & Smith 1985). It is clearly legitimate, of course, to view the descendants of initially massive stars as part of the massive star phenomenon.

As has been noted by many authors, an understanding of massive star formation is hindered by the effects of the stars on their nascent environment. The Kelvin-Helmholtz time t_K of a star of mass M_* and radius R_* is given by

$$t_K = \frac{GM_*^2}{R_* L_*} \simeq 3\,10^7 \left(\frac{M_*}{M_\odot}\right)^2 \left(\frac{R_*}{R_\odot}\right)^{-1} \left(\frac{L_*}{L_\odot}\right)^{-1} \text{yr} . \quad (1)$$

The stellar luminosity increases strongly with stellar mass ($L_* \propto M_*^{3.5}$ on the upper main sequence) and t_K decreases quite rapidly with M_*. The time to assemble a star from a cloud initially marginally supported against its self-gravity by its internal pressure is $t_{\text{acc}} \approx GM_*^2/a^3$ (Shu, Adams & Lizano 1987), where a is the effective sound speed in the cloud. Thus

$$t_{\text{acc}} \approx 4.10^4 \left(\frac{M_*}{M_\odot}\right)^2 \left(\frac{a}{1\ \text{km}\,\text{s}^{-1}}\right)^{-3} \text{yr} . \quad (2)$$

On the upper main sequence, $t_{\rm acc}/t_{KH}$ is generally greater than unity and massive stars are still accreting when they commence nuclear burning. As a result, these stars are initially embedded in dusty accretion envelopes which are optically invisible, but strong infrared sources. Their winds and radiation fields interact with the accretion flow and affect the mass of the star finally formed (e.g. Kahn 1974). Once such influences spread outside the accretion flow, they cause considerable modification to the stellar environment and the resulting conditions there do not reflect the conditions when star formation was initiated. A major goal in the study of massive star formation is to determine the positive and negative effects of such modifications on the star formation process.

In order to study some of the consequences of star formation, we first review the relevant hydrodynamics.

2 Shocks and Ionization Fronts

2.1 Basic Hydrodynamics

Heyvaerts (1991) has given an excellent overview of basic gas dynamics which includes the derivation of the relevant equations. Neglecting mass sources, mass sinks and forces other than pressure forces, the basic equations are:

Mass conservation

$$\frac{\partial \rho}{\partial t} + \underset{\sim}{u} \cdot \nabla \rho + \rho \nabla \cdot \underset{\sim}{u} = 0 \ . \tag{3}$$

Momentum conservation

$$\frac{\partial \underset{\sim}{u}}{\partial t} + \underset{\sim}{u} \cdot \nabla \cdot \underset{\sim}{u} = -\frac{1}{\rho} \nabla P \ . \tag{4}$$

Energy conservation

$$\frac{\partial \epsilon}{\partial t} + \underset{\sim}{u} \cdot \nabla \epsilon + P \left[\frac{\partial}{\partial t} \left(\frac{1}{\rho} \right) + \underset{\sim}{u} \cdot \nabla \left(\frac{1}{\rho} \right) \right] = Q \ . \tag{5}$$

In Eqs (3)–(5), $\underset{\sim}{u}$, ρ and P are the fluid velocity, density and pressure respectively; $\epsilon (\equiv P/(\gamma - 1)\rho)$ is the specific internal energy; γ is the ratio of principal specific heats; Q is the net rate of energy gain per unit mass. Quite generally, Q is a complex function which may involve additional equations describing ionization balance, radiation transfer, etc.

Shock fronts have thicknesses of the order of some appropriate collision mean-free-path or Larmor radius. Ionization fronts have thicknesses of the order of the mean-free-path of an 'average' ionizing photon in neutral gas. Consequently, under practically all circumstances, shocks and ionization fronts are extremely thin compared to the scale sizes of flows where they occur. The conditions on either side of these discontinuities can be related using the [1D] time independent forms of Eqs (3), (4) and (5), i.e.

$$\frac{d}{dx}(\rho u) = 0 \tag{6}$$

$$\frac{d}{dx}(\rho u^2) + \frac{dP}{dx} = 0 \tag{7}$$

$$\frac{d}{dx}\left[\rho u \left(\frac{\gamma}{\gamma - 1}\right)\frac{P}{\rho} + \frac{1}{2}u^2\right] = Q \ . \tag{8}$$

2.2 Shock Fronts

The appropriate equations are (6), (7) and (8), with $Q = 0$. They integrate to give

$$\rho u = \Phi \tag{9}$$

$$P + \rho u^2 = \Pi \tag{10}$$

$$\frac{1}{2}u^2 + \left(\frac{\gamma}{\gamma - 1}\right)\frac{P}{\rho} = \mathcal{E} \ . \tag{11}$$

All three quantities (Φ, Π, $\mathcal{E}$) are unchanged across a shock front. Eqs (9)–(11) can be used to relate upstream (subscript 0) to downstream (subscript 1) conditions and in this form are known as the Rankine-Hugoniot (R-H) conditions. The direction of the jump is determined by the requirement that the specific entropy increases across the shock. The downstream and upstream conditions are related by

$$\frac{\rho_1}{\rho_0} \equiv \frac{u_0}{u_1} = \frac{(\gamma + 1)M_0^2}{2 + (\gamma - 1)M_0^2} \tag{12}$$

$$\frac{P_1}{P_0} = \frac{2\gamma M_0^2 - (\gamma - 1)}{(\gamma - 1)} \tag{13}$$

$$M_1^2 = \frac{2 + (\gamma - 1)M_0^2}{2\gamma M_0^2 - (\gamma - 1)} \tag{14}$$

where the Mach number is defined by

$$M = u/a; \ \ a^2 = \gamma P/\rho \ . \tag{15}$$

Inspection of these equations shows that $M_0 > 1$ and $M_1 < 1$. Shocks are generated by the propagation of supersonic disturbances through gases.

It is usually sufficient to utilize the strong shock ($M_0 \gg 1$) forms of Eqs (12)–(14). For the usual case ($\gamma = 5/3$), these are: $\rho_1/\rho_0 \equiv u_0/u_1 = 4$; $P_1/P_0 = 5M_0^2/4$; $M_1 = 1/\sqrt{5}$; $P_1 = 3\rho_0 u_0^2/4$. These velocities are measured in the shock frame; in a fixed frame where the shock velocity is V_s and the upstream and downstream gas velocities are v_0 and v_1 respectively, $v_1 \simeq (3/4)\,V_s$ for the usual case $|\,V_s\,| >> |\,v_0\,|$.

Under many circumstances of astrophysical interest, the shocked gas cools rapidly. If this gas cools to an (isothermal) sound speed c_2 ($\equiv \sqrt{P_2/\rho_2}$), where $c_2 < c_1$, the strong shock versions of the R-H conditions give

$$u_2 \simeq c_2^2/u_0; \ \ \rho_2/\rho_0 \simeq u_0^2/c_2^2 \ . \tag{16}$$

Often (e.g. when photoionization determines conditions in the upstream and downstream gas), $c_2 \simeq c_0$ and then

$$\rho_2/\rho_0 \simeq u_0^2/c_0^2 = M_0^2 \ . \tag{17}$$

In the fixed frame (if $| V_s | >> | v_0 |$),

$$v_2 = V_s(1 - c_2^2/u_0^2) \simeq V_s; \ P_2 = \rho_0 v_s^2 \ . \tag{18}$$

In general then the post-shock pressure $P = \alpha \rho_0 V_s^2$, $v_s = \alpha V_s$ where $\alpha = 3/4$ (adiabatic) or $\alpha = 1$ (well cooled).

2.3 Ionization Fronts

The basic equations (6)–(8) now retain Q since the process of photoionization injects energy and the ions and electrons produced can produce radiative energy loss in a variety of ways. In general Q is a function of position in the ionization front (IF) and is extremely complex to calculate properly (Mason 1975). Fortunately, the basic principles can be established without doing this. We describe the variation of flow quantities within the IF using the flow equations, which implies that the thicknesses of IFs are much greater than collision mean-free-paths. This is generally well satisfied. The corollary is that shock fronts can occur *within* IFs. Because IFs involve radiative energy input and radiative loses, the entropy increase condition utilized for adiabatic shocks no longer holds and we therefore anticipate a greater range of transition possibilities than for shocks.

Defining the isothermal sound speed as c and Mach number M by $c^2 = P/p$, $M = u/c$ respectively, Eqs. (9) and (10) give

$$\frac{c^2}{u} + u = \frac{\Pi}{\Phi} \ . \tag{19}$$

The stagnation enthalpy $\mathcal{E}$ defined by Eq. (11) can be written

$$\mathcal{E} = \frac{5}{2}c^2 + \frac{1}{2}u^2 = \frac{5}{2}\left(\frac{\Pi}{\Phi}\right) u - 2u^2 \ . \tag{20}$$

Using Eq. (19) the Mach number is given by

$$M = \left(\frac{\Pi}{\Phi u} - 1\right)^{-\frac{1}{2}} \ . \tag{21}$$

There are three fiducial velocities associated with Eqs. (19)–(21).

1. The sonic point ($M = 1$) is realized when $u \equiv u_s = \Pi/2\Phi$. Clearly, subsonic and supersonic flow correspond to $u < u_s$ and $u > u_s$ respectively. It is readily verified that c reaches its maximum value $c_{\rm max}$ at $u = u_s$.
2. At $u \equiv u_m = \Pi/\Phi$, $c = 0$ and $M \to \infty$. The stagnation enthalpy is finite, $\mathcal{E} \equiv \mathcal{E}_0 = \pi^2/2\Phi$.
3. $\mathcal{E}$ reaches a maximum value $\mathcal{E}_{\rm max} = 25\Pi^2/32\Phi^2$ when $u \equiv u_* = 5\Pi/8\Phi$.

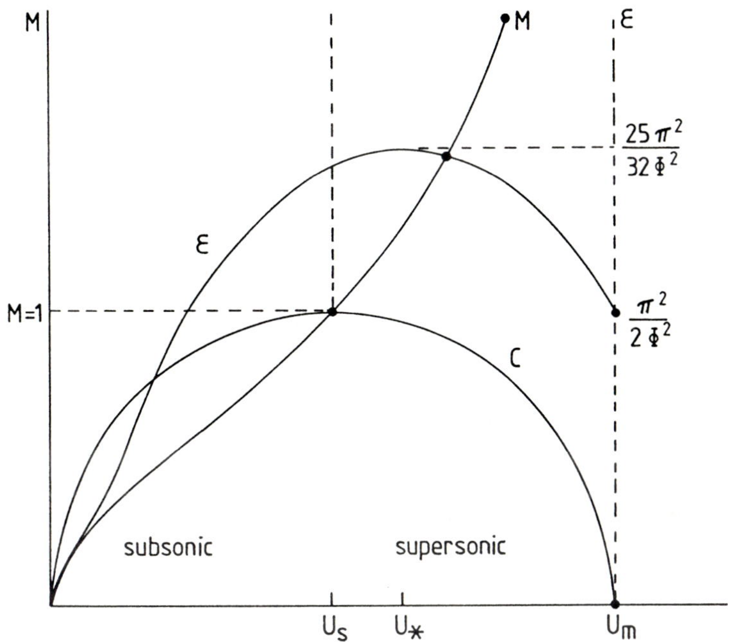

Fig. 1. The internal structure of an ionization front in velocity space.

Figure 1 sketches the variation of $c, \mathcal{E}$ and M through an IF; the gas inside the IF must always sit on the curves. When gas enters an IF, it is heated by photoionization and $\mathcal{E}$ increases. Detailed calculations show that in general, $\mathcal{E}$ reaches a maximum and then decreases eventually to a steady value where radiative heating and cooling balance (Fig. 2). $\mathcal{E}$ can never exceed $\mathcal{E}_{\max}$ and we discuss later what happens if the gas in an IF tries to do this.

In order to accommodate the increase in $\mathcal{E}$ as the gas flows through the IF, Fig. 1 shows that the upstream (neutral) gas can have velocity $u = u_0 \approx 0$ (subsonic upstream flow) or $u = u_0 \approx u_m$ (supersonic upstream flow).

Equation (19) can be written as

$$(c_0^2 + u_0^2)/u_0 = (c^2 + u^2)/u \tag{22}$$

where u and c are the velocity and sound speed respectively at a general point within the IF. Equation (22) has the solutions

$$\delta \equiv \frac{u}{u_0} = \frac{1}{2}\left[\frac{(c_0^2 + u_0^2) \pm \sqrt{(c_0^2 + u_0^2)^2 - 4c^2u_0^2}}{u_0^2}\right] . \tag{23}$$

There are two cases which allow real δ.

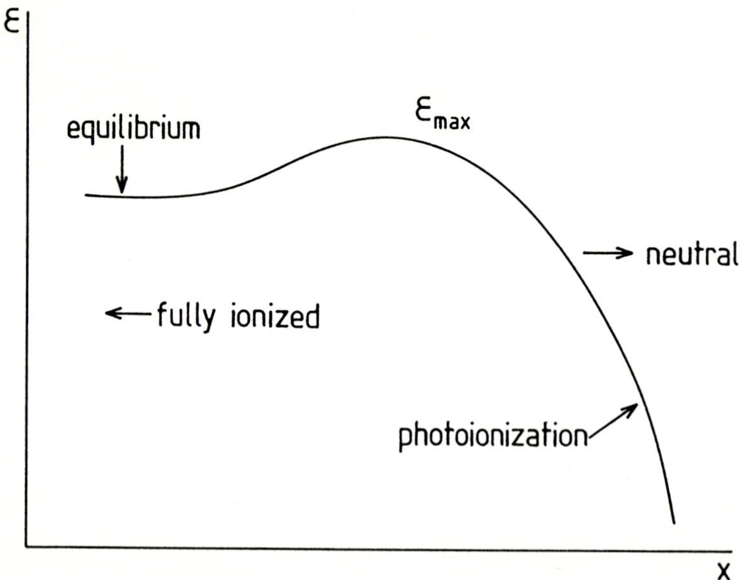

Fig. 2. The spatial variation of stagnation enthalpy $\mathcal{E}$ through an ionization front.

(a)

$$c_0^2 + u_0^2 \geq 2cu_0 \tag{24}$$

i.e.,

$$u_0 \geq u_R = c + (c^2 - c_0^2)^{\frac{1}{2}} \ . \tag{25}$$

Since $c >> c_0$ (because of heating), $u_0 \geq u_R = 2c_{\rm max}$. Far downstream in the ionized gas, $c = c_1 = c_i$ (≈ 12 km s^{-1}). Because of the overshoot in $\mathcal{E}$ (Fig. 2), $c_{\rm max} > c_i$ but in general the difference is small and a reasonable approximation is $u_R \approx 2c_i$ (≈ 25 km s^{-1}). Clearly, $u_0 \gg c_0$ and Eq. (23) has two solutions:-

(i) $\delta \simeq 1$, $u_0 \approx u_1 \geq u_R$. The density change across the IF ($= \delta^{-1}$) is negligible and the downstream velocity is supersonic. The pressure jump across the IF is $P_1/P_0 \approx c_i^2/c_0^2$, i.e., large – Weak-$R$ Front.

(ii) $\delta \approx c_i^2/u_0^2$, $u_1 \lesssim c_i/2$. The density increases across the IF and the downstream velocity is subsonic. The pressure jump $P_1/P_0 \approx 4c_i^2/c_0^2$, i.e. again large – Strong-R Front.

(b) Alternatively,

$$-\,(c_0^2 + u_0^2) \leq -2u_0 c \tag{26}$$

i.e.,

$$u_0 \leq u_D = c - (c - c_0^2)^{\frac{1}{2}} \simeq c_0^2/2c_{\rm max} \ . \tag{27}$$

As before, a good approximation is $u_D \approx c_0^2/2c_i$, and there are two cases:

(i) $\delta \simeq u_0^2/c_0^2$, $u_1 \approx c_0^2/u_0 \gtrsim 2c_i$. The density drops by a large factor across the IF and the downstream velocity is supersonic. The pressure jump is $P_1/P_0 \lesssim 1/4$, i.e. roughly isobaric – Strong-D Front.

(ii) $\delta = c_0^2/c_i^2$, $u_1 \leq c_i/2$. The density again drops by a large factor but the downstream velocity is subsonic, $P_1 \approx P_0$ and the transition takes place at constant pressure – Weak-D Front.

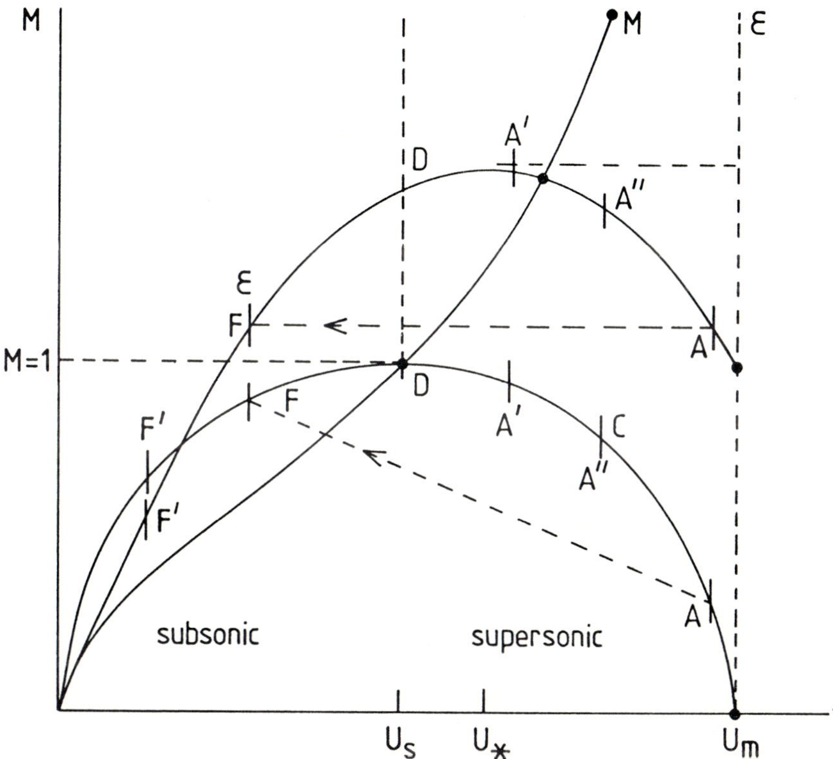

Fig. 3. Internal structures of R type ionization fronts. The path A→A′ →A″ gives a weak-R front. Possible strong-R fronts are A→D→F′ and A→F→ F′. The second path includes a shock within the ionization front.

Figures 3 and 4 show possible internal structures for the R type and D type IFs respectively. Since shock waves can go only in the direction supersonic → subsonic, it is clearly not possible to set up a strong D type front without passing through the point $(\mathcal{E}_{\max}, u_*)$ on the $\mathcal{E}$ curve. This implies that the front structure exactly adjusts to this, i.e. there is an internal constraint on these IFs.

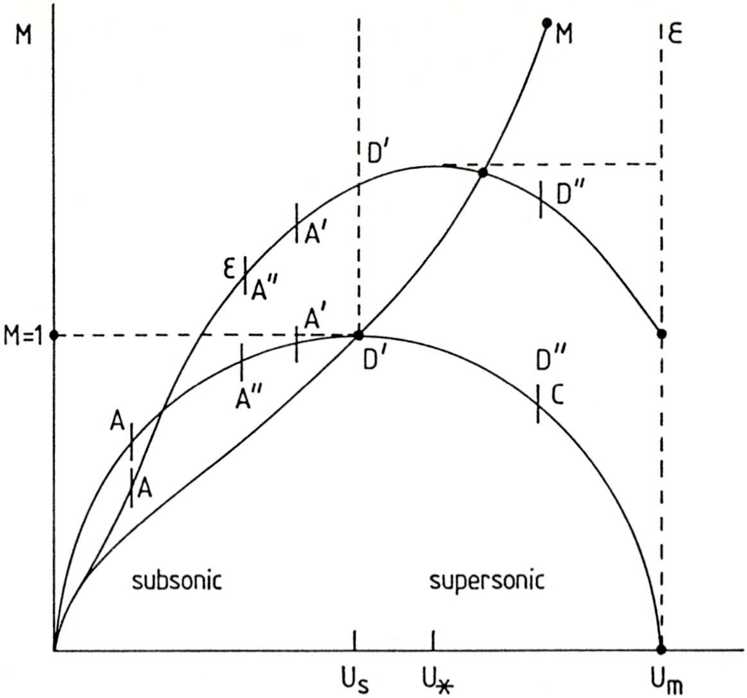

Fig. 4. Internal structures of D type ionization fronts. Weak-D fronts follow paths such as A→A′ →A″. Strong-D fronts *must* follow a path such as A→D′ →D″.

The velocity, V_I, of an IF relative to neutral gas is given by

$$V_I = \frac{J}{n_0} \tag{28}$$

where J is the normal ionizing photon flux incident at the IF and n_0 is the neutral gas density. Equation (28) assumes no recombination inside the IF (generally true) and effectively 100% ionization of H in the ionized gas. The velocity restrictions $V_I \geq u_R$ or $V_I \leq u_D$ can be expressed in terms of density restrictions, $n_0 \leq n_R$ or $n_0 \geq n_D$ for any given J. In the case $n_D > n_0 > n_R$, conditions ahead of the IF must adjust to satisfy the above constraints. Physically, the attempt to set up an IF in the forbidden range is equivalent to forcing $\mathcal{E}$ to exceed $\mathcal{E}_{\max}$. Since $\mathcal{E}_{\max} \propto \Pi^2$, its value can be increased by increasing Π, i.e. by a pressure wave moving ahead of the IF. The only possibility is to have a shock wave heading a D-type IF (since a supersonic front would overtake any disturbance). The initial shock velocity is such as to make $V_I = u_D$, where u_D is measured relative to the shocked gas. This initial configuration is sketched in Fig. 5. The initial velocity of the shock V_{SI} can be derived assuming the shock

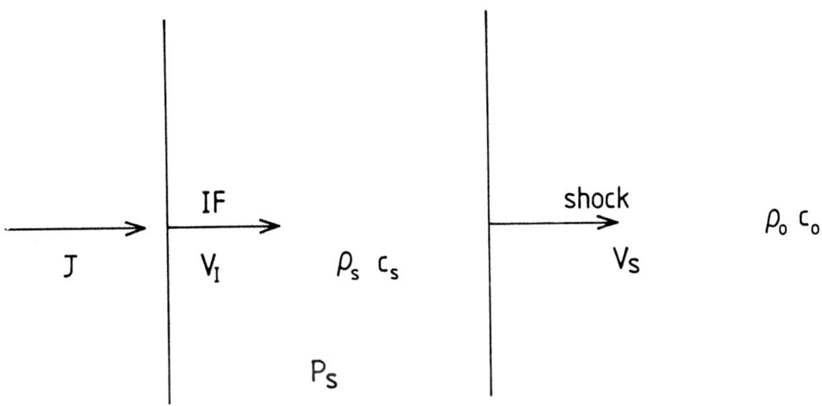

Fig. 5. The initial combination of shock and ionization fronts set up when the density lies in the forbidden range $n_D > n_0 > n_R$.

is strong and the post-shock gas cools well. The post shock pressure is

$$P_s = \rho_s c_s^2 = \rho_0 V_{SI}^2 \tag{29}$$

where subscript s refers to conditions in the shocked gas. Since

$$V_I = u_D = c_s^2 / 2c_i \tag{30}$$

$$\frac{J}{n_s} = \frac{c_s^2}{2c_i} \ . \tag{31}$$

Hence

$$V_{SI} = \left(\frac{2Jc_i}{n_0} \right)^{\frac{1}{2}} \ . \tag{32}$$

By definition, $(c_0^2/2c_i) \leq J/n_0 \leq 2c_i$ and therefore as would be anticipated, $u_D \leq V_{SI} \leq u_R$.

The determinacy of an IF depends on its type. If it moves supersonically into a medium, no disturbance can precede it; conversely, if it moves subsonically, an arbitrary disturbance can affect conditions ahead of it. Fronts moving supersonically with respect to downstream gas are not influenced by downstream conditions, whereas fronts moving subsonically need the specification of the downstream conditions. Taking the various types of IF in order;

(i) Strong-R fronts are analogous to shock fronts. If the up- and down-stream conditions are specified, there is no indeterminacy.

(ii) Since the downstream conditions cannot affect the propagation of a weak-R front there is one degree of indeterminacy associated with it.

(iii) Conditions ahead of a strong-D front can be affected by an arbitrary disturbance, but the downstream conditions cannot affect its propagation. There are then two degrees of indeterminacy.

(iv) Again, an arbitrary disturbance can move ahead of a weak-D front. There is one degree of indeterminacy associated with it.

The velocities of all IFs are given by Eq. (28) which proves one external constraint, thus removing one degree of indeterminacy. Consequently, weak-R and weak-D type fronts are completely specified. Strong-R type fronts are overdetermined and require very special conditions for their existence (cf. Goldsworthy 1961) and consequently, they are of little practical significance. The remaining degree of indeterminacy of strong-D type fronts is removed by the requirement that the flow passes through the point ($\mathcal{E}_{\max}$, u_*).

Quite generally, IF problems are time dependent and IFs can naturally evolve from one type to another. As a useful guide, there are several common situations where specific types of IFs occur:

1. Weak-R – early stages of evolution of HII regions when $V_I >> u_R$.
2. Weak-D – situations where the downstream pressure is kept high, e.g. the later stages of evolution of spherically symmetric HII regions, or when IFs are trapped in gas bounded by shock fronts (stellar wind bubbles).
3. Strong-D – flows off externally photoionized neutral globules where the pressure behind the IF is much greater than that in any surrounding medium.

2.4 Stellar Wind Interactions

Continuous mass loss from stars is ubiquitous in the H-R diagram. The mass loss rates $\dot{M}_*$ range from the insignificant $\sim 10^{-14} M_\odot$ yr^{-1} for the Sun, to perhaps as much as 10^{-3}–$10^{-4} M_\odot$ yr^{-1} in some red giants and supergiants. For any given star, $\dot{M}_*$ may dramatically change during the star's evolution (e.g. Maeder 1984). The wind velocity V_* ranges from $\sim$ 10 km s^{-1} for red supergiants to perhaps as high as 4000 km s^{-1} for some planetary nebulae nuclei. This range reflects different escape velocities and ejection mechanisms, and can again vary during the evolution of a given star.

Winds from early type stars are driven by the momentum in the stellar radiation field and move hypersonically with typical Mach numbers $\sim$ 100. The mass loss rates are characteristically $\sim 10^{-6} M_\odot$ yr^{-1} and the mechanical luminosity $\dot{E}_* (\equiv \dot{M}_* V_*^2/2) \sim 10^{36}$ erg s^{-1}. The impact of these winds sets up

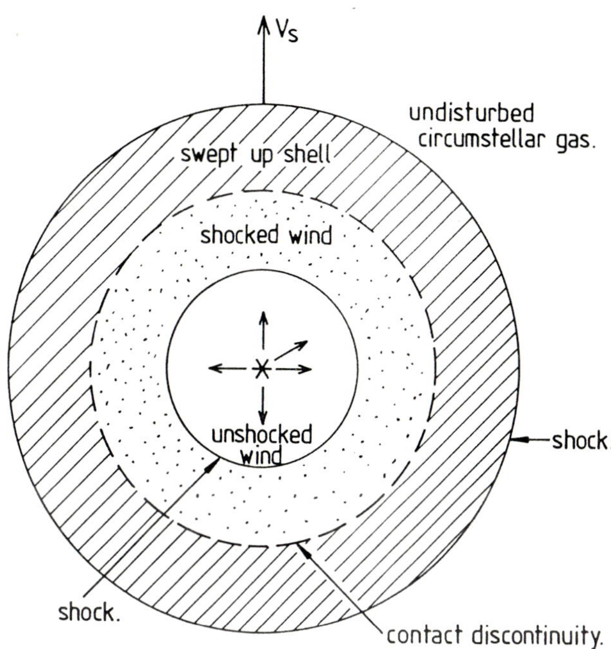

Fig. 6. The 'classical' two shock flow pattern set up when a hypersonic wind impacts on surrounding gas.

the characteristic 'two-shock' flow pattern (Fig. 6, e.g. Dyson & Williams 1980) once the wind has swept up its own mass in circumstellar material. This occurs at a time $t_0 \approx 3(\dot{M}_6/V_2^3 n_0)$ yr and radius $R_0 \simeq 0.06\,(\dot{M}_6/V_2 n_0)^{1/2}$ pc, i.e. essentially instantaneously. In these formulae, $\dot{M}_6 \equiv \dot{M}_*/10^{-6} M_\odot$ yr^{-1} and $V_2 \equiv V_*/2000$ km s^{-1}. The thermal behaviour of the shocked stellar wind governs the flow dynamics. If it cools on a timescale greater than the dynamic timescale, the shocked wind expands adiabatically and its pressure drives the swept-up shell—an 'energy' or 'pressure' driven flow. At the other extreme, the cooling time maybe much smaller than the dynamic timescale. The shell is driven by wind momentum—a 'momentum' driven flow. The transition from one regime to another depends critically on the wind speed (Dyson 1984). Energy driven flows occur when the wind velocity $V_* > V_*^c \simeq 100\,(n_0 \dot{M}_6)^{1/9}$ km s^{-1}; momentum driven flows occur if $V_* < V_*^c$. The transition is extremely sharp (Dyson 1984). Flows driven by early type stars then would be energy driven. It should, however, be stressed that this is true only if the circumstellar material is smoothly distributed. If the medium is very clumpy, enhanced radiative losses can occur in the interfaces between clumps (Hartquist & Dyson 1993) and the flowing shocked stellar wind, and the flow can then be better described as momentum driven.

Neglecting the thermal energy content of the swept up gas, dimensional arguments show that the shell radius $R_s \propto (\dot{E}_* \rho_0^{-1})^{1/5} t^{3/5}$ for a steady wind blowing into circumstellar material of constant density ρ_0. Generalizing to a medium with a radial density distribution $\rho(r) = \rho_* r^\beta$ gives (e.g. Dyson 1984)

$$R_s = \phi(\beta) \left(\frac{2\dot{E}_*}{\rho_*} \right)^{1/(5+\beta)} t^{3/(5+\beta)} \tag{33}$$

where

$$\phi(\beta) = \frac{(3+\beta)(5+\beta)^3}{12\pi(11+\beta)(7+2\beta)} \ . \tag{34}$$

The shell velocity, V_s, and radius, R_s, are related by

$$V_s = \frac{(3+\beta)\, R_s}{5t} \ . \tag{35}$$

A number of interesting implications follow from these equations.

(i) If $\beta = -2$, V_s is constant. This is relevant to the expansion of shells driven by the winds from planetary nebulae nuclei into slow asymptotic giant branch stellar ejecta.

(ii) If $\beta < -2$, the shells accelerate. The dense shell of swept up gas ($\approx 10^4$K) is pushed by tenuous hot ($T \gtrsim 10^7$ K) stellar wind. Consequently, the contact discontinuity may become Rayleigh-Taylor unstable. The maximum size of the fragments produced is about equal to the shell thickness which is usually small because the compression is high if the swept up gas cools well. This type of instability has been suggested as a mechanism for producing clumps in diffuse nebulae (Schneps, Ho & Barratt 1980) and planetary nebulae (e.g. Kahn & Breitschwerdt 1990).

(iii) The optical depth in the Lyman continuum of the shell varies with time in a way determined by the external density distribution. The recombination rate in the shell $\dot{\mathcal{N}}_R \propto n_s^2 R_s^2 \Delta R_s$, where ΔR_s and n_s are respectively the shell thickness and density. The equation of continuity shows that $\Delta R_s \propto R_s^{(1+\beta)}/\rho_s$ and if the outer shock is strong, $\rho_s \propto R_s^\beta V_s^2$. Hence $\dot{\mathcal{N}}_R \propto R_s^{(3+2\beta)} V_s^2 \propto t^{(5+4\beta)/(5+\beta)}$. Two special cases are those for a uniform density where $\dot{\mathcal{N}}_R \propto t$ and for $\beta = -2$ where $\dot{\mathcal{N}}_R \propto t^{-1}$. The uniform density case is most relevant to diffuse nebulae and here the IF becomes trapped in the shell when $\dot{\mathcal{N}}_R = S_*$, where S_* is the stellar ionizing photon output rate. It is readily shown (e.g. Dyson 1981) that this occurs at a time $t = t_T \simeq 10^6 S_{48} \dot{E}_{36}^{-1} n_0^{-1}$ yr, where $S_{48} \equiv S_*/10^{48}$ s^{-1} and $\dot{E}_{36} \equiv \dot{E}_*/10^{36}$ erg s^{-1}. This is relevant only if the shell velocity has not dropped to the sound speed in the ionized gas before t_T and this demands (Dyson 1981) $n_0 \gtrsim n_{\rm CRIT} \simeq S_{48}^2/\dot{E}_{36}^3$ cm^{-3}. As noted in §2, the trapped IF is weak-D type since effectively the pressure is uniform across the shell.

(iv) Flows can evolve from initially momentum driven to energy driven and vice-versa, depending on the external density distribution (Dyson 1984). A very comprehensive discussion of this including the effects of time dependent mechanical luminosity is given by Koo & McKee (1992a,b).

3 Ultra Compact HII Regions

3.1 Introduction to Ultra Compact HII Regions

Ultra compact HII regions (UCHII) are of considerable interest for several reasons. They are associated with OB stars and have often very high associated optical extinction which shows that they are embedded deep within molecular clouds. This has the important inference that at least some massive stars are formed within clouds in distinction to formation at cloud edges (e.g. Elmegreen 1992). They clearly are produced by the interaction of relatively young OB stars with their natal cloud surroundings. Churchwell (1990) has given an excellent overview of the subject and we only very briefly note some of the salient details.

UCHII have high densities ($< n_e^2 >^{1/2} \gtrsim 10^5$ cm^{-3}), small scale sizes ($L \lesssim 0.03$ pc) and high emission measures ($< n_e^2 > L \gtrsim 10^7$ cm^{-6} pc). Because of the high visual extinction and emission measures, most information is derived from high frequency ($\nu \gtrsim 10$ GHz) radio data. The necessary photon production rates necessary to keep them ionized are in the range 10^{45-49} s^{-1}, corresponding approximately to ZAMS spectral types B1–O5. Churchwell (1990) notes that 10–20% of OB stars are associated with UCHII, implying that they have lifetimes $\sim 10^5 - 10^6$ yr.

The morphologies of UCHII are varied. Churchwell (1990) divides them into 5 main morphological types; cometary ($\sim$ 20%), core-halo ($\sim$ 16%), shell ($\sim$ 4%), irregular or multiply peaked ($\sim$ 17%) and spherical or unresolved ($\sim$ 43%). Most attention has been given to cometary objects which are discussed later. Since they provide an interesting test-bed for studying the star–molecular cloud interactions, we first discuss some other possibilities.

3.2 UCHII as 'Classical' HII Regions

We first ignore the presence of stellar winds (which is probably reasonable for the later spectral types) and treat their formation as the ionization of cloud material by stars which turn on with the full ZAMS ionizing luminosities. We neglect the interaction of the stars with infalling material. A discussion of this earlier stage is given by Kahn (1974). An initial Strömgren sphere is set up by the propagation of a weak-R type IF into the cloud. This takes a time $t_{s0} \approx (n_0\beta_2)^{-1} \simeq 1.6 n_5$ yr^{-1} (e.g. Dyson & Williams 1980), where $n_5 \equiv n_0/10^5$ cm^{-3} and β_2 is the hydrogen recombination rate coefficient to levels $n \geq 2$. This Strömgren sphere has a radius R_{s0} given by the usual formula

$$R_{s0} = (3S_*/4\pi n_0^2\beta_2)^{\frac{1}{3}} \simeq 0.017\, S_{48}^{\frac{1}{3}}\, n_5^{-\frac{2}{3}} \text{ pc} \;. \tag{36}$$

When the IF radius is very close to R_{s0}, the IF velocity drops below $u_R\,(\simeq 2c_i)$ and as discussed in §2, a shock advances into the ambient gas and the resultant compression ensures that the IF changes to weak-D type. The resulting motion of the combined IF-shock is well approximated by (Dyson & Williams 1980)

$$R_s = R_{s0}\left(1 + \frac{7}{4}\frac{c_i t}{R_{s0}}\right)^{\frac{4}{7}} \;. \tag{37}$$

The problem is immediate. If, as an example, we take the lower limit of $\sim 10^5$ years as the life time of a UCHII, then after this time, $R_s \approx 10^{18}$ cm for $S_{48} \approx 1$ and $n_5 = 1 - 10$. Clearly this is far too large. Although there is evidence that very dense molecular clumps ($n \gtrsim 10^7$ cm^{-3}) exist in the hot core of the BN–KL region of the Orion Molecular Cloud, it is hard to argue their densities are generally so high and it is unlikely that UCHII can be satisfactorily modeled as small 'classical' HII regions.

3.3 UCHII as Photoionized Clumpy Clouds

Clumpiness affects strongly the structures of HII regions (Shull *et al* 1985) and wind-blown bubbles (e.g. Hartquist et al 1986). We first discuss the structure of HII regions neglecting winds.

Dense clumps ionized on their outside exist in diffuse nebulae (e.g. Dyson 1968). A detailed discussion of the relevant gas dynamics is given by Dyson (1968); Kahn (1969); Bertoldi (1989); Bertoldi & McKee (1990). Provided the clumps are dense enough, the photoionized gas expands at relatively low Mach numbers into the surrounding lower density material. The IF is strong-D type. Effectively the clumps act as injection sources of material. For simplicity here we assume that the clumps inject material at a constant rate $\dot{q}$ per unit volume with, on average, zero velocity relative to the flow, and that the injected material is the only source of mass. Mass is added everywhere inside the HII region. A flow is set up which remains ionized out to some radius R_R at which all the stellar ionizing photons have been absorbed. This transition region is a recombination front (RF) since the flow is now in the ionized-neutral direction. Formally, they are described by the same equations as IFs (i.e. Eqs. (8)–(10)). However there is a major difference. The heating overshoot which produces the maximum in the $\mathcal{E}/x$ diagram (Fig. 2) no longer occurs. The sound speed decreases monotonically through the RF. The only allowed possibilities are then subsonic-subsonic, supersonic-supersonic or supersonic-subsonic with an internal shock (cf. Mestel 1953). There is no equivalent of a Strong-D type IF.

In a steady state, the mass conservation condition is

$$4\pi r^2 \rho u = \int_0^r 4\pi z^2 \dot{q}\, dz = \frac{4}{3}\pi r^3 \dot{q} \ . \tag{38}$$

Since the flow is isothermal, the momentum equation is

$$u\frac{du}{dr} = -c_i^2 \frac{dln\rho}{dr} - \frac{\dot{q}u}{\rho} \tag{39}$$

where the second term on the RHS of Eq. (39) allows for the momentum 'absorbed' in accelerating injected gas to the local flow velocity. With $M \equiv u/c_i$, Eqs. (38) and (39) give

$$\frac{M}{(3M^2+1)^{2/3}} \approx M = Ar \tag{40}$$

$$\rho = \frac{\dot{q}}{3Ac_i(3M^2+1)^{2/3}} \simeq \frac{\dot{q}}{3Ac_i} \tag{41}$$

where A is a constant of integration and M is assumed small. The position of the RF is fixed by

$$S_* = \int_0^{R_R} 4\pi z^2 n^2 \beta_2 \, dz = \frac{4\pi}{3} \frac{R_R^3 \beta_2}{\overline{m}^2} \left(\frac{\dot{q}}{3Ac_i} \right)^2 . \tag{42}$$

Since the flow is very subsonic, the pressure on the ionized side of the RF is equal to the pressure Π_E on the neutral side. Hence $A = \dot{q}c_i/3\Pi_E$ and Eq. (42) gives

$$S_* \simeq \frac{4\pi}{3} \frac{\beta_2 R_R^3 \Pi_E^2}{\overline{m}^2 c_i^4} . \tag{43}$$

Thus

$$R_R \simeq 2\,10^{12}\, \Pi_E^{-2/3} S_{48}^{1/3} \text{ cm} \tag{44}$$

$$n \simeq 3.5\,10^{11} \Pi_E \text{ cm}^{-3} . \tag{45}$$

If $R_R = 10^{17} R_{17}$ cm, $\Pi_E \simeq 9\,10^{-8} S_{48}^{1/2} R_{17}^{-3/2}$ dyne cm^{-2} and $n \simeq 3\,10^4 R_{17}^{-3/2}$ $S_{48}^{1/2}$ cm^{-3}. The required mass injection rate depends on Mach number in the flow, M_F say, and is $\dot{q} \approx 1.5\,10^{-30} S_{48}^{1/2} R_{17}^{-5/2} M_F$ gm cm^{-3} s^{-1}.

If the external pressure is determined by the *effective* sound speed in the molecular cloud (≈ 3 km s^{-1} c_3 say), the mean ambient density required is $n_a \approx 2.5\,10^5 R_{17}^{-3/2} S_{48}^{1/2} c_3^{-1}$ cm^{-3}, which is high but not implausibly so. We return later to the derived value of $\dot{q}$. Note that we have identified n as the observed density. In fact, emission will come also from the surfaces of photoionized clumps and this contribution should really be taken into account.

3.4 UCHII as Wind Driven Flows in Clumpy Clouds

If we allow for the presence of fast winds, there are two possible processes for mass injection: photoionization of clumps as above or the hydrodynamic ablation of clumps by the wind itself, or—more properly—the mass loaded wind. This second process has been discussed in detail by Hartquist et al (1986) in the context of the hydrodynamics of the Wolf-Rayet Nebula RCW 58. Ablation occurs whatever the Mach number, M_F, of the flow relative to the clumps. If this flow is supersonic, the rear of the clump expands at roughly the clump internal sound speed in the flow. Mass loss occurs from 80% or so of the clump surface and the mass injection rate is independent of M_F. If this flow is subsonic, the small pressure differential between the stagnation pressure at the upstream facing surface of the clump and the pressure at the clump edges leads to ablation via the Bernoulli effect. In this case the clump expands at velocity M_F^2 times the clump internal sound speed, and the mass injection rate is proportional to $M_F^{4/3}$. For the purposes of discussion here though, we will assume a Mach number independent mass injection rate as above.

Mass injection into a high speed wind will clearly slow the flow down. The frictional energy dissipated may be radiated away if mixing of hot and cold gas at interface regions leads to enhanced radiative losses (Hartquist & Dyson 1993). We assume this is the case here and also, for simplicity, that the flow has enough momentum so that it always remains supersonic.

The condition of mass conservation remains Eq. (38). Since we neglect pressure forces, the momentum equation is

$$\frac{d}{dr}(r^2u^2\rho) = 0 \ . \tag{46}$$

So that

$$u = \frac{3\dot{\mu}_*}{4\pi\dot{q}r^3} \ ; \qquad \rho = \frac{4\pi\dot{q}^2r^4}{9\dot{\mu}_*} \tag{47}$$

where $\dot{\mu}_* \equiv \dot{M}_* V_*$.

The recombination front position follows again from Eq. (42), but now with the density distribution of Eq. (47). Hence

$$R_R = \left[\left(\frac{891}{64\pi^3}\right)\left(\frac{\dot{\mu}_*^2\overline{m}^2}{\beta_2\dot{q}^4}\right)S_*\right]^{\frac{1}{11}}$$

$$\simeq 2\,10^6\,\dot{q}^{-4/11}\,S_{48}^{1/11}\,\dot{\mu}_{28}^{2/11}\ \mathrm{cm} \tag{48}$$

where $\dot{\mu}_{28} \equiv (\dot{\mu}_*/10^{28}\ \mathrm{gm\,cm^{-1}\,s^{-2}})$. Again identifying the flow density with the observed rms density—with the same cautionary remarks—the predicted rms density is

$$< n^2 >^{\frac{1}{2}} \simeq 1.3\,10^{20}\dot{q}^{6/11}\,S_{48}^{4/11}\,\dot{\mu}_{28}^{-3/11}\ \mathrm{cm}^{-2} \ . \tag{49}$$

The Mach number just before the RF is

$$M_R \equiv \frac{u_R}{c_i} = \frac{3\dot{\mu}_*}{4\pi\dot{q}R_R^3c_i} \simeq 350\dot{\mu}_{28}^{5/11}\,S_{48}^{-3/11}\,\dot{q}^{1/11} \ . \tag{50}$$

The required mass injection rate is

$$\dot{q} = 3.8\,10^{-30}\,R_{17}^{-11/4}\,S_{48}^{1/4}\,\dot{\mu}_{28}^{1/2}\ \mathrm{gm\,cm^{-3}\,s^{-1}} \tag{51}$$

and so

$$M_R \simeq 0.75\,\dot{\mu}_{28}^{1/2}\,S_{48}^{-1/4}\,R_{17}^{-1/4} \tag{52}$$

$$< n^2 >^{1/2} \simeq 1.1\,10^4\,R_{17}^{-3/2}\,S_{48}^{1/2}\ \mathrm{cm}^{-3} \ . \tag{53}$$

There is reasonable consistency with the basic assumptions. At $r = R_R$, the momentum flux in the flow is $\approx \rho_R u_R^2 \approx 9\,10^{-8}R_{17}^{-2}\dot{\mu}_{28}$ dyne cm^{-2}. If $M_R >$ 1, then either a shock must occur within the RF or, more likely, the neutral downstream flow remains supersonic and eventually shocks somewhere in the cloud. The cloud confining pressure in principle could then be appreciably lower than that calculated for the non-wind flow. Note that, in this case, the UCHII would be surrounded by supersonic HI gas, possibly providing one discriminant for this model.

3.5 Mass Injection and Clump Lifetimes

As noted above, for wind driven UCHII, photoionization and hydrodynamic ablation can both effect mass injection. Photoionization will dominate if the pressure in the ionized gas at the IF dominates the ram pressure in the mass loaded flow. At distance r from the star (assuming the simple momentum conserving flow), the ram pressure is $P_F = \dot{\mu}_*/4\pi r^2$. The pressure at the IF is $P_I \approx n_I c_i^2 \simeq n_0 c_0^2$ where we assume approximate pressure balance across the IF, neutral and ionized gas densities n_I cm^{-3} and n_0 cm^{-3} respectively, and sound speed c_0 in the cloud. In order to compare P_I and P_F, some structure must be assumed for the clumps. For simplicity we take them as approximately self-gravitating isothermal clumps (cf. Dyson 1968), but will not attempt to take their proper structure into account here. In this case, the clump size ℓ_0 and density are related by $\ell_0 \approx c_0 (G\rho_0)^{-1/2}$. The IF on the clump surface must be D-type otherwise the clump will be ionized far too quickly. The clump must self-shield itself against ionization implying (cf. Dyson 1968; Kahn 1969)

$$f\, n_I^2\, \beta_2\, \ell_0 \approx J_\infty \tag{54}$$

where $J_\infty \equiv S_*/4\pi r^2$ is the flux incident on the ionized jacket around the clump and f is a geometrical factor giving the effective thickness of the jacket. Typically, $f \approx 0.1$ (Kahn 1969). The ratio of pressure at the IF to the ram pressure in the flow is therefore

$$\begin{aligned} \frac{P_I}{P_F} &\simeq \left(\frac{4\pi r^2\, G \overline{m}^4\, S_*^2 c_i^8}{f^2\, \beta_2^2\, c_0^4\, \dot{\mu}_*^3} \right)^{\frac{1}{3}} \\ &\simeq 1.8\, S_{48}^{2/3}\, R_{17}^{2/3}\, \dot{\mu}_{28}^{-1} \; . \end{aligned} \tag{55}$$

Hence $P_I > P_F$ for $R_{17} \gtrsim 0.4\, S_{48}^{-1} \dot{\mu}_{28}^{3/2}$, i.e. photoionization dominates towards the outer regions of the UCHII, at least for self-gravitating clumps.

The mass injection must fulfill two important requirements. It must be able to supply the required $\dot{q}$ and the clump lifetimes must be $\gtrsim 10^5$ years (though see below for a caveat on this latter requirement).

The mass loss rate from a photoionized clump is $\dot{M}_c \approx 4\pi \ell_0^2 \rho_0 u_D$ if the IF moves into the clump at the upper velocity limit for a D-type IF. Hence putting $u_D \approx c_0^2/2c_i$ and utilizing the condition $\ell_0 \approx c_0 (G\rho_0)^{-1/2}$, $\dot{M}_c \approx 4\pi c_o^4/c_i$. This remarkable result means that the mass loss rate for a photoionized clump is determined only by the sound speed in the neutral gas ahead of the IF. If $c_0 = 0.3 c_3$ km s^{-1}, $\dot{M}_c \simeq 1.5\, 10^{20} c_3^4$ gm s^{-1}.

If the required injection rate for the pure photoionization region is (Sec. 3.3) $\dot{q}_p$, the number of clumps needed is $N_c \approx \dot{q}_p V/\dot{M}_c$ where $V (\approx 4\, 10^{50} R_{17}^3$ cm$^3)$ is the UCHII volume. Hence $N_c \simeq 4\, R_{17}^{1/2} M_F S_{48}^{1/2} c_3^{-4}$ clumps ($\sim$ few) are required. If photoionization dominates for the wind driven UCHII, then (Sec. 3.4, Eq. (51)), a similar result obtains.

The clump characteristic lifetime is $t_\ell \simeq M_c/\dot{M}_c$, where the mass of a clump is $M_c \approx 4\pi\ell_0^3\rho_0/3 \approx ((4\pi c_0^{14}\epsilon\beta_2)/(3G^5C_i^4 J_\infty \overline{m}^2))^{1/3}$. Thus $t_\ell \simeq 3.6\,10^4 c_3^{2/3}\ R_{17}^{2/3}\ S_{48}^{-1/3}$ yr. This is reasonably consistent with the requirement $t_\ell \gtrsim 10^5$ years.

Alternatively, the clumps need not be gravitationally bound. The expansion time of an unbound clump $t_e \sim \ell_0/c_0 \approx 10^4\ell_{16}/c_3$ yr, where $\ell_{16} \equiv \ell_0/10^{16}$ cm. This is too short for a clump to have a permanent existence in the UCHII. However, cloud velocities in the vicinity of the star (mass M_*) powering the UCHII will be roughly $V_c \approx (2GM_*/r)^{1/2} \approx 2(M_*/20M_\odot)^{1/2}R_{17}^{-1/2}$ km s^{-1} and the crossing time $t_c \lesssim r/V_c \approx 1.6\,10^4\,R_{17}^{3/2}(M_*/20M_\odot)^{-1/2}$ yr (it could be less if a group of stars is present). An influx of new clouds from surrounding regions of the molecular cloud into the UCHII could be the source of mass. Such clouds would have to be dense enough to give IF velocities $\lesssim u_D$. Their densities need to be $n_0 \gtrsim 7\,10^7 S_{48}^{1/2}\ell_{16}^{-1/2}R_{17}^{-1}c_3^{-2}$ cm^{-3}. This is extremely high. Nevertheless, such very high clump densities may be present in some molecular cloud cores.

3.6 Cometary UCHII

By far the most attention has been given to cometary UCHII (Van Buren, MacLow & Wood 1990, MacLow et al 1991, Van Buren & MacLow 1992). In these models, the exciting star has a wind and moves supersonically (at velocity V_0) with respect to the circumstellar material (this is equivalent to a stationary frame with material moving at velocity $-V_0$ into the shock). A cometary shaped bow-shock configuration is set up (Fig. 7). The important point is that it is, as above, a stationary configuration, and although gas streams in and out of the shocked region, it will be maintained as long as there is a stellar wind and circumstellar gas. The cometary UCHII is identified with the ionized inner region of the swept up shell. This type of morphology is not uncommon. Dyson (1977) suggested that the galactic nebula S206 was produced as the wind from an O star interacted with gas streaming off a photoionized molecular cloud. In this case, the IF was not trapped in the shell. Dyson & Ghanbari (1989) proposed a similar explanation for the massive ($\approx 400M_\odot$) cometary Wolf-Rayet Nebula NGC 3199.

In order to set up a flow pattern of this kind, considerable distortion of the usual spherically symmetric pattern (Fig. 6) must occur. As first discussed by Weaver et al (1977), this occurs when the shell velocity drops to a value about equal to that of the star. This condition can be expressed as (cf. Dyson & Ghanbari 1989) $t \gtrsim 1.2\,10^6(\dot{E}_{36}/n_0)^{1/2}V_{20}^{-5/2}$ yr, where V_{20} is the velocity of the star relative to the interstellar medium in units of 20 km s^{-1}. Clearly, at densities characteristic of molecular clouds, this will occur very early on in the lifetime of the star.

The analysis given here is adapted from that of Dyson (1974). The principal assumptions are that the shocked circumstellar and shocked wind material cools to a thin layer and therefore that the shocked sandwich is thin. Secondly, the ionized and neutral components in the shell are assumed to move at the same velocity. The most questionable assumption is that the layer of shocked stellar

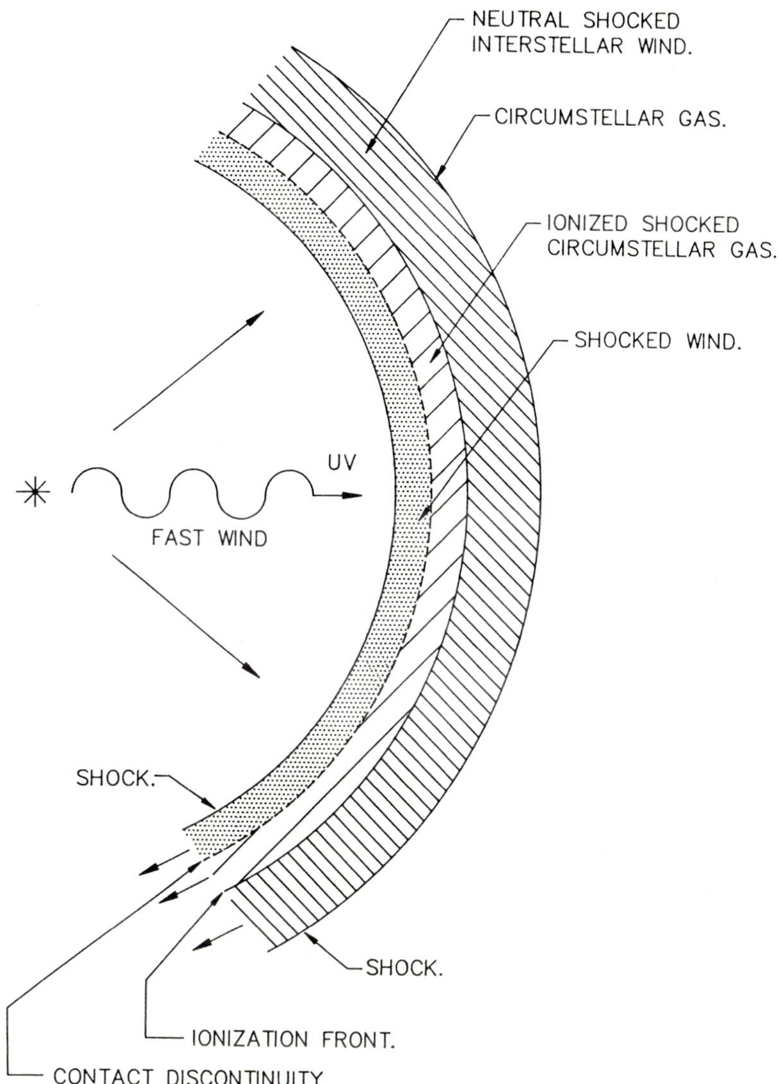

Fig. 7. Sketch of a cometary HII region.

wind is thin, since it is so hot ($T \gtrsim 10^7$ K) and tenuous that it does not cool well. Any mixing between this gas and shocked circumstellar gas which could cause enhanced cooling is likely to be confined to a thin boundary layer. It is possible to correct for this finite thickness of the layer in an approximate fashion (e.g. Dyson 1974), but here we ignore this.

The geometry is sketched in Fig. 8. The surface Σ is determined by the balance of the normal components of the momentum fluxes in the wind and

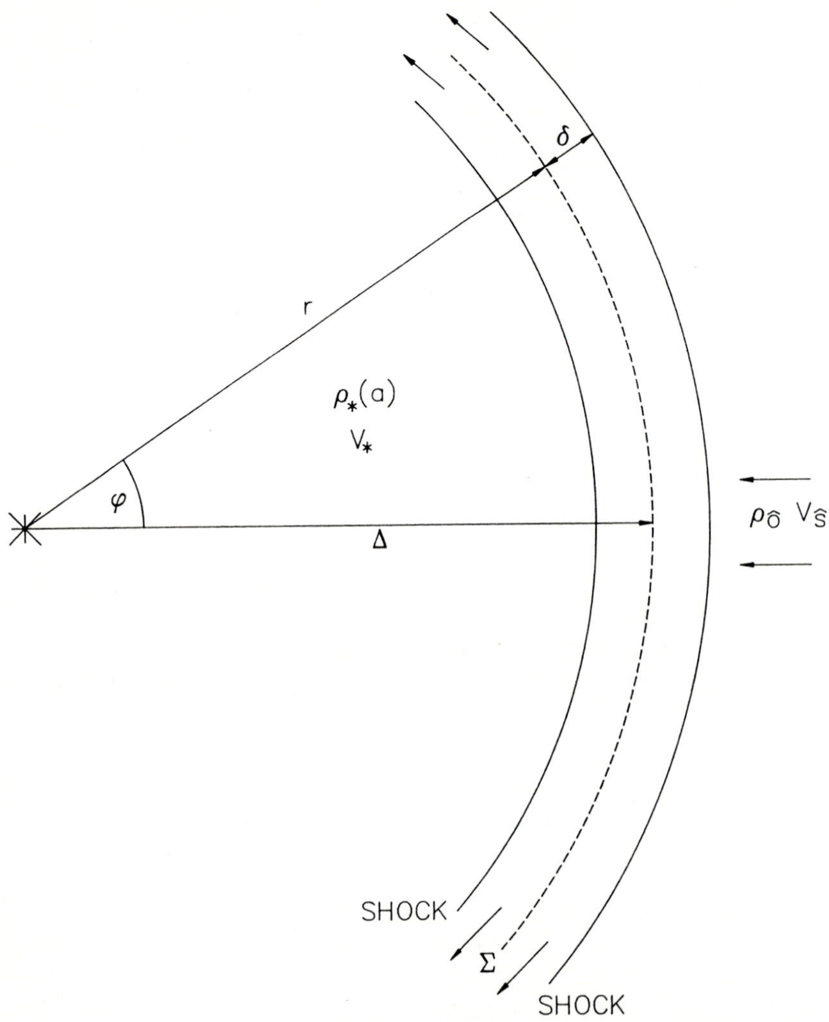

Fig. 8. The geometry for the calculation of the structures of cometary HII regions.

circumstellar gas. (Strictly, since the shocked gas flows along curved trajectories, the centrifugal force should be included in the momentum condition. However (cf. Dyson 1974) the effect is small.) At the stagnation point S, the momentum balance condition is

$$\rho_*(\Delta)V_*^2 = \rho_0 V_0^2 \ . \tag{56}$$

Since the wind flow is ballistic, $\rho_*(\Delta) = \dot{M}_*/4\pi\Delta^2 V_*$ and so

$$\Delta = \left(\frac{\dot{M}_* V_*}{4\pi\rho_0 V_s^2}\right)^{1/2} \simeq 1.2\,10^{17}\dot{E}_{36} n_5^{-1/2} V_5^{-1} V_2^{-1/2}\ \mathrm{cm}\ , \tag{57}$$

where $V_5 \equiv V_0/5\ \mathrm{km\,s^{-1}}$ and V_2 and n_5 are as defined previously. Scale sizes characteristic of UCHII are produced at molecular cloud densities. The shape of Σ is given by the momentum balance condition

$$\rho_*(r)V_{*n}^2 = \rho_0 V_{sn} \tag{58}$$

where V_{*n} and V_{sn} are respectively the normal components of V_* and V_s across Σ. Simple geometry (Dyson 1974) shows that the shape Σ is given by the equation

$$\left(\frac{r}{\Delta}\right)\cos\phi + \frac{d}{d\phi}\left(\frac{r\sin\phi}{\phi}\right) = 1 \tag{59}$$

which has the solution

$$\frac{r}{\Delta} = \frac{\phi}{\sin\phi}\ . \tag{60}$$

The IF must be trapped in the shell in order that the characteristic cometary morphology occurs. The condition for this is that $\Phi \equiv \beta_2 n_s^2\delta \gtrsim J_\infty$, where n_s and δ are respectively the shell density and thickness (assuming that the gas is fully ionized). To obtain δ, we have to solve for the flow in the shell of swept up gas. The mass flux into the shock over the surface with half-angle ϕ (Fig. 8) is

$$F = \pi r^2 \sin^2\phi\,\rho_0 V_s\ . \tag{61}$$

If we neglect pressure gradients in this gas, the velocity V follows from the momentum condition

$$\frac{d(FV)}{dA} = \rho_0\, V_{st}\, V_{sn} \tag{62}$$

where dA is the area element defined by

$$dA = 2\pi r^2 \sin\phi \left[1 + \frac{1}{r^2}\left(\frac{dr}{d\phi}\right)^2\right]^{\frac{1}{2}} d\phi \tag{63}$$

and V_{st} is the tangential component of velocity of the circumstellar gas at Σ. For small ϕ, $V_{sn} \approx V_s$, $V_{st} \approx V_s \sin\phi \approx V_s\phi$, $dA \approx 2\pi r^2\,\phi\,d\phi$. Hence $V \approx V_0\phi(2/3\sqrt{3})$. The surface density $\sigma(\equiv n_s\overline{m}\delta)$ can be found from the continuity condition

$$F \equiv \pi r^2\sin^2\phi\rho_0 V_s \approx \pi r^2\phi^2\rho_0 V_s = 2\pi r\sin\phi\sigma V \simeq 2\pi r\,\phi\,\sigma\,V\ . \tag{64}$$

Since the gas pressure in the ionized gas must balance the ram pressure into the shock, $n_s c_i^2 \overline{m} \simeq \rho_0 V_s^2$ and so $\delta \simeq (3\sqrt{3}/2)\,r\,c_i^2/V_s^2$ and $\Phi \simeq 33 n_5^{1/2}\,\dot{E}_{36}^{3/2}\,V_5^{-1}$ $S_{48}^{-1} V_2^{-3/2}$. Clearly trapping can occur with characteristic parameters.

4 OB Stars and Clumps

OB stars can have positive or negative effects on the clump structures of molecular clouds and this has implications for star formation. Negative effects (photoionization or wind induced ablation) have been introduced in the discussion on UCHII above.

On the positive side, there are several possibilities. Expanding HII regions drive shells of shocked neutral gas ahead of them once the IF velocity has dropped below u_R. Since the shock velocity is always less than the sound speed in ionized gas, swept-up molecules (H_2, etc.) are not dissociated and the shell is molecular. It is easily seen that this shell contains most of the disturbed material. The mass of the neutral shell is $M_n \propto R^3(n_0 - n_i)$, where the neutral shell has radius R and is assumed thin, and n_i ($\approx$ const) is the density in the HII region. To maintain ionization balance, $R^3 n_i^2 = R_{s0}^3 n_0$, and $M_n/M_T \propto 1 - (R_{s0}/R)^{3/2}$, where M_T is the total mass in the expanding system. Clearly $M_n \approx M_T$ once $R >> R_{s0}$.

Small scale inherent clumpiness in the swept up material will tend to be ironed out by the compression once the expansion velocity of the shell is not too great compared to the internal sound speed. In principle, though, the shell could become gravitationally unstable and form clumps. This would occur once the thickness of the shell $\Delta R \approx \ell_J$, where ℓ_J ($\approx c_s(G\rho_s)^{-1/2}$) is the Jeans length in the shell of density ρ_s and sound speed c_s. Magnetic fields have been ignored. The shell thickness $\Delta R \approx (R/3)(c_s^2/\dot{R}^2)$ where $\dot{R} \equiv dR/dt$, and $\rho_s \simeq \rho_0(\dot{R}^2/c_i^2)$, hence $\ell_J \simeq \Delta$ at $t \simeq t_G \approx (G\rho_0)^{-1/2} \approx 2 10^5 n_5^{-1/2}$ yr. The collapse time is a factor $\sim (c_s/\dot{R})$ less. At $t = t_G$, the expansion velocity of the shell follows from (37) and is $\dot{R}(t_G) \approx n_5^{3/14}$ km s^{-1} which will be marginally supersonic.

Wind driven shells dominate over expanding HII regions if the trapping time t_T for the IF in the shell is short enough. Since $t_T \approx 10 S_{48} \dot{E}_{36}^{-1} n_5^{-1}$ yr (§3), this is likely true in molecular clouds. Again, provided most of the mass is in the neutral part of the shell, the time at which clumps are formed is as above. The shell velocity is now $\dot{R}_s(t_G) \approx 3\dot{E}_{36}^{1/5}$ km s^{-1}. (A detailed treatment is given by Elmegreen 1992.)

An alternative possibility is that sudden acceleration of a swept-up shell leads to Rayleigh-Taylor instability. Pressure driven shells can accelerate for two reasons; if the driving pressure suddenly increases, or if there is a decrease in density ahead of the shell. In principle, only small pressure or density variations are required to instigate RT instability. In practice large enough variations are required to make the instability growth rate ($t_g \propto |\ \ddot{R}_s\ |^{-1/2}$) great enough. If the inverse of the growth rate for wavelengths about equal to the shell thickness is comparable to the shell dynamical timescale, the fragments produced have scale sizes about equal to Δ. If the shell gas has sound speed c_s, then $\Delta \approx R_s(c_s^2/3\dot{R}_s^2)$. If, for example, the acceleration of a wind driven shell is invoked to produce neutral fragments (cf. Schneps, Ho & Barratt 1981), $c_s \approx 1$ km s^{-1} and the compression factor is large. Only very small neutral fragments are produced in this way. A more fundamental question is whether shells driven by hot tenuous gas fragment if they accelerate. If a hole appears in the shell, the hot interior gas flows past the clumps and ablation of clump material into the flow occurs.

Meaburn, Hartquist & Dyson (1988) conjectured that the shells might self-seal by this mechanism and some observational support for this view came from their observation of small blister structures on a filament in the Vela supernova remnant. They argued that to obtain genuine shell fragmentation, it was likely that gravitational or thermal instabilities were necessary. (This also applies to the discussion in §2.4.)

A key question is whether or not such induced clumpiness can be distinguished from primordial clumpiness. Primordial is used in the sense of the structure obtaining before the current epoch of star formation. Of course it could retain a memory of previous episodes of star formation. Block, Dyson & Madsen (1992) have suggested that optically visible clumps with scale sizes of 1–2 pc seen in the S-E quadrant of the Rosette Nebula are primordial in the sense defined above. They argued that the optically visible chains of clumps lie well outside the central wind blown cavity of the nebula. Additionally, even if somehow the wind-blown shell could fragment, they noted that the fragments produced would have been much smaller than those observed. They suggested that the clumps observed are produced when stellar UV from the central stars impinges on a primordially clumpy medium and 'etches' the lower density interclump medium away. Provided the clumps are much denser than the interclump medium, the scale size of the etched clumps is $\ell \approx V_I t$, where V_I is the velocity of the IF into the interclump medium and t the time elapsed since ionization began. Since $V_I \approx J/n_I$, where J is the is the incident UV flux and n_I the interclump density, $\ell \approx Jt/n_I \sim 1-2$ pc, using appropriate values for the Rosette Nebula, in good agreement with the clump sizes observed.

5 Effects of Groups of OB Stars

The grouping of OB stars occurs in a wide variety of circumstances with a corresponding range in the scale of the grouping. In the solar neighbourhood, about 16 groups of OB stars, each containing up to a few tens of OB stars, exist in associations within 1.5 kpc of the Sun. Many well known giant diffuse nebulae (e.g. Orion, the Rosette) contain several OB stars. On a larger scale, giant extragalactic HII regions in spiral galaxies such as M101 need $10-10^4$ OB stars to maintain their ionization. Even more extreme are the nuclear regions of starburst galaxies which must contain more than 10^4 OB stars. Finally, more speculatively but with increasingly credibility, active galactic nuclei (and in particular QSOs) may have massive stellar space densities $\gtrsim 10^6$ pc^{-3}.

At its most simple, a sudden localized burst of massive star formation could lead to the following sequence of events: the generation of individual HII regions and wind driven shells, the overlap of these regions and shells and finally a collectively driven 'super' region. The dynamical effects in these various stages are driven by stellar radiation fields and winds, and by supernova explosions. The relative importance of the various phenomena depends on a variety of factors; the age and initial mass function of the cluster, the stellar metallicity and whether the cluster is formed in a sudden burst or if continuous star birth and death occurs.

Detailed studies of the input of mass, momentum and energy into the surroundings of stellar clusters show that OB and Wolf-Rayet stars always dominate the wind power input. In bursts of star formation, stellar winds dominate for times $\lesssim 10^7$ year, after this, supernovae dominate. In the continuous formation case, supernovae dominate provided all stars with initial masses $\gtrsim 8M_\odot$ (which become supernovae) are in equilibrium. Finally, the integrated power injection is constant (to within a factor ~ 10) over timescales $\sim 10^7$ yr.

As far as simple dynamical effects are concerned, all the processes (winds and supernovae) can be lumped together provided all the ejecta involved are thermalized, i.e. they must be able to interact with at least their own mass of surrounding material. In this approach, supernovae injecting energy E_s per event on a characteristic timescale τ_s between events behave rather like a continuous wind of mechanical luminosity $\dot{E} \equiv E_s/\tau_s$. We now briefly discuss four areas where stellar groups play an important role.

5.1 Superbubbles and the Disc-Halo Connection

Considerable evidence points to the existence of large scale structures analogous to wind or supernovae driven bubbles. Such evidence includes the X-ray emitting shell (diameter ≈ 450 kpc) in Cygnus; extended and clearly related HI and Hα loops and X-ray emission in Eridanus; frequently observed HI shells on scale sizes (0.1–3) kpc.

It is easy to estimate the energetics required to produce such super shells. If we combine the winds and supernovae as above, then the results of §3 can be used to show that the shell radius and velocity are respectively

$$R_s = 270(\dot{E}_{38}/n_0)^{\frac{1}{5}} t_7^{\frac{3}{5}} \text{ pc} \tag{65}$$

$$V_s = 16(\dot{E}_{38}/n_0)^{\frac{1}{5}} t_7^{-\frac{2}{5}} \text{ km s}^{-1} \tag{66}$$

where $\dot{E}_{38} = \dot{E}/10^{38}$ erg s^{-1}, $t_7 = t/10^7$ yr. Obviously then, the generation of super-shells requires the energy input of hundreds of OB stars. The interior structure of the bubble depends on whether or not the possibility of mass input from the swept-up shell by conductively driven evaporation is included (MacLow & McCray 1988). If it occurs, the evaporated mass dominates. The viewpoint taken here (cf. Hartquist & Dyson 1988) is that mechanisms which suppress conduction are operative. In that case the interior structure of the shell is that of a hot ($T \approx 5\,10^7 V_2^2$ K) bubble of mean density $< n > \approx 3\,10^{-4} \dot{E}_{38}^{2/5} n_0^{3/5} V_2^{-2} t_7^{-4/5}$ cm^{-3}. (In these expressions V_2 is the typical mass injection velocity in units of 2000 km s^{-1}). Standard theory gives an energy budget of (15/77) of injected energy in shell KE, (35/77) of injected energy in interior thermal energy and the remainder (27/77) used in compressing the shell (and generally radiated away). A caveat to this should be made. The implicit assumption that the bubble is energy driven and the structure is as described above assumes no significant injection of cool material into the bubbles by the ablation of, for example, molecular clouds.

Very important effects arise because the gas density in the galaxy is not homogeneous over the scale sizes of these superbubbles and is stratified perpendicular to the galactic plane. These effects lead to the generation of the 'bubble bath' structure of the galaxy (Brand & Zealey 1975; Dyson 1981) and the 'disc-halo' connection (e.g. MacLow, McCray & Norman 1988; Norman & Ikeuchi 1989).

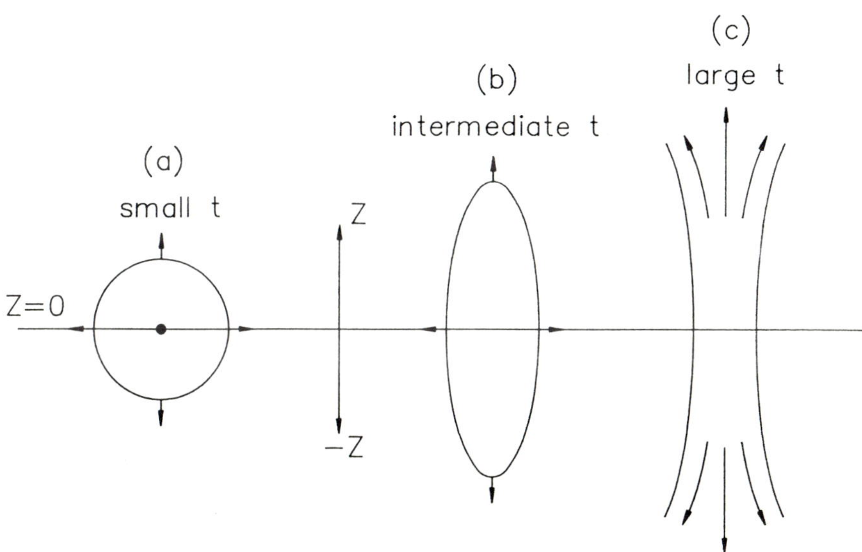

Fig. 9. The time evolution of a bubble in a stratified medium. The initially spherical bubble (a) elongates (b) and eventually blows out (c).

The principles are easily derived. For simplicity we assume energy driven bubbles (although essentially the same conclusions apply to the momentum driven case). Consider a bubble originating in the $z = 0$ plane of symmetry of a medium stratified in the z direction (Fig. 9). The bubble interior is spatially isobaric to a very good approximation since generally the expansion speed of the shell is much less than the sound speed in the hot bubble gas. If R is a characteristic shell dimension, the interior pressure, P, is proportional to the thermal energy per unit volume, i.e. $P \approx \dot{E}\, t/R^3$, for a constant energy injection rate $\dot{E}$. If the ambient density distribution takes the form $\rho = \rho_0\, f(z)$ (where $f(z)$ is symmetric about $z = 0$), the shell section moving in the z direction has velocity $\dot{R}(z) \approx (P/\rho(z))^{1/2}$. Since $t \approx R/\dot{R}(z)$, $\dot{R}(z) \approx (\dot{E}/R^2 \rho_0 f(z))^{1/3}$. If, for example, $f(z)$ has a form such as $f(z) \propto e^{-|z|/h}$, where h is an appropriate scale height, $\dot{R}(z) \propto (e^{|z|/h}/R^2)^{1/3} \sim (e^{|z|/h}/z^2)^{1/3}$. Along the plane of symmetry, however, $\dot{R}(0) \propto R^{-2/3}$, i.e. the standard uniform density behaviour. At small t, (Fig. 9(a)), the bubble is initially spherically symmetric, but clearly eventually $\dot{R}(z) > \dot{R}(0)$ and the bubble elongates along the z axis (Fig. 9(b)).

The velocity in the z direction initially decreases with time, but at $|z| \approx h$, the shell starts to accelerate and in this approximation, the velocity can in principle become very high. This phenomenon is called 'blow out'. Since the dense shell is accelerated by hot tenuous gas, the contact discontinuity is Rayleigh-Taylor unstable and the shell breaks up (Fig. 9(c)). Hot gas and shell fragments can flow into the galactic halo—the so-called 'disc-halo connection' (e.g. Norman & Ikeuchi 1989). Since at blow out, $\dot{R}(z) \approx (\dot{E}/h^2\rho_0)^{1/3}$, the blow-out time is roughly $t_b \approx (h^5\rho_0/\dot{E})^{1/3} \approx 10^6 h_{100}^{5/3} n_0^{1/3}/, \dot{E}_{38}^{-1/3}$ yr, where $h_{100} \equiv h/100$ pc. The velocity is $\dot{R}(h) \approx 50 \dot{E}_{38}^{1/3} n_0^{-1/3} h_{100}^{-2/3}$ km s^{-1}. So, for example, in the solar neighbourhood with $n_0 \approx 1$ cm^{-3}, $h_{100} \approx 1.5$, and assuming activity in an OB association lasts for a time $\approx 10^7$ years, blow-out occurs for $\dot{E}_{38} \approx 0.01$, i.e. only a small number of active stars is required in principle (though gravity has been ignored). However, the velocity predicted at blow-out is then ~ 10 km s^{-1} and genuine shell fragmentation may not occur (cf. remarks in §4). If thermal instability is needed, a blow-out velocity $\dot{R}(h) \approx 100$ km s^{-1} (Gaetz, Edgar & Chevalier 1988) is necessary and a value $\dot{E}_{38} \approx 10$ (i.e. hundreds of OB stars) may be required.

Amongst the many potentially very important consequences of this scenario are the production of high velocity clouds and the production of metal enhanced supernovae ejecta which eventually fall back to the galactic plane (e.g. Norman & Ikeuchi 1989).

5.2 Giant Extra-galactic HII Regions (GEHR)

A comprehensive overview is given by Kennicutt (1991) and only a brief resumé of salient points is given here. Typical parameters of GEHR are diameters $\sim (10^2 - 10^3)$ pc, ionizing OB stars $\sim (10 - 10^4)$ and ionized masses $M_I \sim (10^3 - 10^7) M_\odot$. (As a galactic comparison, the Orion Nebula has diameter ≈ 1 pc, contains 6 OB stars and has an ionized mass of several hundred $M_\odot$). Kennicutt (1991) notes three main reasons for their importance. The stellar content can be studied directly and used in direct comparison with theoretical evolutionary models of stellar groups. Their properties can be used to study the formation of stars in groups. Finally, the feedback into the ISM from star formation can be studied. These studies have particular relevance to starburst galaxies and active galactic nuclei (e.g. Perry 1993; Williams & Perry 1993).

A number of interesting general results have been obtained which have important implications for theories of star formation. For example, the initial mass functions may be appreciably flatter than in the solar neighbourhood. The upper mass limit may also vary with metallicity.

The kinematics are complex. Nearby GEHR are characterised by discrete shells, $V \sim 10 - 30$ km s^{-1} and the largest shells are associated with OB associations. Several GEHR (e.g. 30 Dor) contain high velocity shells ($V \approx 100 - 300$ km s^{-1}) which coincide with X-ray or non-thermal radio sources. Quite generally, gas velocities are highly supersonic with the consequence that continuous energy input from supernovae and stellar winds is required to balance strong post-shock radiative energy dissipation.

Several interesting features are thrown up by models of GEHRs. Analysis of the motions in the GEHR of M33 and M101 (Dyson 1978) shows that shells are driven by groups of stars which must have turned on simultaneously over relatively short timescales ($\sim 10^5 - 10^6$ yr). The shell lifetimes are comparable and re-energisation of shells must occur. This could be effected through supernovae events or stars increasing their wind output by becoming WR stars. This then implies that continuous star formation is ongoing and presumably these complexes must still contain molecular clouds. Very fast shells must be related to recent events such as sudden localized bursts of star formation or very recent supernovae explosions. It may be significant that the highest velocity shells have velocities of about 300 km s^{-1} which is when typical supernovae remnants become radiative. One interesting problem is, given the level of activity, why interstellar material remains around the sites of such supernovae. Possibilities include the movement of the presupernova star into a region where ISM remains or that the supernova occurs in circumstellar ejecta (e.g. red supergiant ejecta). In this latter case, the shells could be very short-lived (~ 100 yr) phenomena.

5.3 Starburst Galaxies

The definition of a starburst is one which takes place on a timescale much less than the Hubble time at a rate which could not be sustained for a Hubble time. An excellent overview of starburst properties and the relationship of this phenomenon to the subject of Active Galactic Nuclei is given by Perry (1993). We briefly note a few salient points. Observational signatures of a starburst include nebular emission lines, an OB continuum rising to the blue, very blue UBV colours and radio supernova remnants. The efficiency of star formation may be much higher than in galactic molecular clouds ($\approx 100\%$ as opposed to a few %). However, high levels of star formation activity may eject material from the star formation region and the efficiency (in terms of mass conversion) is hard to quantify. There is some evidence that IMFs are flatter than in the solar neighbourhood or, alternatively, that the low mass cut-off occurs at a higher value. Whether these result from the inhibition of low mass star formation by high mass stars or that conditions favour massive star formation is an open question.

One of the most interesting and important phenomena associated with compact starbursts is that of galactic superwinds (Heckman, Armus & Miley 1990). The lines of evidence pointing to the existence of these winds include large scale ($\gtrsim$ kpc) optical line and X-ray emission perpendicular to galactic discs; line ratios consistent with shock phenomena; double peaked optical emission line profiles extending over kpc which are consistent with expanding bubbles or bipolar flows; line splitting $\sim 200 - 600\ \mathrm{km\,s^{-1}}$.

The energy source for such winds is energy injection from winds and supernovae in the starburst. Provided that the stellar ejecta is thermalized, the situation is somewhat analogous to that described in §5.1, but differ in having a much higher energy (and consequently mass) injection rate. Let us assume kinetic energy and mass injection rates $\dot{E} = 8\,10^{42}$ erg s^{-1} and $\dot{M} = 3\,M_\odot$ yr^{-1} respectively for a starburst bolometric luminosity of $10^{11}\,L_\odot$. If we suppose this is injected in a very small volume, blow-out (cf. Sec. 5.1) would be reached at time $t_b \approx 2\,10^4\,h_{100}^{5/3} n_0^{1/3}$ yr at a velocity $\dot{R}(h) \approx 2\,10^3\,n_0^{-1/3}\,h_{100}^{-2/3}$ km s^{-1}, i.e. very quickly. Shell fragments would be produced with these velocities and the hot gas would escape outwards into the galactic halo. Optical emission would be produced by shocks and/or radiation fields interacting with these fragments. The connection between the optical emission and the wind is then far from simple.

The motion of the hot gas expanding away from the galactic disc has been modeled by Chevalier & Clegg (1985). They assume a spherically symmetric model in which the starburst is localized within a sphere of radius R. Mass and energy injection take place interior to R with respective volume rates $\dot{q}$ and $\dot{e}$. The gas is assumed to flow adiabatically and gravity is neglected. Interior to R, the continuity equations are

$$\rho = \frac{\dot{q}r}{3u} \qquad r < R \tag{67}$$

$$\rho = \rho_R \left(\frac{R^2}{r^2}\right)\left(\frac{u_R}{u}\right) \qquad r > R \tag{68}$$

where subscript R denotes a value at $r = R$. Likewise the momentum equations are

$$u\frac{du}{dz} = -\frac{1}{\rho}\frac{dP}{dz} - \frac{\dot{q}u}{r} \qquad r < R \tag{69}$$

$$u\frac{du}{dz} = -\frac{1}{\rho}\frac{dP}{dz} \qquad r > R\ . \tag{70}$$

Finally, the energy equations are ($\gamma = 5/3$)

$$\rho u \left(\frac{u^2}{2} + \frac{5}{2}\frac{P}{\rho}\right) = \frac{\dot{e}r}{3} \qquad r < R \tag{71}$$

$$\rho u r^2 \left(\frac{u^2}{2} + \frac{5}{2}\frac{P}{\rho}\right) = \rho_R U_R R^2 \left(\frac{u_R^2}{2} + \frac{5}{2}\frac{P_R}{\rho_R}\right) \qquad r > R\ . \tag{72}$$

It is convenient to use the Mach number $M \equiv u/a$ where $a^2 = \gamma P/\rho$ and rewrite the momentum equations in the form

$$\frac{dM}{dr} = \frac{M(5M^2+1)(\frac{2}{3}M^2+2)}{2r(1-M^2)} \qquad r < R \tag{73}$$

$$\frac{dM}{dr} = \frac{M(\frac{2}{3}M^2+2)}{r(M^2-1)} \qquad r > R\ . \tag{74}$$

The supersonic branch of (69) gives $P \to 0$ as $r \to \infty$; and inspection of (68) and (69) gives $M = 1$ at $r = R$. It is readily shown that the asymptotic behaviour of the solution ($r >> R$) is $u \to u_\infty \equiv \sqrt{2}(\dot{e}/\dot{q})^{1/2} = \sqrt{2}(\dot{E}/\dot{M})^{1/2} \simeq 3000$ km s^{-1}. Very approximately, the system will behave like a giant wind driven bubble with $\dot{e} = 3\dot{E}/4\pi R^3$, $\dot{q} = 3\dot{M}/4\pi R^3$ if the supersonically moving gas pushes against ambient gas.

Clumps embedded in the region may have very important effects. In principle, interfaces between the flow and clumps could enhance energy losses and the flow might then become momentum driven (cf. Hartquist & Dyson 1993). Clumps could also act as diagnostics of either the thermal pressure ($r < R$) or ram pressure ($r > R$) since they would be pressurized to these values either by simple pressure waves ($r < R$) or shocks ($r > R$). Heckman et al (1990) have compared pressure profiles predicted ($P \propto r^{-2}$, $r > R$; $P \propto$ const ($r < R$)) with those deduced for M82 and NGC 3256 and obtained reasonable agreement.

The implications of superwind activity are profound. For example, they inject chemically enriched matter, energy and momentum into the intergalactic medium. Shells driven by superwinds may provide an origin for the $z_{\rm abs} \simeq z_{\rm em}$ narrow absorption line systems seen in many QSOs. (This idea has been explored in detail by Falle et al (1981), Dyson et al (1980), before the recognition of the superwind-starburst connection. However, in this work the implications of a central non-thermal continuum source were also investigated.)

5.4 The Starburst-AGN Connection

One of the most important questions is the relationship between active galactic nuclei and starbursts (Heckman 1991; Filippenko 1993). We here briefly discuss whether starbursts or stellar clusters are necessary to account for the mass spectral features of radio quiet QSOs, in particular, the presence of broad emission lines (ubiquitous) and broad absorption lines ($\approx 10\%$ of radio quiet QSOs).

There are basically two schools of thought. Terlevich and collaborators, in a stimulating series of papers, have argued that at least the radio quiet QSOs can be explained purely in terms of starburst activity. In their picture a young stellar cluster is formed with a high metallicity. (This latter requirement is important to produce massive hot stars with the hard photon dominated radiation fields needed to produce the required ionization state.) Stars then evolve to supernovae which, when they explode, are surrounded by dense ($n \sim 10^7$ cm^{-3}) circum- or interstellar-matter. The radiation produced by the thermalization of supernova kinetic energy energizes the dense surrounding matter which at the same time is accelerated to velocities of ~ 1000 km s^{-1}. Terlevich and collaborators identify this accelerated material with the broad emission line region. Clearly, this is also a natural connection with the generation of superwind activity in this model. This model has appealing features, but there are difficulties, such as the production of observed fast ($\sim 10^3$ s) X-ray variability and the need for all supernovae to have very dense environments and, perhaps most importantly, why jet phenomena are observed.

The other approach (Dyson & Perry 1982; Perry & Dyson 1985) retains the dense stellar cluster but adds a black hole and an accretion disc. The central black hole is fueled by material from the accretion disc which is constantly replenished by mass loss (e.g. winds) from the stellar cluster. Part of this mass loss partakes in outflow (a wind) driven either by the central radiation field or by energy input from supernovae. Broad emission lines arise when the wind shocks against supernovae and cools, initially by the Compton scattering of the central radiation field. There is some contribution to the lines from fast moving supernovae ejecta. Supernovae ejecta or swept up ISM may also produce the broad absorption lines (Perry & Dyson 1992).

Acknowledgements I am grateful to the organizers of this EADN School for their invitation to present these lectures.

References

Bertoldi, F.: Astrophys. J. **346** (1989) 735
Bertoldi, F., McKee, C.F.: Astrophys. J. **354** (1990) 529
Block, D.L., Dyson, J.E., Madsen, K.: Astrophys. J. Lett. **390** (1992) L13
Brand, P.W.J.L., Zealey, W.J.: Astron. Astrophys. **38** (1975) 363
Chevalier, R.A., Clegg, A.W.: Nature **317** (1985) 44
Churchwell, E.: Astron. Astrophys. Review **2** (1990) 79
Dyson, J.E.: Astrophys. Space Sci. **1** (1968) 388
Dyson, J.E.: Astrophys. Space Sci. **35** (1975) 299
Dyson, J.E.: Astrophys. Space Sci. **51** (1977) 197
Dyson, J.E.: Astron. Astrophys. **73** (1978) 132
Dyson, J.E.: in *Investigating the Universe*, ed. Kahn, F. D., D. Reidel Publ. Co., Dordrecht (1981) p.125
Dyson, J.E.: Astrophys. Space Sci **106** (1984) 181
Dyson, J.E., Falle, S.A.E.G., Perry, J.J.: Mon. Not. R. astr. Soc. **191** (1980) 785
Dyson, J.E., Ghanbari, J.: Astron. Astrophys. **226** (1989) 270
Dyson, J.E., Perry, J.J.: in Proc. 3rd European IUE Conference, Madrid (1982) p.595
Dyson, J.E., Smith, L.J.: in *Cosmical Gas Dynamics*, ed. Kahn, F. D., VNU Science Press, Utrecht (1985) p.173
Dyson, J.E., Williams, D.A.: *The Physics of the Interstellar Medium* Manchester Univ. Press (1980)
Elmegreen, B.: in *Star Formation in Stellar Systems*, eds. Tenorio-Tagle, G., Prieto, M., Sanchez, F., Cambridge Univ. Press (1992) p.381
Falle, S.A.E.G., Perry, J.J., Dyson, J.E.: Mon. Not. R. Astr. Soc. **195** (1981) 397
Filippenko, A.V.: in *Physics of Active Galactic Nebulae*, eds. Duschl, W.J., Wagner, S.J., Springer-Verlag, Berlin (1993) (in press)
Gaetz, T.J., Edgar, R.J., Chevalier, R.A.: Astrophys. J. **329** (1988) 927

Goldsworthy, F.A.: Phil. Trans. **A253** (1961) 277

Hartquist, T.W., Dyson, J.E.: Astrophys. Space Sci. **144** (1988) 615

Hartquist, T.W., Dyson, J.E.: Quart. J. R. astr. Soc. **34** (1993) 5

Hartquist, T.W., Dyson, J.E., Pettini, M., Smith, L.J.: Mon. Not. R. astr. Soc. **221** (1986) 715

Heckman, T.M.: in *Massive Stars in Starbursts*, eds. Leitherer, C., Walborn, N.E., Heckman, T.M., Norman, C.A. (1991) p.289

Heckman, T.M., Armus, L., Miley, G.K.: Astrophys. J. Suppl. **74** (1990) 833

Heyvaerts, J.: in *Late Stages of Stellar Evolution and Computational Methods in Astrophysical Hydrodynamics*, ed. de Loore, G. B., Springer-Verlag, Berlin (1991) p.313

Kahn, F.D.: Physica **41** (1969) 172

Kahn, F.D.: Astron. Astrophys. **37** (1974) 149

Kahn, F.D., Breitschwerdt, D.: Mon. Not. R. astr. Soc. **244** (1990) 521

Kennicutt, R.C.: in *Massive Stars in Starbursts*, eds. Leitherer, C., Walborn, N.R., Heckman, T.M., Norman, C.A., Cambridge Univ. Press (1991) p.157

Koo, B.-C., McKee, C.F.: Astrophys. J. **388** (1992a) 93

Koo, B.-C., McKee, C.F.: Astrophys. J. **388** (1992b) 103

Leitherer, C.: in *Massive Stars in Starbursts*, eds. Leitherer, C., Walborn, N. R., Heckman, T.M., Norman, C.A., Cambridge Univ. Press (1991) p.1

MacLow, M.-M., Van Buren, D., Wood, D.O.S., Churchwell, E.: Astrophys. J. **369** (1991) 395

MacLow, M.-M., McCray, R.: Astrophys. J. **324** (1988) 776

MacLow, M.-M., McCray, R., Norman, M.L.: Astrophys. J. **337** (1988) 141

Maeder, A.: in *Observational Tests of the Stellar Evolution Theory*, IAU Symposium 105, eds. Maeder, A., Renzini, A., D. Reidel Publ. Co., Dordrecht (1984) p.299

Maeder, A.: Astron. Astrophys. Suppl. **84** (1990) 139

Mason, D.: Ph.D. Thesis, Victoria University of Manchester (1975)

Meaburn, J., Hartquist, T.W., Dyson, J.E.: Mon. Not. R. astr. Soc. **230** (1988) 243

Mestel, L.: Mon. Not. R. astr. Soc. **114** (1953) 437

Norman, C.A., Ikeuchi, S.: Astrophys. J. **345** (1989) 372

Perry, J.J.: in *Central Activity in Galaxies*, Lecture Notes in Physics, Springer-Verlag, Berlin (1993) 413

Perry, J.J., Dyson, J.E.: Mon. Not. R. astr. Soc. **213** (1985) 665

Perry, J.J., Dyson, J.E.: Proc. Baltimore Meet. *Testing the Black Hole Paradigm*, eds. Holt, S.S., Neff, S.G., Orry, C.M., AIP Press, New York (1992) p.553

Schneps, M.H., Ho. P.T.P., Barratt, A.H.: Astrophys. J. **240** (1980) 84

Shu, F.H., Adams, F.C., Lizano, S.: Ann. Rev. Astron. Astrophys. **25** (1987) 84

Shull, P., Dyson, J.E., Kahn, F.D., West, K.A.: Mon. Not. R. astr. Soc. **212** (1985) 799

Van Buren, D., MacLow, M.-M., Wood, D.O.S., Churchwell, E.: Astrophys. J. **353** (1990) 570

Van Buren, D., MacLow, M.-M.: Astrophys J. **394** (1992) 534

Weaver, R., McCray, R., Castor, J., Shapiro, P., Moore, R.: Astrophys. J. **218** (1977) 377
Williams, R.J.R., Perry, J.J.: Mon. Not. R. astr. Soc. (1994) (in press)

Observing Far-Infrared and Submillimeter Continuum Emission

J. P. Emerson

Department of Physics
Queen Mary & Westfield College
University of London
Mile End Road
London E1 4NS
UK

Abstract

Observational techniques in far-IR (space and airborne) and submm (ground based) continuum astronomy are discussed with emphasis on general observational issues rather than on the detailed workings of instruments. Instruments are treated as black boxes whose characteristics must be well known by the observer, or user of data from the instrument. The material is aimed at an astrophysicist unfamiliar with observing in this wavelength region and wanting to understand the operation and limitations of, and so make optimum use of, the capabilities of available instrumentation. The difficulties involved in making such observations are pointed out, and some guidance given on how to deal with them.

1 Introduction & Philosophy

There are various books, conference proceedings and reviews on the topics of far-IR and submm astronomy (e.g. Fazio 1976, Fazio 1977, Shaver 1985, Wolstencroft & Burton 1988, Watt & Webster 1990, Liege 1990). Many workers now routinely use archival IRAS data, and submm continuum observations are currently accessible to many workers through the new large telescopes and the availability of common user instrumentation. I hope in these lectures to fill a possible gap in the literature by concentrating not on the workings of the instrumentation, nor on the results from using it, but rather on the information one should be aware of when defining a far-IR or submm observing programme, on the difficulties in precisely defining the parameters of the system, and on the factors that must be taken into account in observing and in data reduction.

In these lectures I will adopt the following definitions of the terms given in the title.

Far-Infrared is $\lambda > 25\mu$m beyond which the atmosphere becomes opaque from the ground (apart from a poor window at 35μm) until $\sim 350\mu$m where the submm begins.

Submillimetre is strictly $\lambda < 1000\mu$m but here I shall include wavelengths up to 2000μm since some submm continuum instruments work up to 2mm.

Continuum implies low spectral resolution, insufficient to study lines.

In an observational approach to an astrophysical problem one may proceed by asking the following questions:

1. What is known?
2. What can be measured for testing theory?
3. How does one obtain the observations?
4. What do the results mean?

To find the answers one can:

1. Consult books, reviews, papers in journals, telescope archives.
2. Be aware of capabilities of existing and planned telescopes and instrumentation.
3. Consult manuals, previous users, telescope staff, write a proposal, prepare fully, observe very carefully, and reduce the data well.
4. Think and model and write. This is the fun part!

In all cases it is important to appreciate, *even for theorists*, the nature of the detection process and in particular its limitations. The nature of the observation must be precisely defined and the associated uncertainties given.

Here I will discuss techniques of observation. I am deliberately not discussing the detailed workings of individual instruments because these details are well treated elsewhere, and because the lifetime of front line instrumentation is often short, due to the rapid pace of advance in instrument technology.

2 Measurables: Flux Density & Specific Intensity

I start with a reminder of what one actually measures. In star formation, as in most of astronomy, we are limited to detecting photons (or their energy $E = h\nu$) from remote objects.

2.1 Flux Density

Ultimately we can measure the *Number of photons* N (or their energy $Nh\nu$), passing through an *area* ΔA m^2, in *solid angle* $\Delta\Omega_{\mathrm{b}}$ ster (of defined shape and orientation), centered on a particular direction *direction* (RA,δ), in *time* interval Δt sec centered on time t, in *frequency* interval $\Delta\nu$ Hz centered on frequency ν Hz and which are *polarized* in a given direction.

From this we can deduce the *Flux density*, S_ν, of the incident radiation arriving in solid angle $\Delta\Omega_{\mathrm{b}}$ centered on direction (RA,δ), at time t, and at frequency ν, which are polarized in a given direction, as

$$S_\nu = \frac{Nh\nu}{\Delta A \Delta t \Delta\nu} \mathrm{Watts\ m^{-2}Hz^{-1}}.$$

Astronomer's use the unit of the Jansky, Jy, where a flux density of 1 Jy = 10^{-26}W m^{-2}Hz^{-1}. Magnitudes are rarely used in far-IR or submm work.

We must define the shape profiles of the solid angle $\Delta\Omega_{\mathrm{b}}$, and of the frequency range $\Delta\nu$. These are not likely to be given by a top hat function as the above has implied for simplicity. In the case of a Gaussian profile we use the Full Width at Half Maximum (FWHM) of the Gaussian.

By considering different directions of polarization we can also determine the percentage linear polarization, p, the position angle of linear polarization θ, measured increasing from N to E (anticlockwise) on the sky and, by also using a phase plate, the degree of circular polarization q.

We wish to determine how the flux density S_ν, and its associated polarization parameters (p, θ and q) vary with frequency ν, time t, and sky position with as fine a resolution as possible in all parameters. The electromagnetic radiation reaching us contains no more information than this.

2.2 Polarization

Although electromagnetic radiation carries polarization information many far-IR/submm continuum systems used to ignore this and measured only total flux density. Amongst the (not necessarily good) reasons for the relative lack of work in this field were that if one is to determine p, θ and q then S_ν must anyway be measured. The extra optical components needed, the necessity of measuring only a fraction of the total flux density, and the extra signal-to-noise needed to determine small differences, meant that only the brighter objects could be measured. Also if objects are unpolarized, or are not perceived as likely to be polarized, little is gained. Although to save space I will not associate a p, θ, and q with every S_ν it should be remembered that strictly one should do so as polarization is important. The observational situation is happily now changing (Gonatas et al 1990, Platt et al 1991, Clemens et al 1990, Leach et al 1991) and common-user polarimeters are becoming available (Flett & Murray 1991). There is much to be learnt from far-IR/submm polarization (Hildebrand 1988), and it should *not* be ignored.

2.3 Specific Intensity

If the flux density is assumed to be distributed uniformly over the solid angle $\Delta\Omega_{\mathrm{b}}$ (and this is probably a fair assumption if the solid angle of the telescope beam $\Delta\Omega_{\mathrm{b}} << \Delta\Omega_{\mathrm{obj}}$, the solid angle subtended by the object being observed, and the object is fairly uniform) we can deduce the *specific intensity* I_ν where

$$I_\nu = \frac{S_\nu}{\Delta\Omega_{\mathrm{b}}} \mathrm{Jy\ ster}^{-1}.$$

Note that in the absence of absorption, emission or scattering, the specific intensity is constant along a ray path, so if we have determined the specific intensity of the radiation from an object, and the ray has not been affected by intervening material, we have determined an intrinsic property of the object. In the case of blackbody radiation the specific intensity emitted by a surface depends only on frequency and temperature so we could deduce the temperature of the object $T_{\rm obj}$.

Taking into account the shape profile of $\Delta\Omega_{\rm b}$ we relate the flux density to the specific intensity in general by

$$S_\nu = \int I_\nu \cos\theta d\Omega.$$

I_ν is physically the more fundamental quantity. S_ν is observationally the more relevant quantity.

If the solid angle of the telescope beam, $\Delta\Omega_{\rm b} >> \Delta\Omega_{\rm obj}$, the solid angle subtended by the object being studied, we must use $\Delta\Omega_{\rm obj}$ to relate S_ν to I_ν. However in this case we usually do not know $\Delta\Omega_{\rm obj}$. For this situation the object is called a point source (to this particular telescope beam) and it is customary to report results in terms of Flux density. For an extended object (resolved by this particular telescope beam) it is customary to report results in terms of specific intensity.

2.4 Using νS_ν to Present Results

Some workers, usually in IR for μm, or in spectroscopy for cm^{-1}, prefer to normalize the W m^{-2} (or W m^{-2} ster^{-1}) to wavelength interval, μm, or to wavenumber interval, cm^{-1}, rather than to frequency interval, Hz.

A certain amount of energy is detected per m^2 and is independent of the units being used so

$$S_\nu d\nu = S_\lambda d\lambda = S_{\bar\nu} d\bar\nu$$

$$d\lambda = \frac{c}{\nu^2} d\nu \rightarrow S_\lambda = \frac{\nu^2}{c} S_\nu = \frac{c}{\lambda^2} S_\nu$$

$$d\bar\nu = \frac{1}{c} d\nu \rightarrow S_{\bar\nu} = c S_\nu .$$

So that putting in numerical values and using the (unfortunately) common units of cm^{-2} instead of m^{-2}

$$\left[\frac{S_\lambda}{\text{W cm}^{-2}\mu\text{m}^{-1}}\right] = 3 \times 10^{-16} \left[\frac{\lambda}{\mu\text{m}}\right]^{-2} \left[\frac{S_\nu}{\text{Jy}}\right]$$

$$\left[\frac{S_{\bar\nu}}{\text{W cm}^{-2}(\text{cm}^{-1})^{-1}}\right] = 3 \times 10^{-20} \left[\frac{S_\nu}{Jy}\right].$$

Note that as long as we keep units in m^{-2} in both quantities

$$S_\lambda = \frac{c}{\lambda^2} S_\nu \rightarrow \lambda S_\lambda = \frac{c}{\lambda} S_\nu \rightarrow \lambda S_\lambda = \nu S_\nu .$$

So, even though some people prefer W m^{-2} Hz^{-1} and some W m$^{-2}\mu$m^{-1}, (and a very few W m^{-2} (cm^{-1})$^{-1}$) as a way of expressing results, this simple conversion allows an easy intercomparison and $\nu S_\nu (= \lambda S_\lambda)$ has other advantages as a way of presenting results as we now see.

The incident flux, or power per unit area, F (W m^{-2}) measured over some frequency range and over some solid angle, is related to the flux density S_ν by $F = \int S_\nu d\nu$ where S_ν is the power per unit area per (linear) frequency interval. Now

$$F = \int S_\nu d\nu = \int \nu S_\nu \frac{d\nu}{\nu} = \int \nu S_\nu d \ln \nu$$

and we see that νS_ν represents the power per unit area per *natural logarithmic* frequency interval. It is now also clearer why $\nu S_\nu = \lambda S_\lambda$ since both are expressing power per unit per natural logarithmic frequency interval.

When plotting Spectral Energy Distributions (SEDs) it is anyway often necessary, for practical reasons, to plot ν (or λ) logarithmically along the horizontal axis, and if we do this the plots are also simple to interpret if we plot νS_ν, or its logarithm, along the vertical axis. (When doing this do not forget the difference between natural and base 10 logarithms for the horizontal axis, as $\ln \nu = \ln 10 \times \log_{10} \nu = 2.30 \log_{10} \nu$.) If we plot S_ν (or S_λ) against $\log \nu$ it is *not* a very good indication of where the power is radiated. However plotting νS_ν against $\log \nu$ *does* give an excellent indication of where the power is received as the area under a plot of νS_ν vs $\log \nu$, between two frequencies, is proportional to the energy received between those frequencies. Also as $\nu S_\nu = \lambda S_\lambda$ such plots are more easily assimilated by different workers who may think in terms of S_ν or in terms of S_λ. In practise it will usually be necessary to plot $\log \nu S_\nu$ versus $\log \nu$ due to the large range in νS_ν, but this representation still gives a more physically useful plot (see Fig 1) than other varieties.

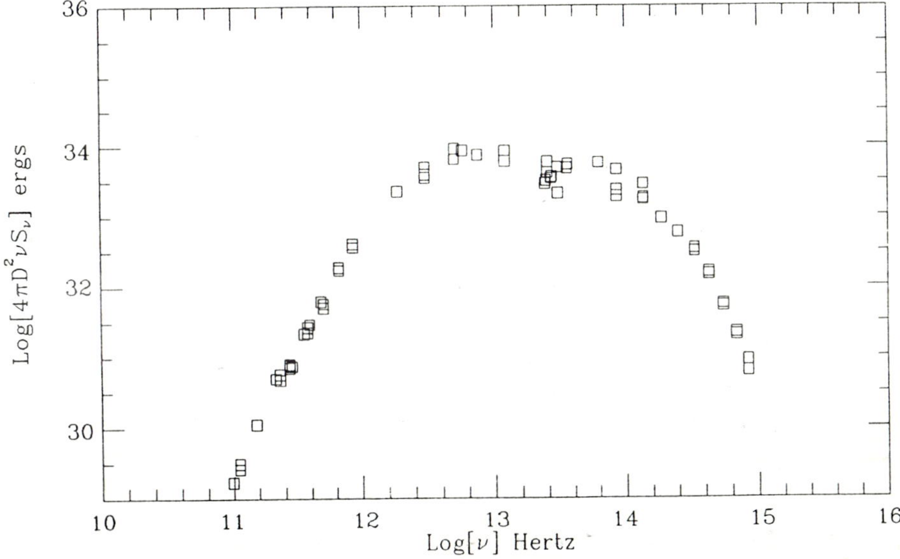

Fig. 1. Photometry of HL Tau from the literature.

A further great advantage of νS_ν arises when interpreting these plots in terms of blackbody, or modified blackbody, curves. Plots of B_ν vs ν or $\log \nu$ *have a different shape* for different temperature blackbodies, where B_ν represents the Planck function. However if one plots νB_ν or $\log \nu B_\nu$ vs. $\log \nu$ then blackbody curves at all temperatures *have the same shape* – they just shift in position about the diagram (see Fig 2).

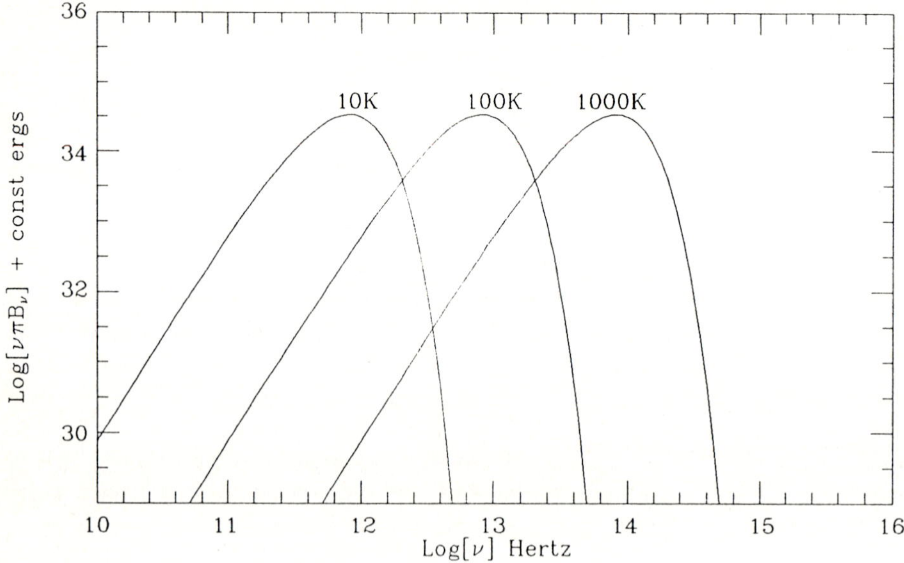

Fig. 2. Blackbody curves for 3 temperatures

Note that the total blackbody power (integrated over frequency) and the blackbody temperature can be readily deduced from such plots using the relations

$$\int_0^\infty B_\nu d\nu = 1.3586[\nu B_\nu]\mathrm{max} \text{ ergs s}^{-1}\text{cm}^{-2}\text{ster}^{-1}$$

$$\log \nu_{\mathrm{max}} = 10.91 + \log T$$

where ν_{max} is in Hz and T in Kelvin. There are similar simple relations for some types of modified blackbody radiation (Emerson 1988a). Similar statements apply to plots of λB_λ (or $\log \lambda B_\lambda$) vs $\log \lambda$, which just look like reflected versions of νB_ν (or $\log \nu B_\nu$) vs. $\log_\nu$ plots. I conclude that we should all use plots of $\log \nu S_\nu$ vs $\log \nu$ (or $\log \lambda S_\lambda$ vs $\log \lambda$) to display SEDs.

3 Observing Frequency and Bandpass

The effective frequency, ν_{eff}, at which a detection system operates is determined by:

1. the system transmission, $t_s(\nu)$, (excluding the atmosphere) where

$$t_s(\nu) = \epsilon_t(\nu) t_o(\nu) t_f(\nu) \epsilon_d(\nu)$$

and the relevant transmissions and efficiencies are:

the telescope efficiency	$\epsilon_t(\nu)$
the transmission of the optics	$t_o(\nu)$
the transmission of the filters	$t_f(\nu)$
the (quantum) efficiency of the detector	$\epsilon_d(\nu)$

2. the intrinsic spectrum, S_ν, of the object being observed,
3. and in the case of observations made through the atmosphere by the atmospheric transmission $t_A(\nu)$.

We define the effective frequency $\nu_{\rm eff}$ as

$$\nu_{\rm eff} = \frac{\int_0^\infty \nu S_\nu t_A(\nu) t_s(\nu) d\nu}{\int_0^\infty S_\nu t_A(\nu) t_s(\nu) d\nu}.$$

Note that if we specified that S_ν = constant, the same $\nu_{\rm eff}$ would be obtained if we defined it, independently of the nature of the object being observed, as

$$\nu_{\rm eff} = \frac{\int_0^\infty \nu t_A(\nu) t_s(\nu) d\nu}{\int_0^\infty t_A(\nu) t_s(\nu) d\nu}$$

but that for other spectral dependences than S_ν = constant such an object independent definition of $\nu_{\rm eff}$ is *not* appropriate. In general the effective frequency of a system should be quoted along with the SED (form of S_ν) and atmospheric transmission (often given in terms of amount of precipitable water vapour above the telescope) to which that effective frequency refers. The uncertainty is defining $\nu_{\rm eff}$ increases with the frequency range over which radiation is detected, which in continuum detectors can be $\Delta\nu \sim \nu$.

As well as the effective frequency of a system one should define its effective bandwidth $\Delta\nu_{\rm eff}$. Often the precise meaning of this term is not clearly defined in the literature. Possible definitions include:

1. The full bandwidth at the points where $t_s(\nu)$ reaches half its maximum value.
2. The full bandwidth at the points where the product $t_A(\nu)\, t_s(\nu)$ reaches half its maximum value.
3. The (top-hat) bandwidth that would contribute the actual power received for an object with the same SED as that assumed for the object observed i.e.

$$\Delta\nu_{\rm eff} = \frac{\int_0^\infty S_\nu t_A(\nu) t_s(\nu) d\nu}{S_{\nu_{\rm eff}}}.$$

In case 2) $\Delta\nu_{\text{eff}}$ would be dependent on atmospheric conditions and in (the preferable) case 3) it depends also on the spectrum S_ν of the object being observed. As we measure S_ν, use of ν_{eff} and $\Delta\nu_{\text{eff}}$ are logical, but frequent use is made of the equivalent λ_{eff} and $\Delta\lambda_{\text{eff}}$ whose definitions are analogous.

A system must be carefully designed so that the product $t_s(\nu) \times t_a(\nu)$ is high (ideally 1) in the regions we wish to study and low (ideally 0) in the regions we wish to reject. It is *convenient* to assign a unique effective frequency to a detection system and express results in terms of flux density at that frequency. However *in reality we always* measure a finite band pass and *many different energy distributions could produce the same signal at the output of the system* i.e.

$$\text{Received Signal} \propto \left[\frac{P}{\text{W m}^{-2}}\right] = \int S_\nu t_A(\nu) t_s(\nu) d\nu$$

so there is often insufficient information to uniquely determine S_ν for continuum detectors.

3.1 Example with No Atmosphere: IRAS

IRAS (the Infrared Astronomical Satellite), launched in 1983, performed an all sky IR survey in 4 broad wavelength bands at 12μm, 25μm, 60μm & 100μm, from above the atmosphere. The "IRAS Catalogs and Atlases, Volume 1, Explanatory Supplement" (IRAS 1988, hereafter Exp Supp) provides an excellent example of a case where all the necessary information for understanding an instrument, and the meaning of its results, is given.

Table II C.4 of the Exp Supp gives the $\Delta\lambda_{\text{eff}}$s of 7.0$\mu$m, 11.15$\mu$m, 32.5$\mu$m, & 31.5 μm respectively and the magnitude of out of band leaks (or upper limits to them). The system transmission, $t_s(\nu)$, of the 4 survey bands is plotted and tabulated in Fig II.C.9 and Table II.C.5 of the Exp Supp (IRAS 1988).

The λ_{eff} of 12μm, 25μm, 60μm & 100μm were assigned as round number values and the effective bandpasses were calculated for a source spectral energy distribution, SED, νS_ν = constant. Thus the IRAS Flux densities are the flux densities at the effective wavelengths (12, 25, 60 and 100μm) that would yield the observed in band signals (in W m^{-2}) if the source spectrum is $S_\nu \propto \nu^{-1}$.

The IRAS Science Team could have chosen different λ_{eff} or/and a different source spectrum in which case the quoted flux densities would have been different. Of course the satellite measured (within the experimental uncertainties) a perfectly well defined amount of energy. The choice comes in what assumptions to make to convert the flux measured over a broad band to a monochromatic flux density.

If the true form of S_ν is known, it is possible to perform a color-correction (see Exp Supp VI 26-28) to determine the true flux densities. Another (less useful) way of thinking about it is that the quoted flux densities refer to different effective wavelengths for spectra not of the form νS_ν = constant. Table VI.C.6 in the back cover of each IRAS Catalog volume, tabulates colour correction values K, where

$$S_{\text{true}} = \frac{S_{\text{catalog}}}{K}$$

for S_{true} (& K) appropriate to black-bodies and power laws. For other shapes the user must make the appropriate calculation of K using the information on pages VI-27 and II-18 of the Exp Supp. The IRAS approach should ideally be taken for other broad band ($\frac{\nu_{\text{eff}}}{\Delta\nu_{\text{eff}}} \sim 2 - 10$) photometric systems, given all the necessary information.

In determining the nature of objects it is often useful to calculate the far-IR luminosity using the distance and IRAS Fluxes (in W m^{-2}). As the intrinsic shape of S_ν is not a priori known, guessing a blackbody temperature (or other spectral form), colour correcting, plotting the SED and integrating under it, may not even give the correct IRAS Fluxes (as true S_ν will not be known) and is very tedious. Recalling that IRAS actually determined in-band fluxes (W m^{-2}) and, for convenience, merely converted them to flux densities at effective wavelengths for a particular spectrum, it is clearly easier just to invert that process. It has been shown (Emerson 1988b) that the actual IRAS pass bands can be approximated by 4 contiguous, top-hat shaped pass bands whose lower and upper wavelengths and effective frequency bandpasses are given in Table 1.

Table 1. Contiguous synthetic square bands for deriving in band fluxes from IRAS survey flux densities

Nominal $\lambda(\mu m)$	Lower $\lambda(\mu m)$	Upper $\lambda(\mu m)$	Effective $\Delta\nu_{\text{l-u}} \times 10^{12}$Hz
12	7	16	20.653
25	16	30	7.538
60	30	75	4.578
100	75	135	1.762

so that, for example 75–135μm fluxes (and hence given a distance, luminosities) are calculated as

$$\left[\frac{F(75 - 135\mu\text{m})}{10^{-14}\ \text{W m}^{-2}}\right] = 1.762 \left[\frac{S(100\mu\text{m})}{\text{Jy}}\right].$$

This prescription makes optimum use of the data, minimizing assumptions, as well as being simple. It illustrates that broad band photometers really measure W m^{-2} over a frequency range and that conversion from flux to flux density, S_ν, involves assumptions.

4 Allowing for Atmospheric Transmission

Discrete lines and bands of various molecules lie throughout the far IR/submm region. The opacity of the atmosphere is mainly determined by H_2O, but under dry conditions CO_2, O_3 and O_2 can also be important. The amount of H_2O in the atmosphere above sea level is $\sim$ 1 cm of PreciPitable Water Vapour (ppwv).

The opacity, τ, in clearer spectral regions is dominated by pressure induced line wings for which $\tau \propto WP^{0.5}$, where $W =$ is the amount of the absorber and

P the pressure. As we go up in the atmosphere τ drops as we get above some W, and as P drops. At high-dry mountain sites (Mauna Kea, 4.2 km) ppwv $\sim$ 1 mm. At the tropopause ($\sim$ 12 km) a temperature minimum is reached ($\sim$ 218 K), and the densities of H_2O and CO_2 become low enough for the atmosphere to radiate directly into space and convection ceases. Above the tropopause ppwv drops (there is no convection to bring water up from below, and it is also frozen below) and the transmission improves. The 0.9m KAO (Kuiper Airborne Observatory) flies at $\sim$ 12.5 km where ppwv $\sim 10\mu$m and work in the $30 - 300\mu$m region is possible. Even at airplane altitudes ozone is little reduced as it lies at heights of 10–35km where UV radiation can make it (CFCs permitting!). Whilst in the seventies much far-IR astronomy was done from balloons at higher altitudes, the advent of the KAO and IRAS data, and various financial considerations, have caused a drop in the amount of balloon far-IR work done.

The atmospheric transmission can be calculated for assumed height distributions of the densities of the absorbers and of the temperature and pressure of the atmosphere, using line lists (McClatchey et al 1973). This has been done by several groups (Kyle & Goodman 1987, Traub & Steir 1976, Marten et al 1977) and, using an updated database, at 350 & 450μm (Naylor et al 1991) and at 800μm (Danese & Partridge 1989). Transmissions can be measured using spectra of bright celestial objects of known spectrum (Sun, Moon).

In general $t_A(\nu)$is a function of time (clouds, air current, atmospheric motions, weather) which implies that one must be careful that the filtering etc. prevents these changes from significantly modifying $\nu_{\rm eff}$ and $\Delta\nu_{\rm eff}$. This implies that the system transmission $t_s(\nu)$ not the atmospheric transmission $t_A(\nu)$ should control the pass band and hence determine $\nu_{\rm eff}$ and $\Delta\nu_{\rm eff}$.

The atmospheric transmission $t_A(\nu)$ is very sensitive to ppwv especially at $350/450\mu$m where, with 2mm ppwv, observations become virtually impossible. Accurate statistics are surprisingly hard to come by but it seems that only $\sim$ 30% of the time, even on Mauna Kea, does the ppwv drop below 1mm. Such moments are of course very valuable for submm work at the shorter wavelengths of $350/450\mu$m, and some telescopes try to actively adapt their schedules to take advantage of this.

In attempts to maximize system transmission some submm systems have let the atmosphere do the short and/or long wavelength blocking. This can be a dangerous approach as, although the extra bandwidth may produce a higher signal to noise detection, the extra uncertainty in the true effective frequency and bandwidth, and the fact they will vary in time, can negate the signal to noise gain because it becomes unclear what exactly is being measured! The measured flux density may not be at the frequency the observers imagined!

4.1 Example with Atmosphere: UKT14 on Mauna Kea

In the IRAS case there was no atmosphere to contend with and so the bandpasses of the filters could be designed subject to the constraints of available materials for optics, filters and detectors. For observations below the atmosphere, this is a far more difficult task as we want the filters, not the (variable) atmosphere, to define the band passes. Even were the atmosphere to remain completely stable, as an object rises and sets we look through different path lengths of atmosphere so $\nu_{\rm eff}$ and $\Delta\nu_{\rm eff}$ could potentially change with source elevation.

For 1mm ppwv (e.g. Mauna Kea), at the zenith the optical depth τ and atmospheric transmission t_A are given in Table 2 for some atmospheric windows. The wavelengths and frequencies quoted are to label the windows, and should not be taken as effective wavelengths or frequencies. At 1000 μm a simple formula

$$\tau(1\text{mm}, 287\ \text{GHz}) = 0.0126 + 0.06 \left[\frac{\text{ppwv}}{\text{mm}}\right]$$

allows calculation of the optical depth of the atmosphere at 1mm as a function of precipitable water vapour (Danese & Partridge 1989). This formula predicts that for 1mm ppwv $\tau(1\ \text{mm}) = 0.1$. Calculations of transmission for ppwv of 0.5mm, 1mm and 3mm for Mauna Kea are shown in Fig 3.

Table 2. Atmospheric transmission above Mauna Kea for 1mm precipitable water vapour

Wavelength region	Frequency	Optical depth	Transmission
λ (μm)	ν (GHz)	τ	t_A (%)
350	850	1.4	25
450	690	1.4	25
600	460	0.7	50
850	345	0.22	80
1300	230	0.10	90

UKT14 is the submm continuum receiver on the 15m James Clerk Maxwell Telescope (hereafter JCMT) on Mauna Kea and acquired its name as it was originally the 14th cryogenic system made for UKIRT, the UK IR Telescope (3.8m), but was later transferred to the JCMT. UKT14 and its frequency response is well described in the literature (Duncan et al 1990). The filters are constructed (Ade et al 1984) to ensure that the effective frequencies are relatively insensitive to both the prevailing atmospheric conditions and to the spectral index of the source. Their transmission characteristics are shown in Fig 4 and the resulting $\nu_{\rm eff}$ and $\Delta\nu_{\rm eff}$ are tabulated in Table 3 varying by only a few % for a wide range of conditions (Duncan et al 1990).

We now understand what S_ν, ν and $\Delta\nu$ mean for our observation (henceforth I drop the $_{\rm eff}$ subscript), and something of the assumptions and uncertainties involved, although we have not yet discussed the calibration process for S_ν. Always be aware of the $\nu_{\rm eff}$, $\Delta\nu_{\rm eff}$ and the out of band rejection characteristics of a system to assess the interpretation of its S_ν! It is also crucial to know the atmospheric transmission as a function of airmass and time, to properly calibrate ground-based submm observations. This is discussed below.

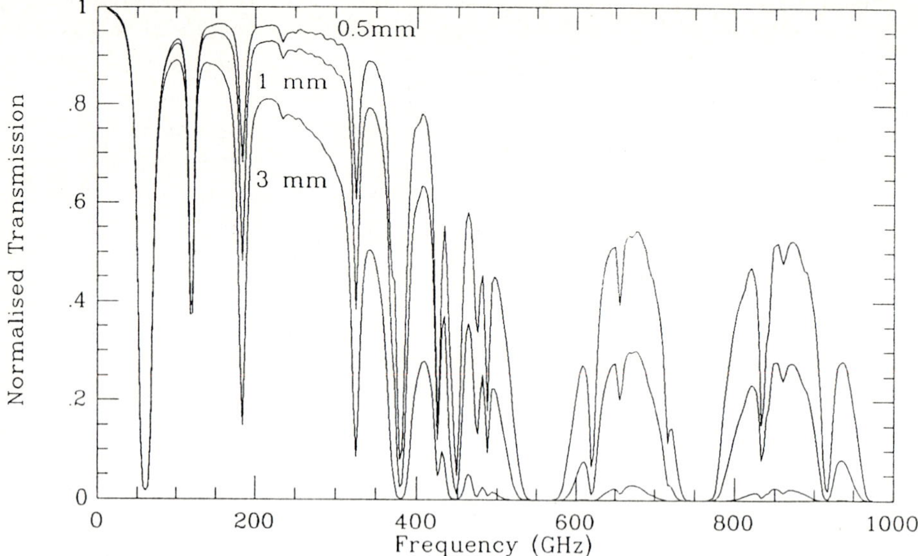

Fig. 3. Calculated atmospheric transmission at Mauna Kea for 0.5, 1 & 3mm of precipitable water vapour

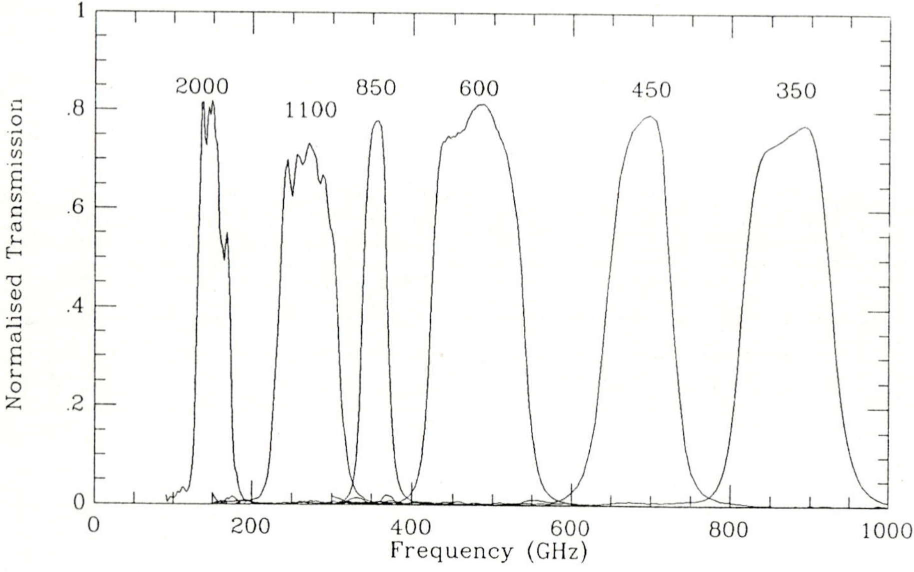

Fig. 4. UKT14 filter characteristics

Table 3. UKT14 effective frequencies for power law SED with spectral index n and various water vapour contents (Duncan et al 1990)

Filter name	Centre ν GHz	$\Delta\nu$ GHz	Effective frequency ppwv = 1mm			power law n = 2			
			n=0	n=2	n=4	ppwv=0.5	ppwv=1	ppwv=2	ppwv=4
2.0mm	149	40	149.1	155.6	177.6	156.7	155.6	154.3	153.0
1.3mm	233	64	233.2	237.9	243.8	238.3	237.9	237.5	237.1
1.1mm	266	74	265.9	271.6	278.1	272.3	271.6	270.6	269.3
WBMM	290	220	291.7	316.9	342.6	323.1	316.9	308.1	296.0
850μm	354	30	351.0	353.0	354.3	353.3	353.0	352.6	351.9
800μm	384	101	372.9	379.5	385.9	381.7	379.5	375.9	370.1
600μm	480	112	457.8	463.0	467.8	469.8	463.0	451.5	434.3
450μm	682	84	673.7	675.8	677.8	677.1	675.9	674.4	672.9
350μm	864	113	850.8	854.9	858.3	856.3	854.9	854.0	854.0

5 Calibration & Determination of Atmospheric Transmission

We have now discussed some of the potential pitfalls in describing the results of an observation in terms of a flux density, S_ν, measured at an effective frequency ν, in pass band $\Delta\nu$, within a solid angle $\Delta\Omega_{\rm b}$, at position RA,δ, at time t, without polarization information. For observation from the ground we need to know the atmospheric transmission, which may (and usually will!) vary with time. We now discuss how the atmospheric transmission is determined.

An ideal calibrator is bright, point-like and has the same spectrum as the object being studied, is observed at the same time and airmass as the object, is very close to it in the sky, and has a well determined flux density! Under these circumstances calibration is simple - just the take ratio of output signals and multiply by the Known S_ν of the calibrator. However one still has to assign $\nu_{\rm eff}$ and $\Delta\nu_{\rm eff}$. Also one must be aware of the effects of beam shape and sidelobe responses, especially on extended sources, or if there are other objects nearby.

In practise, especially in the submm, the calibrator will probably not be point like, will be observed through a different atmosphere (time, airmass, sky position) and will have a different spectral shape from the object. If the atmospheric transmission is known for each observation the problem is tractable and the procedures can be deduced from our previous discussion.

We must thus know $t_A(\nu)$ for both calibration and object observations and in the submm (ground based) this is equivalent to determining the ppwv – the precipitable water vapour content of the atmosphere overhead.

Possible methods of determining the atmospheric transmission, or equivalently the atmospheric optical depth, are:

1. For a calibration object of known spectrum and a system of known relative transmission at two or more frequencies the ratio of the signals at the two frequencies will depend on, and hence can give, the atmospheric water vapour

– though models of this system or tables are needed. For other objects this approach is not so good as the spectrum may not be known (one is trying to measure it!) and signal to noise may not be high.

2. *If* the response of the system (in milliVolts) to an incident flux density of one Jy (incident on the telescope, not the top of the atmosphere) is known and stable then observations of objects of known flux density can be used to deduce the atmospheric transmission (e.g Adams et al 1990).
3. If a night is very stable, plots of the log of the signal versus airmass for a single object will be straight lines the slopes of which give the atmospheric optical depth at the zenith. This method does not work if the atmosphere varies a lot.
4. Measure the sky emission as a function of airmass ("skydips") deduce the emissivity at the zenith and hence deduce the ppwv and atmospheric optical depth. By Kirchoff's Law the emissivity, ϵ_ν, and the absorptivity, a_ν are related:

$$\epsilon_\nu = a_\nu = \frac{I_\nu^{\text{abs}}}{I_\nu^{\text{obj}}}$$

but

$$t = \frac{\text{transmitted}}{\text{incident}} = \frac{I_\nu^{\text{obj}} - I_\nu^{\text{abs}}}{I_\nu^{\text{obj}}} = 1 - a_\nu = 1 - \epsilon_\nu$$

and hence the transmission (and optical depth) can be deduced from the skydip measurements of ϵ_ν. Therefore do skydips regularly to calibrate the atmospheric transmission. On Mauna Kea the Caltech Submillimeter Observatory (hereafter CSO) maintain a continuous (every 10 min) monitor which automatically logs conditions at 1300μm and is remotely accessible. Use it and your skydips to help reduce your data.

Often the atmospheric optical depth, τ, is determined by one of the above methods in one part of the sky and then, after a correction for differing airmass, applied to an object in a different part of the sky. If there are substantial time or sky position dependent changes in ppwv (as is not uncommon) this is a problem. Calibrate frequently and near to your object of interest even though this requires discipline because it is so time consuming! You must do your own skydips in the direction of your object as the CSO will not usually happen to be doing its regular dips in that direction. Then knowing how the atmospheric transmission varies with time and sky position you can properly calibrate your S_ν, ν_{eff} and $\Delta\nu_{\text{eff}}$.

6 Cancellation of Sky & Telescope Emission

We have noted that the atmosphere emits – indeed it will contribute many more photons than a typical point source being observed because sky photons will arrive into all of the telescope beam solid angle $\Omega_{\rm b}$, whereas object photons will only arrive from the much smaller solid angle subtended by the object $\Omega_{\rm obj}$. Furthermore the atmosphere is at $\simeq 250 - 280$ K and in star forming regions the object is often much colder – the main effect however is the solid angle. Furthermore the telescope, optics etc. will radiate as a blackbody at ~ 280 K providing more background. One therefore tries to minimize the beam size, but there are practical limits set by the diffraction limit, and by the magnitude of the telescope pointing and tracking uncertainties. A beam of zero size will detect no flux and will be hard to put accurately on a point!

IR/submm telescopes and detection systems are therefore carefully designed to have low emissivity and to see only the sky, and not high emissivity bits of telescope or dome. For example a baffle which may be good for optical work is bad for IR work as it will have emissivity $\epsilon \sim 1$, compared to $\epsilon \sim 0.1$ for the sky. IR telescopes are made with undersized secondaries to define the effective aperture, avoiding the emission from high emissivity mirror edges. IR/submm telescope design is discussed in Low & Rieke (1974) and in Soifer & Pipher (1978). The aim in the thermal IR ($\lambda > 3\mu$m) is to see the sky ($\epsilon \sim 0.1$) rather than the telescope etc. ($\epsilon \sim 0.5$–1). Whilst careful design may minimize the telescope background we still have the sky (+ residual telescope) backgrounds as well as the photons from the object of interest. Clearly we must subtract the sky background. This is usually done by chopping and nodding. Note also that the background from telescope and sky will fluctuate and is therefore an additional source of noise. The correct choices of parameters for chopping and nodding can considerably improve the quality of background subtraction and hence signal to noise ratio.

6.1 Chopping

The chopping procedure is to look at the object (signal Obj) and the sky in the same direction (signal $\rm Sky^1$) to get a total signal ($\rm Obj + Sky^1$) and then at a nearby (hopefully empty of sources) piece of sky (signal $\rm Sky^2$). If the positions are close and the time difference is small presumably the signals from the sky at the two positions are similar ($\rm Sky^1 \simeq Sky^2$) so the signal form the object, which is what is wanted, is $\rm Obj = (Obj + Sky^1) - (Sky^2)$. But typically Sky >> Obj and it is difficult to determine a small number accurately as the difference of two (noisy) big numbers.

The solution is to continuously (at a fixed frequency) switch between ON ($\rm Obj + Sky^1$) and OFF ($\rm Sky^2$) source sky positions at frequency $f_{\rm chop}$ and to make the electronics respond only to signals in a narrow bandwidth around the modulation frequency $f_{\rm chop}$ and in phase with the modulation. This is known as Phase Sensitive Detection, the modulation as chopping. The beam movement is usually achieved by moving the secondary mirror in an (approximation to) a square wave. The amplitude of the chopper throw – the distance between the object and the reference OFF source position – is an important piece of information. The system is sensitive to *gradients* on that size scale and is insensitive to structure and backgrounds on larger scales.

If the object is extended on the scale of the chop throw or larger only the gradient, not the absolute value, of the object's flux density will be measured. For an extended source with respect to the chopper throw the signal will be a function of chopper throw. Due regard for this fact is crucial for studying extended objects.

Note that it is always possible that there will be (un)serendipitously an (unrelated) object in the reference (OFF) beam. Ultimately unrelated objects in the ON and OFF beams (i.e. background stars or galaxies) may constrain our ability to correctly subtract backgrounds leading to "confusion noise".

Scanning a telescope along the chop direction will produce both positive and negative going double sided signals, and techniques have been developed, particularly for mapping purposes, to recover maps from such data (Emerson et al 1979, Richer 1992) but it is very important that both beams should, at the ends of the scans, fall fully off the source if the maps are to be fully restored.

The design of a chopping secondary may give you no choice of chop direction if it is single axis. For a polar mounted telescope chopping in RA or Dec is natural and for an Az-El mount (JCMT) Az or El is natural. JCMT allows a continuous variation of chop direction so RA or Dec, or any position angle, chopping is possible. Over typical ($<$ 1 arc min) chop throws variation of airmass is insignificant so El chopping is as good as Az or RA, Dec. Chopping at constant position angle in RA, Dec is however to be preferred to Az, El chopping to avoid background (confusing) objects passing through the reference beams due to field rotation in long integrations. Chose the position angle to avoid any objects in the reference beam and optimize its amplitude with respect to any structure in the object.

The throw must always be at least a beam diameter or the object signal will appear in both beams and the object will not be correctly measured (flux density will be lower limit). Typical chop throws are between one and a few beam diameters. If the object is extended the throw should be greater than the object extent to maximize signal to noise ratio. However the larger the throw the greater the likelihood of large sky gradients and nodding failing to remove these. For point sources a chop throw slightly more than a beam diameter is optimum. For extended sources a large throw will tend to maximize the signal and the noise. The effect on the signal to noise ratio depends on the source gradient and the noise. If there is a source in either of the reference (OFF source) beams the chop throw should be adjusted to avoid it.

There will be mechanical constraints on the throw amplitude. The duty cycle (ratio of time spent in a beam to total time including time to switch beams) will depend on chop amplitude and frequency. One wants a square chop cycle (minimum time spent in switching) for efficiency.

The duty cycle and chop amplitude considerations just mentioned suggest low frequencies are required. The timescale of atmospheric fluctuations suggests higher frequencies. The response time of the detector system will also be a factor. An optimum value will be determined experimentally, and may vary with sky conditions, but it is typically a few ($\sim$ 10) Hz.

6.2 Nodding

Chopping alone will work if the sky and telescope backgrounds are the same ON and OFF source. This may not be exactly so due to different optical paths of the two beams and gradients in sky emission. To compensate for this potential (inevitable) beam imbalance we use two OFF positions either side of the ON position.

Call the beam in which the object produces a positive signal the Right beam and that in which it produces a negative signal the Left beam. Let the object contribute signal Obj and the sky at that position Sky^1. Let the sky one chop throw away to the right contribute $\mathrm{Sky}^2 = \mathrm{Sky}^1 + \Delta\ \mathrm{Sky}$ and the sky one chop throw away to the left $\mathrm{Sky}^3 = \mathrm{Sky}^1 - \Delta\ \mathrm{Sky}$ (we are assuming a sky gradient of Δ Sky /chop-throw).

Put the Right beam on the source:

$\mathrm{Right} = (\mathrm{Obj} + \mathrm{Sky}^1) - (\mathrm{Sky}^2)$

$\mathrm{Right} = (\mathrm{Obj} + \mathrm{Sky}^1) - (\mathrm{Sky}^1 + \Delta\ \mathrm{Sky}) = \mathrm{Obj} - \Delta\ \mathrm{Sky}$

Put the Left beam on the source:

$\mathrm{Left} = (\mathrm{Sky}^3) - (\mathrm{Obj} + \mathrm{Sky}^1)$

$\mathrm{Left} = (\mathrm{Sky}^1 - \Delta\ \mathrm{Sky}) - (\mathrm{Obj} + \mathrm{Sky}^1) = -\Delta\ \mathrm{Sky} - \mathrm{Obj}$

Then $\mathrm{Obj} = (\mathrm{Right} - \mathrm{Left})/2$

The process of measuring in the Right beam, then moving the object into the Left beam and subtracting the results removes gradients in sky emission and is known as nodding. Chopping removes a level, the additional process of nodding removes a gradient. If there is sky emission structure at higher spatial frequencies chopping and nodding will fail to remove its effects. Structure in backgrounds leads to confusion noise.

IRAS had none of these problems because it was above the atmosphere and because it was cooled to $\sim$ 20K so that the thermal emission from the telescope was negligible at 100μm. There was thus no need for a chopper and it measured levels, not gradient, although zero point drifts are still a problem. Of course we cannot cool ground based telescopes because the atmosphere would freeze on them!

Chopping at (say) 10Hz involves moving the relatively light secondary. Nodding must clearly be done at a lower frequency and involves slewing the relatively massive complete telescope. The duty cycle of these slews will be an important factor in determining observing efficiency. (Note that for nodding to work well the telescope must be able to make these slews accurately and reproducibly. If the nod amplitude differs from the chop amplitude things get nasty!).

Typical rates are < 0.1 Hz – a 10 sec integration in chopping mode is carried out followed by a nod, 2 x 10 sec cycles in the Left beam, back to right, 2 x 10 sec in Right etc., making sure at the end of the integrations that equal number of Left and Rights have been done. Slower nod rates are more efficient but may not be as effective if sky conditions are changing rapidly.

6.3 Sky Noise

The sky emission is a source of photon/wave noise and also fluctuates due to clouds, weather, wind etc. The performance of the chopping and nodding strongly influence the performance of the system and its ability to be sensitive despite the presence of sky noise. The residual sky noise from a poorly chopping and nodding system can be disastrous. The ability to cancel the background accurately limits the performance of ground based telescopes (Low & Reike 1974, Sollner 1977). Note that reducing the beam size will usually reduce the sky noise and, even though the object signal will be reduced, the signal to noise ratio may well improve (Ade et al 1984).

The ability to cancel sky emission accurately currently provides the practical limitation to the sensitivity of ground based submm systems. If perfect cancellation were achieved signal to noise ratio should increase as $\sqrt{\text{integration time}}$. In practise it is found, with both the UKT14 (JCMT) and MpIfR (IRAM) bolometer/submm systems that this occurs only down to a certain point after which further integration does not reduce the noise. For IRAM the limiting rms point source sensitivity is ~ 2 mJy, reached after integration times of 30 – 60 mins. Longer integration does *not* improve sensitivity. Therefore, when observing, one should monitor the reduction of standard error of the integration with $\sqrt{\text{integration time}}$ and when it stops decreasing as expected consider terminating the observation.

The limit seems to arise through imperfections in the ability to difference the total-power sky contribution due to the different beams sampling different columns of atmosphere. For example the limit on IRAM of 2 mJy is caused by the fractional precision of sky subtraction being only (!!) $\sim 10^{-5}$! Even such a small residual fraction of the background power sets the practical sensitivity limit (Duncan et al 1990, Ade et al 1984). Improved chopping techniques are required to push these limits lower.

7 Spatial structures

7.1 Spatial Resolution

Let us now consider the beam solid angle/shape/size. Fraunhofer diffraction theory at a circular aperture tells us we have a central maximum with surrounding dark and bright rings - the Airy pattern (Airy 1835). A Gaussian fits such a function quite well and the width of the gaussian is described by the Full Width at Half Maximum which, knowing the first minimum is at $1.22\frac{\lambda}{D}$, is $\sim \frac{\lambda}{D}$ where D is the telescope diameter. In convenient units

$$\left[\frac{\theta_{\text{fwhm}}}{\text{arcsec}}\right] = 0.2 \left[\frac{\lambda}{\mu\text{m}}\right] \left[\frac{D}{\text{m}}\right]^{-1}.$$

In practise due to various imperfections in the optics we will find $\theta_{\rm fwhm}$ to be somewhat bigger. An aperture is chosen that will admit all of the central image ($\sim$ 83% of the energy) or sometimes the first bright ring as well ($\sim$ 94% of the energy) at the expense of some loss of resolution.

However just because authors quote a FWHM for their beam it does not necessarily mean that the beam is truly Gaussian. You must read the paper carefully to find out what the beam profile really is if you wish to use it. Imperfections in the telescope surface can degrade the beam and the telescope radiation can be poorly coupled to the detector (e.g. JCMT at 2000μm). Other possible problems include: anomalous refraction/radio seeing can distort submm beam shapes (Zylka et al 1992). This effect is particularly noticeable around sunset on Mauna Kea (JCMT) (Duncan et al 1990). Sidelobe response can be significant (Hills 1992) and as planetary calibrators are not point like this poses a calibration problem. If a point source is not isolated power may be picked up in the telescope beam side lobes. Great care is needed for photometry of extended objects. Both the object and the beam should always be mapped to deal with side lobes. Pointing fluctuations (& chopping) will tend to broaden the beams. When observing one should always check beam profiles or map the beam on a point source to check the beam profile and determine $\Delta\Omega_{\rm b}$.

In the far-infrared the biggest aperture currently available is the 0.9m Kuiper Airborne Observatory, so at 100μm $\theta \geq 22''$. From the ground the 30m IRAM is the biggest 1.3mm dish, $\theta \geq 9''$, and the 15m JCMT is the biggest 350μm dish where $\theta \geq 5''$. In practise the IRAM beam is $11''$ and the JCMT beam $7''$ at these respective frequencies. Table 4 gives some beam sizes at other frequencies.

Table 4. Beam sizes available on some far-IR/submm telescopes

Wavelength μm	2000	1300	1100	850	800	600	450	350	100	60	50	25
Telescope & size	Beam size arcsec											
IRAM 30m		11		8								
JCMT 15m	28	21	19	16	14	9	7	6				
CSO 10m			22		13	13						
KAO 0.9m Univ Texas array									14x28		8x16	
IRAS 0.6m survey									180x300	90x285		46x280
IRAS 0.6m DSD									180x150	90x77		46x140

It might be thought that bigger telescopes could be built to get better spatial resolution but practical difficulties due to thermal and gravitational effects in these large structures result in surface irregularities which destroy the needed accuracy (to 0.1λ) of the dish surface and hence its effectiveness.

One solution is to use arrays and perform interferometric aperture synthesis observations. Of the four mm arrays currently operational (the Nobeyama Millimeter Array (NMA), the Owens Valley Radio Observatory (OVRO), the Berkeley Illinois Massachusetts Array (BIMA), and the Institut Radio Astronomie Millimetrique (IRAM) interferometer) only one, OVRO, operates at down to 1.4mm and it has synthesized a beam of 3.3″ (Woody et al 1989). All four arrays are moving towards operation at 1mm. No space, balloon, rocket, or airplane borne far-IR interferometers have been used. OVRO has achieved a sensitivity of 0.1 Jy/beam (5σ) at 1.4mm. Several reviews describe mm-arrays (Welch 1988, Welch 1990). The Smithsonian Astrophysical Observatory are carrying out a design study for a 6 antenna array to operate down to 350μm on Mauna Kea, and NRAO is seeking funding for an array operating down to 800μm.

7.2 Source Sizes from Beam/Source Profiles

If a Gaussian shaped object of FWHM $\theta_{\rm obj}$ is observed with a Gaussian shaped beam of FWHM $\theta_{\rm beam}$ the resulting profile, when the telescope is scanned across the object, will have a Gaussian profile of FWHM $\theta_{\rm obs} = \sqrt{\theta^2_{\rm beam} + \theta^2_{\rm obj}}$ by the convolution theorem.

Thus if you believe $\theta_{\rm beam}$ is known to 1%, or to 10%, or to 20% respectively then $\theta_{\rm obj}$ can only be reliably determined if $\theta_{\rm obj}/\theta_{\rm beam} > 0.14, > 0.46$, or > 0.66 respectively. Adopting the philosophy that nothing is ever measured to better than 10% we see that sizes can only be determined for objects with size $\geq 0.5\theta_{\rm beam}$. Objects smaller than this will be hard to resolve with this beam size $\theta_{\rm beam}$. This is further emphasized in Table 5. Thus $\theta_{\rm beam}$ and $\theta_{\rm obj}$ must be determined with very high accuracy to extract sizes confidently, unless $\theta_{\rm obj} \sim \theta_{\rm beam}$.

Table 5. Object size as fraction of beam size, resulting fractional beam broadening, and accuracy needed to determine object size

$\theta_{\rm obj}/\theta_{\rm beam}$	$\theta_{\rm obs}/\theta_{\rm beam}$	Accuracy required
0.1	1.005	0.5%
0.2	1.02	2%
0.3	1.04	4%
0.4	1.08	8%
0.5	1.12	12%
0.6	1.17	17%
0.7	1.22	22%
0.8	1.28	28%
0.9	1.35	35%
1.0	1.41	41%

The distance to the nearest star forming regions is $\sim 0.3 \times 10^8$ AU (140pc) & the far edge of the galaxy is at $\sim 0.5 \times 10^{10}$ AU (25 kpc). To put the distances and the sizes of Table 4 in some sort of context the approximate distances to which a 1″ beam would resolve various galactic star formation regions are

Planets	10^{-7}pc
Sun	10^{-2}pc
Planetary system (Saturn)	10pc
Prestellar disks	$10 - 100$pc
Ultra Compact HII region	10^4pc
Fragmenting/collapsing protostellar cloud	$10^2 - 10^4$pc
Outflow	10^5pc
Cores/Protocluster in GMCs (B335)	$10^4 - 10^6$pc
GMCs	$10^7 - 10^8$pc

7.3 Super-resolution

Most beam used are circular, but not all are, for example the IRAS survey array beams. Under favorable conditions image processing techniques can be used to achieve higher spatial resolution than implied by the (Gaussian) FWHM and/or to achieve different beam shapes.

For example IRAS had rectangular beams because of the focal plane geometry. Coaddition of survey data (COADDs produced by the IRAS Processing and Analysis Center (IPAC) at Cal Tech) at high ecliptic latitude produces a cross-like beam shape due to adding observations by rectangular detectors scanning the sky in various directions during the IRAS mission. The IRAS edge detectors were smaller cross-scan than the survey array and were used to make some "DSD" Additional Observations (see e.g. Marston 1989).

Also IRAS's two Chopped Photometric Channels gave circular apertures of 60″ at 50 & 100 μm for some "CPC" Additional Observations, but hysteresis response problems plagued early versions of this data and it is not much used.

The multiple coverages of IRAS scans allows extraction of more spatial detail than the survey beam sizes would suggest, especially in the long detector dimension, if the signal to noise ratio in the scans is high. In favorable cases the resolution achievable is more $\simeq$ scan separation than to the detector width. There are two methods in use to exploit this.

1. The Maximum Correlation Method (Aumann et al 1990) which is available from IPAC as "Hi-Res".
2. The Maximum Entropy Method (MEM) (Bontekoe et al 1991) available on UK Starlink and at Groningen & Wyoming.

Both work remarkably well. In favorable cases one can achieve almost circular beams with FWHM ~ 0.5 of the in-scan value, for data with good signal to noise ratio. (It is very important that backgrounds be carefully removed before attempting this process).

MEM is also used by the University of Texas group who have been operating a high spatial resolution multichannel photometer with slit shaped detectors for mapping/scanning across objects at 50 & 100 μm on the KAO. The current system has 20 detectors each $\sim \frac{\lambda}{2D} \times \frac{\lambda}{D}$ (8×16 arcsec at 50μm, and 14×28 arcsec at 100μm). Multiple scans are made with frequent sampling to get high signal to noise ratio, and the slit scans are then deconvolved with the beam profile using the Maximum Entropy Method. Upper limits to 50 and 100 source μm sizes of $3''$ and $6''$ respectively have been achieved (Lester et al 1986, Butner et al 1991). These sizes are $\sim 0.25\frac{\lambda}{D}$ indicating that MEM has improved angular resolution by a factor of $\sim$ 2. Thus for bright objects this technique can yield spatial information on scales of $3''$ at 50μm and $6''$ at 100μm from the KAO.

MEM should be more widely used on high signal to noise data.

7.4 Pointing, Tracking, Peaking Up

Pointing a massive structure like JCMT at a point on the sky and expecting an object to appear at the centre of a $6''$ beam at 350μm (i.e. pointing within $1''$) is a difficult feat. It works amazingly well but, wherever possible, one should peak-up (using for example a quick 5 or 9 point map) on the object of interest after a large slew (for example to a calibration object) or after a longish time (30 mins) to ensure the object is centred in the beam. If the object itself is not bright enough then peak up on a nearby bright object (of which there are not enough in the submm) and then go to the object. Relative pointing over a few degrees is likely to be much better than absolute pointing.

Once the object is acquired tracking should be good but the object will tend to drift out of the beam on some timescale. If it is bright this will be obvious from the stripchart or individual Right-Left pairs. If it is faint it will not be obvious. An hours integration will be misleading if the object leaves the beam after 20 minutes!

Check pointing frequently (the frequency will depend on the telescope and its current performance) to ensure the object is in the beam. The pointing and tracking performance of large telescopes may vary from observing session to observing session, particularly when maintenance or "improvements" are being made during the day.

7.5 Mapping

In principle, mapping is no more than repetition of point photometry at a number of points. This may, however, be inefficient and if the objects mapped are found to be extended data reduction is quite different.

"On-the-fly" mapping, as used at the JCMT, and by IRAS, involves continuously scanning the telescope across the source at a slow rate with (for JCMT but not IRAS) chopping. The map thus made may be repeated many times to build up signal to noise. This method (with no nodding) is a more efficient use of time in many situations, and individual maps can be normalized before coaddition to cope with transmission or gain changes. The point-by-point method is much more susceptible to gain changes across the map. Scan rate must be determined depending on brightness of object and area to be mapped (Padman & Prestage 1992).

Note, that, because of chopping the map will have positive and negative parts. It must be restored to a positive only form. An algorithm to do this efficiently is implemented in the "NOD2" software in use at several telescopes (Emerson et al 1979). More recently a MEM deconvolution program is proving very effective (Richer 1992).

During 1993 a submm Continuum Bolometer Array (SCUBA) with 91 detectors at $450/350\mu$m and 37 at $850/750/600\mu$m covering a $2.3'$ field of view with each element 10 times as sensitive as UKT 14 will be commissioned on JCMT. This camera will revolutionize submm mapping.

8 Sensitivity & How Long to Integrate

A crucial factor in planning observations has not yet been addressed. For a given telescope and receiver system how long must one integrate for to achieve a specified signal to noise ratio on a source of flux density S_ν?

The sensitivity is usually expressed as the flux density that would produce a 1 σ signal in 1 second of on source integration time. This is called the Noise Equivalent Flux Density (NEFD). Statistics tells us that this will scale as (integration time)$^{-0.5}$ and, for historical reasons, the NEFD is expressed in units of Jy Hz$^{-0.5}$, although Jy sec$^{0.5}$ might seem more obvious. Due to the Fourier relation between frequency and time an integration time of 0.5 sec correspond to a bandwidth of 1 Hz (Kraus 1966). The use of Hz$^{-0.5}$ thus really implies the NEFD in an integration time of 0.5 sec. However, as half the time is spent off source – in the chopped reference beam – the NEFD does indeed refer to 1 sec of elapsed time on the sky.

What limits sensitivity? If there are fundamental physical limits and our systems reach these, how can we proceed?

8.1 Photon/Wave Noise

Consider first an ideal photon counting detector which counts every incident photon and is noiseless.

The photon arrival rate at the detector per Hz of bandwidth of incident radiation $N = N_o + N_b$, where N_o is the rate of arrival of photons from the object, and N_b is the rate of arrival of photons from the background. Let the received bandwidth of the incident radiation be $\Delta\nu$. Then, if the photon/quantum picture is good, the arrival of individual photons are uncorrelated and we have Poisson statistics where the arrival rate is $N\Delta\nu \pm \sqrt{N\Delta\nu}$

However on the photon/wave picture the arrival of the peak of a wave correlates with the previous/next arrival so we get bunching or wave noise (Purcell 1956). In time interval τ and frequency interval $\Delta\nu$ Bose-Einstein statistics tells us (Mandl 1988, Kittel 1969) that the rms fluctuation in the number of photons is

$$\left[N\Delta\nu\tau\left(1+\frac{\epsilon f}{e^{\frac{h\nu}{kT}}-1}\right)\right]^{0.5}$$

where ϵ is emissivity of the source of the photons and f is the radiation/detector transfer (quantum) efficiency.

For $h\nu >> kT$ (Wien) $e^{\frac{h\nu}{kT}} >> 1 \Rightarrow$ Photon Noise

For $h\nu << kT$ (Rayleigh-Jeans) $e^{\frac{h\nu}{kT}} << 1 \Rightarrow$ Wave Noise

In the far-IR/submm we are not quite in either regime so we must use the full expression.

Suppose N_o is small compared to N_b then the noise will be limited by the background photons. (The object photons can only increase the statistical fluctuations.) The signal and noise are (respectively) given by:

$$S = N_o\Delta\nu\tau$$

$$N = (N_b\Delta\nu\tau)^{0.5}(1+\epsilon fb)^{0.5}$$

where

$$b = \frac{1}{e^{\frac{h\nu}{kT}}-1}.$$

Therefore

$$S/N = \frac{N_o\Delta\nu^{0.5}}{(N_b(1+\epsilon fb))^{0.5}}\ \tau^{0.5}$$

Now in terms of the power onto the detector (not that above the atmosphere) the S/N is 1 in 0.5 sec if the Noise Equivalent Power (NEP) falls onto the detector. NEP is commonly used as a detector figure of merit.

$$\text{NEP} = h\nu N_o\Delta\nu = \left[\frac{2N_b(1+\epsilon fb)}{\Delta\nu}\right]^{0.5} h\nu\Delta\nu$$

where the square rooted term is N_o for $S/N = 1$ in 0.5 seconds and as the background power

$$P_b = N_b h\nu\Delta\nu$$

therefore

$$\text{NEP}_{\text{limit}} = \left[\frac{2P_b(1+\epsilon fb)}{h\nu\Delta\nu\Delta\nu}\right]^{0.5} h\nu\Delta\nu$$

so

$$\text{NEP}_{\text{limit}} = [2P_b h\nu(1+\epsilon fb)]^{0.5}.$$

In the Rayleigh-Jeans region $b \sim \frac{kT_b}{h\nu}$, where T_b is the temperature of the background.

$$\text{So}\quad \text{NEP}_{\text{limit}} = [2P_b\epsilon fkT_b)]^{0.5}$$

where wave noise dominates.

Thus the background noise limit in the Rayleigh-Jeans region (which applies for ground based work at $\lambda > 350\mu$m) is proportional to the square root of the power in the background, P_b (Lewis 1947, van Vliet 1967). For the background behaving like a blackbody $P_b = tB_\nu(T_b)A\Omega_b\Delta\nu$ where Ω_b is the solid angle of the beam, and t is the effective transmission of the optics. The product $A\Omega_b$ is known as the throughput and must be conserved through the optics leading to design problems on A (as $\Omega_b < 2\pi$) (Hildebrand 1986). For diffraction limited ($\theta_{FWHM} = \frac{\lambda}{D}$) operation $A\Omega = \left(\frac{\pi}{4}\right)^2 \lambda^2$ and so we cannot reduce P_b and we are stuck with a certain noise from the background. Once our detector NEP reaches the background induced NEP limit there is little point improving the detector further as noise sources add quadratically. Thus when a system is background limited it is as good as it can be, until the background is somehow reduced (e.g. by going into space and cooling the telescope as IRAS did).

Note that on the ground the background limit is due to the atmosphere and telescope. For a cold telescope in orbit (e.g. IRAS) these background sources can become negligible.Then other celestial backgrounds, such as that from zodiacal light, interstellar dust/cirrus, or even the cosmic background radiation can limit the sensitivity. Further limits can be imposed by confusion noise arising from cirrus (Gautier et al 1992), background galaxies etc (Helou & Beichman 1990).

9 Detection systems

9.1 Bolometers

Bolometers absorb incident radiation, warm up, change their resistance and hence can produce an electrical signal proportional to the incident power. By careful design and cooling they can reach background limited IR performance in the airborne (KAO) and ground-based (submm) regions. Bolometer NEPs are limited by the statistical fluctuations of phonons involved in heat flow from element to heat sink (cf photon fluctuations). At low (zero) background these will dominate. Thus bolometers are not so suitable for very low (space) background environments. There are several papers describing how bolometers work (Mather 1982, Low 1961, Haller 1985) and comparing coherent and incoherent detectors (White 1988,Phillips 1988). Table 6 gives some state of the art values for bolometer NEPs.

Table 6. State of the art bolometer NEPs at various temperatures

Coolant	Operating Temperature K	NEP (W $Hz^{-0.5}$)
He^4	4.2	5×10^{-13}
He^4 at 1mb	1.5	2×10^{-14}
He^3 at 1μb	0.32	5×10^{-16}
ADR	0.10	5×10^{-17}

9.2 Photodetectors

In low backgrounds (e.g. a cooled telescope in space such as IRAS or ISO) better NEPs can be achieved with photodetectors which do not suffer from phonon noise. They just count photons energetic enough to excite electrons into conduction bands in doped semiconductors. They have a long wavelength energy cut off and are described in various places (Willardson & Beer 1970, Willardson & Beer 1977, Haller 1985, Wynn-Williams & Becklin 1987) and in various recent meetings relating to ISO and in the IRAS Exp Supp. In the low background situation of ISO at 100μm their NEP is 4×10^{-19} W $\mathrm{Hz}^{-0.5}$, the state of the art. At high backgrounds they tend to saturate and noise is dominated by the signal current. In practise great care is needed to minimize other forms of noise in the electronics etc. Several texts discuss photoconductors and natural limits (Arams 1973, Keyes 1977, Haller 1985).

9.3 Optics/Filters/Cooling/Systems

The optics of the system have to match the telescope beam to the detector preserving the throughput $A\Omega_b$. In systems with bolometers, Winston cones (Winston 1978,Keene et al 1978) are usually used to couple the radiation to the bolometer.

In addition to the filtering effects of cryostat window materials, interference filters (Whitcomb & Keene 1980) are often used. Care must be taken to block long and short wavelength leaks in the overall transmission.

Cooling the system both reduces the emission from the instrument optics (but not the telescope or atmosphere unfortunately) and is necessary to attain good bolometer NEP performance (as NEP $\propto T^{5/2}$). Operation at 100 mK is being pursued.

General descriptions of far-IR/submm bolometer systems and descriptions of some systems used on some major facilities including the KAO, IRTF, IRAM & SEST, and JCMT can be found in the literature (e.g. Hildebrand 1986, Harvey 1979, Whitcomb et al 1980, Kreysa 1990, Duncan et al 1990 respectively), and there are many briefer descriptions in papers reporting astrophysical results.

10 Observational Capabilities

The main sub-mm observatories are JCMT & CSO on Mauna Kea, the Swedish-ESO Submm Telescope (SEST) in Chile, the IRAM 30m MST in Spain, and the 12m US National Radio Astronomy Observatory dish which has worked down to $\sim$ 1mm. Attempts to summarize current capabilities are invariably out of date because of technical advances. Nevertheless in Table 7 below I give some parameters as derived from an incomplete look at the literature. They are intended to give some idea of what is currently possible, rather than to be definitive. Consultation with those closely involved with particular instruments is recommended to get up to date (& no doubt more correct) information.

For the far-IR the only currently available observatory is NASA's Kuiper Airborne Observatory (KAO), although archival IRAS data is available, and heavily used. Archival IRAS data may be obtained from IPAC, who issue a regular Newsletter describing the products available. NASA Ames Research Center distribute an annual cumulative publication list for its airborne observatories (ie mostly KAO),which now runs to some 200 pages and is a good way of finding what has been done with the KAO. There is a proposal for a more advanced airborne observatory (Erickson 1992).

Plans for further ground based submm telescopes include: a University of Arizona/Max Planck Institut fur Radioastronomie 10m on Mt Graham, Arizona operating down to 350 μm; extending the operating frequency of existing mm arrays to 1mm; and constructing new mm arrays that will operate down to 870 μm. Bolometer arrays are also becoming available, and the SCUBA array on JCMT will be sky noise limited at all wavelengths.

Many future IR/submm space missions have been proposed and are at various stages of design, but budgetary constraints or other delays make it unclear as yet when (if at all) most of them are likely to fly (Erickson & Werner 1992, Hayashi 1992, Kaplan 1992). The Infrared Space Observatory, ISO, is however now due to be launched in late 1995 and will carry four instruments which together will offer imaging and spectroscopic capabilities from the near-IR into the far-IR up to a wavelength of $\sim 200\mu$m. The anticipated system sensitivities and capabilities (which are voluminous) are available in various project books (see also Kessler et al 1992), and will be made widely available when the request for proposals is released in 1994.

11 Conclusions

In these lectures I have attempted to bring out matters relating to observational techniques that are not often discussed in this kind of forum. One should be prepared to discuss the instrument, one is using with the staff at the facility one is observing at to be sure to get the best out of the instrument. Ultimately when one publishes one must clearly present the results, the parameters of the observational system and the uncertainties, together with a description of the procedures used, so that the reader will know what was done, and will have confidence in your data. In the submm and far-IR this task is not as simple as one might at first sight hope!

There is a wealth of further astrophysics to be uncovered as new advances in instrumentation allow us to probe with ever increasing sensitivity and resolution into the far-IR and sub-mm. We can expect much further information on star-formation to be uncovered over the next few years.

Table 7. Capabilities of some current (& archival) far-IR & Submm Observatories

λ μm	ν GHz	$\Delta\nu$ GHz	θ_{FWHM} arcsec	1 σ limit mJy	NEFD mJy $Hz^{-0.5}$
MRT	D = 30m	IRAM	h= 2.9 km		MPIfR bolometer
1300	240	<100	12.4	2	90
870	345		8		380
JCMT	D = 15m	JAC	h = 4.2 km		UKT14 bolometer
2000	150	40	28	30	2000 (2000-3000)
1300	233	64	21	18	300 (300- 500)
1100	264	75	19	8	300 (200- 700)
850	354	30	16	9	800 (700-5000)
800	394	103	14	9	700 (500-5000)
600	480	114	9	30	
450	685	84	7	70	6000 (4000-∞)
350	870	249	6	180	12000 (8000-∞)
SEST	D = 15 m	ESO	h = 2.3 km		MPIfR bolometer
1300	240	<100	23		200
870	345		16		400
NRAO	D = 12m	NRAO			Coherent
1300	232	0.6	30		
CSO	D = 10.4m	CIT	h = 4.1 km		Bolometer
1100	284	67	22	16	1000
800	390	96	17	30	3500
600	480	52	13	70	
IRAS	D = 0.6m	archival (IPAC)	h = orbital		Survey array[1]
100	3000	1760	180 x 300		> 1.5 (PSC2), 1.0 (FSS)
60	5000	4600	90 x 285		> 0.5 (PSC2), 0.2 (FSS)
KAO	D = 0.9 m		h = 14 km		U.Texas array
100	3000	2000	14 x 28		30000
50	6000	2900	8 x 16		30000
KAO	D = 0.9 m		h = 14 km		U. Chicago array
190	1400	300	45		
160	1700	520	45		
100	3100	1300	45		
60	5000	2250	33		
40	7500	3900	33		

[1] In the IRAS case the values quoted under NEFD are not NEFDs but the smallest flux densities listed in the Point Source Catalog version 2 and the Faint Source Survey whose Explanatory Supplements should be consulted for more details.

References

Adams, F.C., Emerson, J.P. & Fuller, G.A.: Submillimeter photometry and disk masses of T Tauri disk systems. ApJ **357** (1990) 606–620

Ade, P.A.R., Griffin, M.J., Cunningham, C.T., Radostitz, J.V., Predko, S. & Nolt, I.G.: The Queen Mary College/University of Oregon photometer for submillimetre continuum observations. Infrared Phys. **24** (1984) 403–415

Airy, G.B.: Trans. Camb. Phil. Soc. **5** (1835) 283–

Arams, F.R.: Infrared-to-millimeter wavelength detectors, Artech House (1973)

Aumann, H.H., Fowler, J.W. & Melnyk, M.: A maximum correlation method for image construction of IRAS survey data. AJ **99** (1990) 1674–1681

Bontekoe, T.R., Kester, D.J.M., Price, S.D., de Jonge, A.R.W. & Wesselius, P.R.: Image construction from the IRAS survey. A&A **248** (1991) 328–336

Butner, H.M., Evans, N.J., Lester, D.F., Levreault, R.M. & Strom, S.E.: Testing models of Low-Mass Star Formation: High-resolution far-infrared observations of L1551 IRS 5. ApJ **376** (1991) 636–653

Clemens, D.P., Leach, R.W., Barvainis, R. & Kane, B,D.: Millipol, a Millimeter & Submillimeter wavelength Polarimeter: Instrument, Operation and Calibration. PASP **102** (1990) 1064–1076

Danese, L., & Partridge R.B.: Atmospheric Emission Models: Confrontation between Observational Data and Predictions in the 2.5 – 300 GHz range. ApJ **342** (1989) 604–615

Duncan, W.D., Robson, E.I., Ade, P.A.R., Griffin, M.J. & Sandell, G.: A millimeter/submillimetre common user photometer for the James Clerk Maxwell Telescope. MNRAS **243** (1990) 126–132

Emerson, D.T., Klein, U. & Haslam, C.G.T.: A Multiple Beam Technique for Overcoming Atmospheric Limitations to Single-Dish Observations of Extended Sources. A&A **76** (1979) 92–105

Emerson, J.P.: Infrared Emission Processes. In: Dupree A.K. & Lago M.T.V.T. (eds.) Formation & Evolution of Low Mass Stars Kluwer (1988a) 21–44

Emerson, J.P.: IRAS Observations. In: Dupree A.K. & Lago M.T.V.T. (eds.) Formation & Evolution of Low Mass Stars Kluwer (1988b) 193–207

Erickson, E.: SOFIA: Stratospheric Observatory for Infrared Astronomy, Sp.Sci.Rev. **61** (1992) 61–68

Erickson, E., & Werner, M.: SIRTF: Space Infrared Telescope Facility, Sp.Sci.Rev. **61** (1992) 95–98

Fazio, G.: Infrared Astronomy. In: Avrett, E.H. (ed.) Frontiers of Astrophysics. Harvard Univ Press (1976) 203–258

Fazio, G., ed.: Infrared & Submillimeter Astronomy. Astrophysics & Space Science Library **63**, Reidel (1977)

Flett, A.M., & Murray, A.G.: First results from a submillimetre polarimeter on the James Clerk Maxwell Telescope. MNRAS **249** (1991) 4P–6P

Gautier, T.N., Boulanger, F., Perault, M. & Puget, J.L.: A calculation of confusion noise due to infrared cirrus. AJ 103 (1992) 1313–1324

Gonatas, D.P, Engargiola, G.A., Hildebrand, R.H., Platt, S.R., Wu, X.D., Davidson, J.A., Novak, G., Aitken, D.K. & Smith C.: The far-Infrared Polarization of the Orion Nebula. ApJ **357** (1990) 132–137

Haller, E.E.: Physics and Design of Advanced IR Bolometers and Photoconductors. Infrared Physics. **25** (1985) 257–266

Harvey, P.M.: A Far-Infrared Photometer for the Kuiper Airborne Observatory. PASP **91** (1979) 143—148

Hayashi, M.: Plans for Submillimetre and Infrared Satellites in Japan, Sp.Sci.Rev. **61** (1992) 99–102

Helou, G. & Beichman C.A.: The confusion limits to the sensitivity of submillimeter telescopes. In: 29th Liege Ap Colloquium, From Ground Based to Space Borne submm astronomy. European Space Agency, ESA SP-314 (1990)

Hildebrand, R.H.: Focal plane optics in far-infrared and submillimetre astronomy. Optical Engineering **25** (1986) 323–329

Hildebrand, R.H.: Magnetic Fields and Stardust. QJRAS **29** (1988) 327–351

Hills, R.: Jupiter's ring. JCMT-UKIRT newsletter No. **4** (1992) 24–26

IRAS Science team, Beichman, C.A., Neugebauer, G., Habing, H.J., Clegg, P.E. & Chester, T.J. eds.: IRAS Catalogs and Atlases, **1**, Explanatory Supplement. NASA–RP–1190 (1988)

Kaplan, M.: NASA's plans for Space Astronomy & Astrophysics, Sp.Sci.Rev. **61** (1992) 103–112

Keene, J., Hildebrand, R.H., Whitcomb, S.E. & Winston, R.: Compact infrared heat trap field optics. Applied Optics **17** (1978) 1107–1109

Kessler, M.F., Metcalfe, L., & Salama,A.: The Infrared Space Observatory, Sp Sci.Rev. **61** (1992) 45–60

Keyes, R.J. ed.: Optical and Infrared Detectors, Springer (1977)

Kittel, C.: Photon Fluctuations. In: Thermal Physics John Wiley & sons 1st. edn. (1969) 260–262

Kraus, J.D.: Radio Astronomy. McGraw Hill (1966) 244–246

Kreysa, E.: In: 29th Liege Ap Colloquium, From Ground Based to Space Borne submm astronomy. European Space Agency, ESA SP-314 (1990)

Kyle, T.G., & Goodman A.: Atlas of Computed Infrared Atmospheric Absorption Spectra, National Center for Atmospheric Research, NCAR-TN/STR-112 (1975)

Leach, R.W, Clemens, D.P., Kane B.D., & Barvainis R.: Polarimetric mapping of Orion using Millipol: Magnetic Activity in BN/KL. ApJ **370** (1991) 257–262

Lester, D.F., Harvey, P.M. & Joy, M.: The spatial structure of IRC+10216 and NGC7027 in the Far-Infrared. ApJ **304** (1986) 623–633

Lewis, W.B.: Fluctuations in streams of thermal radiation. Proc. Phys. Soc. (London) **59** (1947) 34–40

Liege, 29th Ap Colloquium, From Ground Based to Space Borne submm astronomy. European Space Agency, ESA SP-314 (1990)

Low, F.J.: Low-Temperature Germanium Bolometer, J.Opt.Soc.Am. **51** (1961) 1300–1304

Low, F. and Rieke G.H.: The Instrumentation and Techniques of Infrared Photometry. In: N. Carleton (ed.), Methods of Experimental Physics,**12**, Astrophysics, Part A, Optical and Infrared. Academic Press (1974) 415–462

Mandl, F: Statistical Physics, John Wiley 2nd Edn. (1988) p. 271 and 312

Marston, A.P.: High-Resolution Far-Infrared Images of M83, AJ **98** (1989) 1572–1580

Marten, A., Baluteau, J.P. & Bussolletti, E.: High Resolution Infrared Spectra of the Earth's Atmosphere – I. Numerical Simulation of Atmospheric Spectra. Infrared Physics **17** (1977) 197–209

Mather, J.C.: Bolometer noise: nonequilibrium theory. Applied Optics **21** (1982) 1125–1129

McClatchey, R.A., Benedict, W.S., Clough,S.A., Burch, D.E., Calfee, R.F., Fox, K., Rothman, L.S., & Garing, J.S.: AFCRL Atmospheric Absorption Line Parameters Compilation. Air Force Cambridge Research Laboratories AFCRL-TR-73-0096 (1973)

Naylor, D.A., Clark, T.A., Schultz, A.A. & Davis, G.R.: Atmospheric transmission at submillimete wavelengths from Mauna Kea. MNRAS **251** (1991) 199–202

Padman, R. & Prestage R. Digital demodulation and fast sampling with UKT14. JCMT-UKIRT newsletter No. **3** (1992) 19–22

Phillips, T.G.: Techniques of submillimeter astronomy. In: Wolstencroft R.D. & Burton W.B. (eds.), Millimetre and Submillimetre Astronomy , Kluwer (1988) 1–25

Platt, S.R., Hildebrand, R.H., Pernic, R.J., Davidson, J.A. & Novak, G.: 100μm Array Polarimetry from the Kuiper Airborne Observatory: Instrumentation, Techniques, and First Results. PASP **103** (1991) 1193–1210

Purcell, E.M.: Nature **178** (1956) 1449–1450

Richer, J.S.: Dual beam mapping with a maximum entropy algorithm MNRAS **254** (1992) 165–176

Shaver, P.A. ed.: ESO-IRAM-Onsala workshop on (Sub)Millimetre astronomy. ESO Conference and Workshop Proceedings **22** (1985)

Soifer, B.T. & Pipher, J.L.: Instrumentation for Infrared Astronomy. ARA&A **16** (1978) 335–369

Sollner, G.: Frequency Spectrum of Fluctuations in Submillimetre Sky Emission and Absorption. A&A **55** (1977) 361–368

Traub, W.A., & Stier, M.T.: Theoretical atmospheric transmission in the mid- and far-infrared at four altitudes. Applied Optics **15** (1976) 364–377

van Vliet, K.M.: Noise limitations in solid state photodetectors: Applied Optics **6** (1967) 1145–1168

Watt, G.D. & Webster A.S. eds.: Submillimetre Astronomy. Kluwer (1990)

Welch, W.J.: Techniques and Results of Millimeter Interferometry. In: Wolstencroft R.D. & Burton W.B. (eds.), Millimetre and Submillimetre Astronomy , Kluwer (1988) 95–116

Welch, W.J.: Millimeter and Submillimeter Interferometry. In: Watt G.D. & Webster A.S. (eds.) Submillimetre Astronomy, Kluwer (1990) 81–86

Whitcomb, S.E., Hildebrand, R.H. & Keene J.: An f/35 submillimeter photometer for the NASA Infrared Telescope Facility. PASP **92** (1980) 863–869

Whitcomb, S.E., & Keene J.: Low-pass interference filters for submillimeter astronomy. Applied Optics **19** (1980) 197–198

White, G.J.: Receiver Technology. in Millimetre and Submillimetre Astronomy , eds. Wolstencroft R.D. & Burton W.B., Kluwer (1988) 27–94

Willardson, R.K., & Beer A.C. eds.: Semiconductors and Metals **5** IR Detectors, Academic Press (1970)

Willardson, R.K., & Beer A.C. eds.: Semiconductors and Metals **12** IR Detectors II, Academic Press (1977)

Winston, R.: Cone collectors for finite sources. Applied Optics **17** (1978) 688–689

Wolstencroft, R. & Burton W.B. eds: Millimetre and Submillimetre Astronomy. Kluwer (1988)

Woody, D.P., Scott, S.L., Scoville, N.S., Mundy, L.G., Sargent, A.I., Padin, S., Tinney, C.G. & Wilson, C.D.: Interferometric Observations of 1.4mm continuum sources. ApJ **337** (1989) L41–L44

Wynn-Williams, C.G., & Becklin, E.E. (eds.): Infrared Astronomy with Arrays. Univ. of Hawaii, Institute for Astronomy (1987).

Zylka, R., Mezger, P.G. & Leasch, H.: Anatomy of the Sagittarius A complex: II $\lambda 1300\mu m$ and $\lambda 870\mu m$ continuum observations of Sgr A_* and its submm/IR spectrum. A&A (1993) in press

Near Infrared Techniques for Studies of Star Formation

Steven V. W. Beckwith

Max-Planck-Institut für Astronomie, Heidelberg, Germany

1 Infrared Appearances of Young Stars

1.1 Stages of Star Formation

Stars are born in clouds of gas and dust, and they are most probably born in the centers of particularly cool and dense regions, sometimes referred to as *cores*, shown in Fig. 1.1a. These cores are quasi-stable. They may survive for many times their free-fall collapse timescales, but at some time they begin to fall under the force of their self gravity and collapse to create new stars (Fig. 1.1b). The way they are supported is as yet unknown as is the trigger for collapse, although current thinking holds that clouds are supported in some fashion by magnetic fields (Shu et al. 1992).

A well-known theoretical principle is that the cores collapse from the inside out; that is, because they are densest in the centers, and the free fall time is proportional to the inverse square root of the density, the inner parts fall much faster than the outer parts, and a luminous object is born inside a core long before the outer parts have reached the center (Larson 1973). Small enhancements in the density of different regions can also create fragmentation and runaway collapse of subregions. It is a fact that most nearby stars are members of binary or multiple star systems (Abt and Levy 1976, Abt 1983) as are young, pre-main sequence stars (Leinert et al. 1993; Ghez et al. 1993), a probable consequence of this tendency towards fragmentation during "inside-out" collapse.

In its gestation, a star shines from within a collapsing core which is itself inside a molecular cloud. It should be no surprise that these young stars are invisible in the optical band owing to strong extinction by the dust grains in the cores and clouds. Typically, the extinction to a very young star is many tens to many hundreds of visual magnitudes.

The stellar luminosity originally comes from latent energy released by the central collapse. The young star is too cold to burn hydrogen; it will burn deuterium, if only briefly, but it must get rid of the energy from the collapse (Palla and Stahler 1990). This energy, given off at the near-infrared ($1 - 3\,\mu$m) and optical wavelengths ($0.3 - 1\,\mu$m), is absorbed by the infalling dust and reradiated at longer wavelengths in the thermal (3 – 30 μm) to far infrared (30 – 300 μm). Extinction at far infrared wavelengths is generally quite small even in dense cores, so this radiation escapes the cloud. Figure 1.2 shows the radial structure of a protostar calculated by Stahler, Shu, and Taam (1980) with the dust "blanket" surrounding the young object.

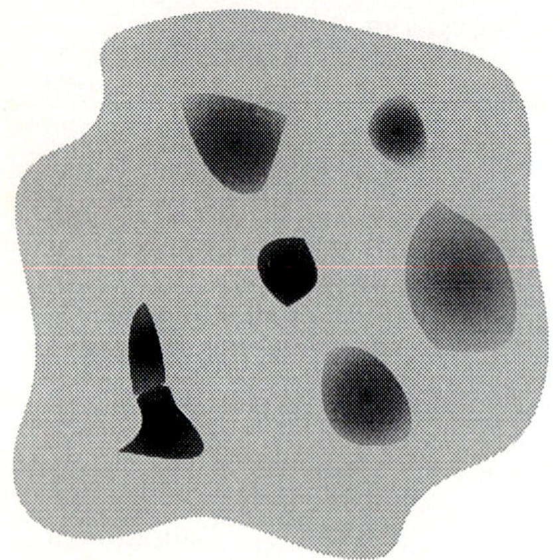

Fig. 1.1a. A molecular cloud with a series of loosely bound cores. The cores are the seeds for new stars.

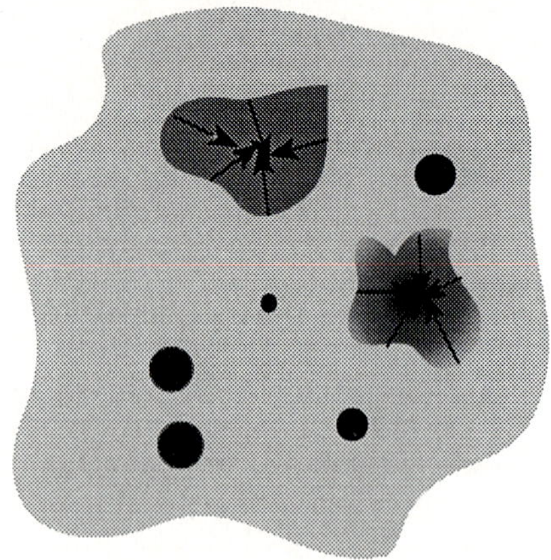

Fig. 1.1b. Several cores are in the process of collapse, and there are three new stars within cores that have already collapsed.

There are, then, two reasons why star formation is observed primarily at infrared wavelengths. The first is because extinction by interstellar dust is a strong function of wavelength, λ, and the extinction is usually too large at optical wavelengths to allow meaningful observations; the opacity falls nearly as $\lambda^{-1.9}$ from 0.5 to 10 μm. Ten visual magnitudes of extinction ($A_V = 10$) corresponds to only 1 magnitude of extinction at 2.2 μm and much less in the thermal infrared. A region which is completely opaque optically may be relatively transparent at longer wavelengths. So the infrared is important simply to *see into* the clouds and observe the processes taking place in the early life of a star.

The second reason is because the bulk of the luminosity from young stars will emerge at infrared wavelengths in the early stages owing primarily to radiative transfer in the surrounding dust blankets. To locate young stars in molecular clouds and to measure their luminosity, it is advantageous to observe at the wavelengths of peak power; indeed, no survey of young stars in a dark cloud is complete without knowledge of the infrared radiation from the cloud.

Other wavelengths serve, too, as important diagnostics of the way in which stars form. The radio (HII regions and stellar winds), millimeter and submillimeter (molecular emission from clouds), far infrared (dust envelopes), and x-ray (stellar chromospheres) observations all play an important role in the study of young stars. These observations and the accompanying techniques are discussed in other chapters. But the stars themselves shine predominantly in the optical and near-infrared bands, meaning that infrared observations are always desirable and often necessary to understand the early evolution of young star forming regions.

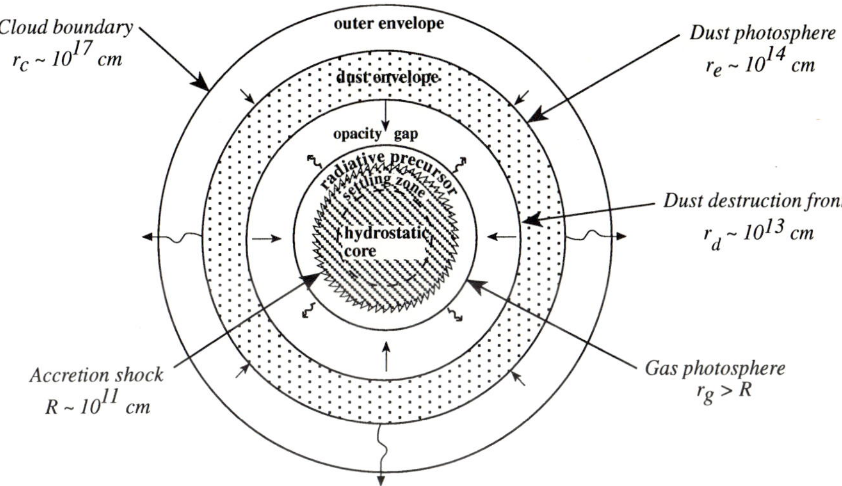

Fig. 1.2. The radial structure of a few $M_\odot$ protostar with a dust envelope surrounding the luminous object (after Stahler, Shu, and Taam 1980). These authors calculated the structure assuming spherical collapse from an initially stable cloud with a mixture of gas and dust (100 to 1 by mass). The *opacity gap* develops because the high temperature radiation from the accretion shock at 10^{11} cm – effectively the stellar surface, R_* – melts the dust out to 100 R_*, 10^{13} cm. Beyond about 1000 R_*, the dust emission is optically thin in the infrared, creating a "dust photosphere" at this radius. The cloud extends to at least 10^6 R_* in this model.

Stars of 2 $M_\odot$ or less have relatively long pre-main sequence lifetimes, long enough to disperse their cores and shine visibly well before they become main-sequence stars (*e.g.*, Cohen & Kuhi 1979). During this time they will often be encircled by disks of gas and dust (Beckwith and Sargent 1992), perhaps the precursors to planetary systems, and they lose mass at prodigious rates in oppositely collimated *jets* (Mundt 1988; Edwards, Ray, & Mundt 1993). The disks emit over a broad spectrum from the ultraviolet to the submillimeter (Bertout 1989; Strom et al. 1989; Beckwith et al. 1990). The jets shock the material around them, stimulating strong optical and near-infrared line emission, and they gather up gas from the surrounding clouds which may be seen in millimeter-wavelength rotational emission from heavy molecules (cf. chapters by S. Cabrit in this volume).

The high mass stars probably do much the same thing but so quickly that it is difficult to catch them in the act. In any case, they stay immersed in cloud material until after they become main sequence stars. Their entire pre-main sequence life is shielded from direct optical observation by the clouds.

These chapters concern the observations of young stars at near and thermal infrared wavelengths and concentrate on the methods by which the observations are gathered. This chapter gives an overview of the most general principles concerning the production of infrared light and its use for the observation of newly born stars. It is brief and meant only to motivate potential uses of these techniques. The second, third, and fourth chapters discuss the observational techniques themselves. The fifth chapter describes a few of the areas in which new infrared technology will have an immediate impact. The emphasis here is on general principles, since the detailed applications change rapidly.

1.2 Extinction by Dust Grains

The interaction of small particles with radiation is a rich and difficult subject, far larger than we can possibly cover here (Spitzer 1978, Bohren & Huffman 1983). For the purposes of these lectures, it is necessary only to recall a few basic principles:

a. Most dust grains are small compared to near infrared wavelengths, much less than 1 μm in size.

b. For our purposes, spherical grains represent a reasonable approximation to the particle shapes. Therefore, absorption and scattering may be treated by the Mie theory.

c. As a consequence of *a* and *b*, both the absorption and scattering cross sections decrease as the wavelength of the radiation increases.

If the grain radius is a, the absorption and scattering cross sections, σ_a and σ_s, are usually written in terms of πa^2, the geometrical cross section: $\sigma = Q(\lambda)\pi a^2$, with $Q_a(\lambda)$, $Q_s(\lambda)$, and $Q_e(\lambda)$ representing the respective efficiencies of absorption, scattering, and extinction. Along the line of sight, $Q_e = Q_a + Q_s$.

For simple substances, $Q_s(\lambda) \sim \lambda^{-4}$ and $Q_a(\lambda) \sim \lambda^{-1}$. Therefore, scattering plays a negligible role in the extinction of almost all infrared radiation; the wavelengths are sufficiently long that $Q_s \ll Q_a$. Moreover, radiation which is scattered should be bluer in the sense that it is dominated by the shortest wavelengths in any spectral band. The sky is blue, because it is just scattered sunlight.

The actual extinction cross sections in the near infrared are approximately proportional to $\lambda^{-1.9}$. The optical depth, defined as the natural logarithm of the factor by which radiation is decreased, is unity for almost exactly one magnitude of extinction ($10^{-\frac{1}{2.5}} = 0.4$; $e^{-1} = 0.37$), so we can use magnitudes and optical depth interchangeably in the discussion of extinction or absorption. The usual rule of thumb is that 10 visual magnitudes ($A_V = 10$) corresponds to 1 magnitude of extinction at 2.2 μm. The infrared extinction curves are discussed in detail by Becklin et al. (1978) and Rieke and Lebofsky (1985).

1.3 Infrared Continuum Emission

Of the various processes that produce radiation over a wide range of wavelengths, only thermal emission is important for consideration here. Synchrotron emission and bremsstrahlung (free-free) radiation are rarely strong enough near young stars to be important in the infrared.

Thermal emission from a single object at a uniform temperature, T, depends only on the area of the object, A, the emissivity as a function of frequency, $\epsilon(\nu)$, and T. The spectral energy distribution is a Planck function modified by the surface emissivity function:

$$\nu F_\nu = \epsilon(\nu) \frac{A}{D^2} \frac{2h\nu^4}{c^2} \frac{1}{e^{\frac{h\nu}{kT}} - 1}, \tag{1.1}$$

where ν is the observing frequency, F_ν is the flux density (units of power per area per bandwidth), D is the distance from the source to the observer, and the physical constants take on their usual definitions. We will switch between frequency ν and wavelength λ when convenient. If the emissivity is unity at all frequencies, the source is a *blackbody* and has maximum emission at a wavelength:

$$\begin{aligned} \lambda_{\max} &= 0.255 \frac{hc}{kT} \\ &= \left(\frac{3674 \text{ K}}{T}\right) \quad \mu\text{m}. \end{aligned} \tag{1.2}$$

Peak emission in the infrared means temperatures of a few hundred to a few thousand K. Stellar photospheres, even for protostars, are usually more than about 2500 K.[1] Dust envelopes are much cooler.

A small particle at a distance r from an object of luminosity L will be heated and cooled by the absorption and emission of radiation. If the particle is a spherical blackbody of radius a and rapidly spinning (to equalize the temperature throughout), the equilibrium temperature is:

$$\begin{aligned} T_{\text{eq}} &= \left(\frac{L}{16\pi\sigma_b}\right)^{-\frac{1}{4}} r^{-\frac{1}{2}}, \\ &\sim 500 \text{ K} \left(\frac{L}{10 L_\odot}\right)^{-\frac{1}{4}} \left(\frac{r}{1 \text{ AU}}\right)^{-\frac{1}{2}} \end{aligned} \tag{1.3}$$

Small particles which absorb well at short wavelengths but radiate poorly at long wavelengths will be somewhat hotter (cf. Spitzer 1978), but (1.3) gives one a general idea of the temperatures in the vicinity of young stars. It is useful to note that carbon particles (graphite in various forms) melt at about 1200 K, and silicate particles melt at about 1600 K, so one does not expect to see any dust at temperatures more than about 2000 K. Therefore, all radiative transfer involving dust shifts the energy spectrum into the infrared (compare with (1.2)),

[1] we will not speculate on the appearance of brown dwarf stars

and between the photosphere of a young star and its inner dust boundary there is generally a gap in which there are no solid particles (the "opacity gap" in Fig. 1.2).

1.4 Infrared Line Emission

Ions, atoms, and molecules all have energy transitions in the infrared. Usually, the important transitions are those for which the level-excitation energy is less than 1 eV. In atoms and ions, this limit means the normal electronic transitions are usually *not* important in the infrared, since the energies are similar to the binding energies of outer electrons which are a few eV or more except for rare, heavy elements. Figure 1.3 shows energy level diagrams for neutral hydrogen and oxygen.

One exception to this general rule occurs for the upper electronic transitions of hydrogen. The Lyman and Balmer series ($n = m \rightarrow 1$ and $n = m \rightarrow 2$) are at ultraviolet and optical wavelengths, but the Paschen and Brackett series are in the infrared ($n = m \rightarrow 3$ and $n = m \rightarrow 4$, respectively). Hydrogen is sufficiently abundant to make these series relatively strong when the gas is ionized, so that radiative recombinations populate the upper levels. Near infrared observations of these lines are often helpful to study ionized regions buried in dense molecular clouds.

Among the more important of the various atomic and ionic lines are:

- the Brackett series to detect ionized gas and estimate the extinction to the plasma from the observed line ratios: Br_α (4.05 μm) & Br_γ (2.16 μm),
- HeI (2.06 μm) to trace very hot plasma,
- FeII (1.64 μm) for abundance and ionization estimates,
- NeII at 12 μm for abundance estimates and to trace gas velocities in the vicinity of young stars,
- a series of fine structure lines from heavy elements used for abundance studies: e.g. SIV (10 μm), ArIII (9 μm), SiII (35 μm), etc.

Of these, the Br_γ and Br_α lines have been by far the most important to date.

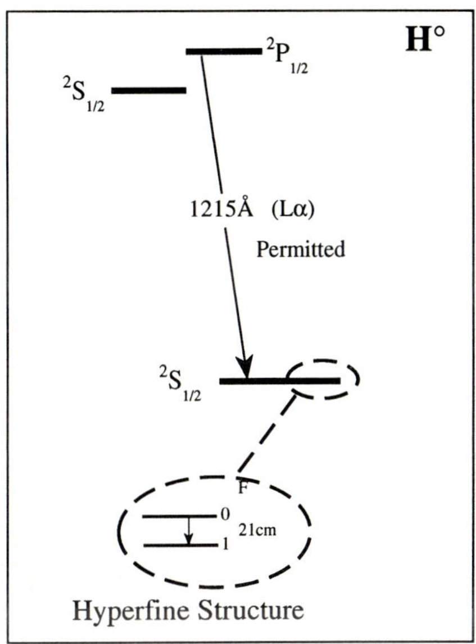

Fig. 1.3a. Energy level diagrams for the hydrogen atom indicating the standard electronic transitions and the hyperfine structure lines.

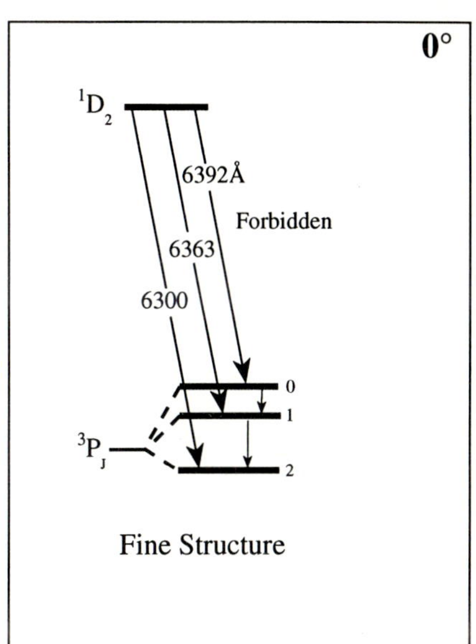

Fig. 1.3b. An energy level diagram for neutral oxygen showing the electronic and fine structure lines.

Molecular energy levels are more complex than the ionic levels owing to the greater number of degrees of freedom available to a molecule. Figure 1.4 shows the three principle types of level structure and the corresponding energies for a simple diatomic molecule such as CO. The electronic levels are similar to those for ions and have about the same energies: the nuclear potential from several nuclei separated by a few angstroms is similar in magnitude to that from a single nucleus, so the electronic binding energies are of order 10 eV or more. The electronic transitions are generally in the ultraviolet and optical portions of the spectrum and need not concern us here (there are high order transitions analogous to the higher order hydrogen transitions discussed above, but they have thus far not been very useful to astronomy).

The vibrational levels come about because the internuclear binding energy is a potential well with quantized levels corresponding classically to the movement of the two nuclei bound by a spring; the wavefunctions are nearly identical those of a simple harmonic oscillator: $E_v = h\nu_0(v + \frac{1}{2})$, where v is the vibrational quantum number. The vibrational energies for most nuclei are of order *tenths* of an eV, and the transitions are in the near infrared. The most important lines to date have been the isolated vibrational lines of molecular hydrogen (e.g. $v = 1 - 0$ $S(1)$ at $2.122\,\mu$m) and the fundamental ($\Delta v = 1$) and overtone ($\Delta v = 2$) lines of CO near 4.6 and $2.4\,\mu$m, respectively. H_2O is an important molecule, but it is almost impossible to observe from the ground, because the Earth's atmosphere absorbs strongly in all transitions. NH_3 and SiO near 10 μm have been observed in the envelopes of late-type stars and could become important for star formation observations as the sensitivity of thermal infrared spectrometers improves.

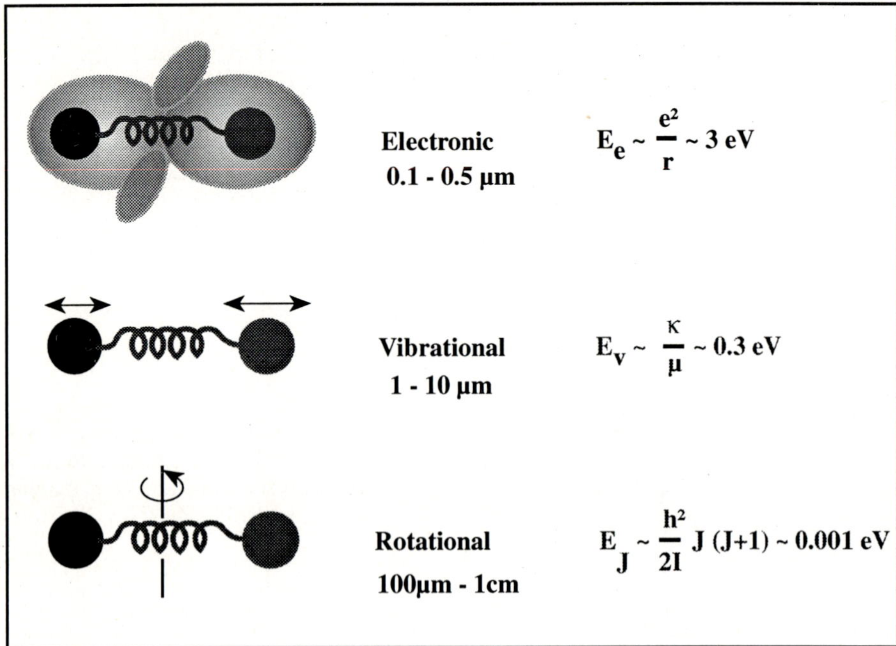

Fig. 1.4. Schematic pictures of a diatomic molecule showing the three principle types of energy levels along with the corresponding wavelengths of the fundamental transitions. Molecules also have even more complex equivalents of fine and hyperfine structure which split the rotational levels.

Quantization of angular momentum means that molecules have rotational energy levels, too. The separation of levels depends only on the moment of inertia, I, of the molecule: $E_J \sim \frac{h^2}{2I} J(J+1)$, where J is the rotational quantum number. Very light molecules, such as H_2 and HD, have the highest energy rotational levels with transitions in the thermal infrared. Heavy molecules, such as CO, have levels starting at millimeter wavelengths. Pure rotational transitions of molecules have been seen near 4 μm (Knacke & Young 1981), and there is some hope that the lowest transition of the hydrogen molecule at 28 μm will become important with the introduction of cooled, space telescopes.

The higher energy levels, such as the vibrational levels, are split by the lower energy levels just as is the case for ions. Thus, each vibrational transition of H_2 is split into rotational sublevels, the equivalent of molecular fine structure. H_2 is light and the different vibration-rotation transitions are well separated: the three lowest transitions in which the vibrational quantum number, v, changes by 1 are the $v = 1 - 0$ $S(0)$ ($2.22\,\mu$m), $v = 1 - 0$ $S(1)$ ($2.12\,\mu$m), and $v = 1 - 0$ $S(2)$ ($2.02\,\mu$m). The vibrational transitions of heavy molecules, such as CO, are split into many closely-spaced lines and constitute distinct *bands* of vibrational transitions. Each band has a rather sharp *bandhead* where the energies of sequential rotational transitions occur at nearly the same wavelength, giving the spectrum a characteristic appearance.

The H_2 lines are mainly important for analyzing the high temperature regions occurring behind shock waves in molecular clouds and the ion-neutral interface at the edge of HII regions. In star formation studies, the shock diagnostics have received the widest attention. CO bandheads are strong in cool stellar atmospheres, such as young, low-mass stars. They are useful to determine effective temperatures and spectral types for embedded stars. They are often seen in emission, probably in hot gas in the inner parts of circumstellar disks, as well.

Herzberg (1950) and Townes and Schalow (1975) give excellent introductions to the physics of molecules and the formation of molecular lines. Descriptions of atomic and ionic lines can be found in many places, for example, in Leighton (1959). Genzel (1992) presents a much more extensive review of infrared atomic and molecular lines as applied to the analysis of molecular clouds.

1.5 Solid State Resonances

Solids normally lack the narrow absorption and emission lines of atoms and molecules. There are, however, cases where particular materials have resonant vibrations over a relatively narrow wavelength region giving rise to *broad* absorption and emission lines with characteristic shapes, depending on the solid. Furthermore, there are molecules that are extremely large, 100 Åsmall to qualify as solid particles. These large molecules normally have strong resonances at several wavelengths, the most famous examples now being the polycyclic aromatic hydrocarbons, or PAHs, for short (Puget & Léger 1989).

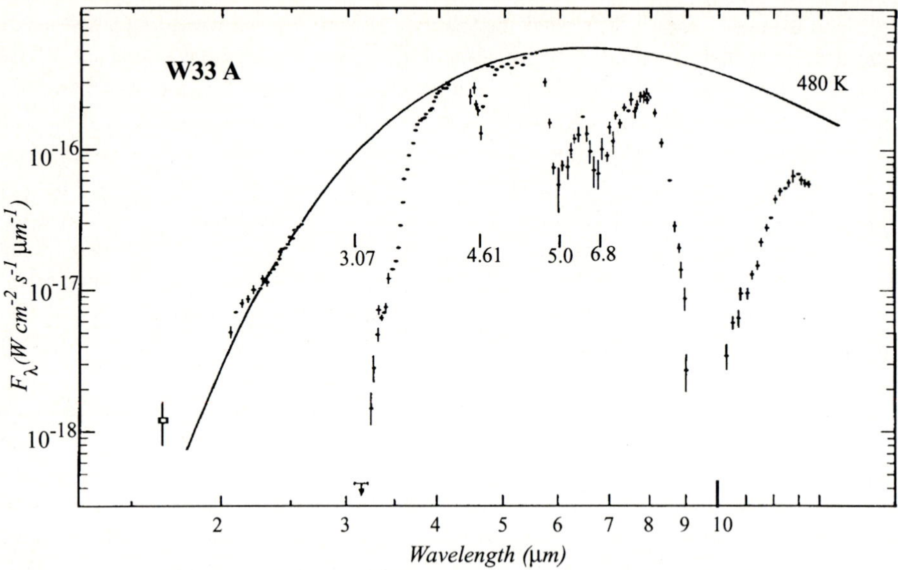

Fig. 1.5. The spectrum of the infrared source W33A from Soifer et al. (1979). It is a heavily extincted young star in a molecular cloud. The strong, broad absorption bands from solid state resonances are evident throughout the spectrum.

Solid state resonances due to ice (3.3 μm), silicon compounds ($\sim$ 10 μm), and various organic materials (5 – 8 μm) were seen in the first infrared spectra of young stars and are ubiquitous throughout the Galaxy. They are seen both in absorption and emission, and they are generally attributed to the constituents of the dust particles surrounding these stars. An array of previously unidentified features in the thermal infrared has now been ascribed to PAHs; PAHs are evidently responsible for unusually hot thermal emission, $T \sim 2000$ K, from small particles in reflection nebulae near pre-main sequence clusters such as the Pleiades (Sellgren 1981, 1984).

The utility of these resonances for understanding star formation has thus far been limited. In fact, there is often enough controversy about the identification of particular resonances to preclude even a definite statement about the composition of the dust in the circumstellar envelopes. When absorption from water-ice is seen, it indicates large extinction ($A_V \geq 10$ mag) and very cool envelopes ($T \leq 200$ K), but it is difficult to be quantitative. Nevertheless, infrared wavelengths are most important for the identification and study of these spectral features (see Fig. 1.5).

1.6 The Orion Star Forming Region

Figure 1.6 shows the Orion Nebula, perhaps the most famous of HII regions, in the visible (Fig. 1.6a) and infrared (Fig. 1.6b) light. The five bright stars which make up the Trapezium – $\theta^1_{\rm C}$ is the brightest – provide essentially all of the ionizing radiation for the nebula. They are young main sequence stars that are still surrounded by some of the gas and dust from which they were born, hence, the HII region.

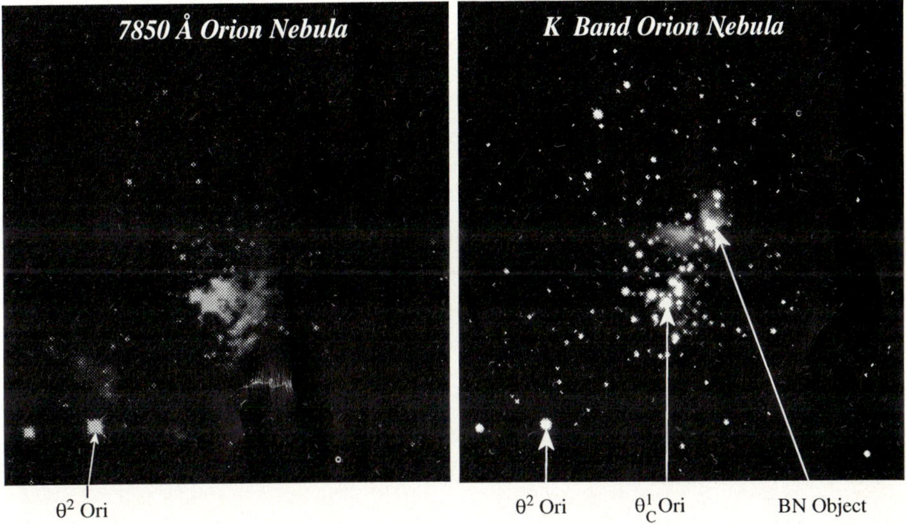

Fig. 1.6a. This red-light photograph of the Orion Nebula with the Trapezium cluster at its center was taken by Herbig (1982). Notice that the star θ^2 is similar brightness as in the infrared image in 1.6b, but there are far fewer stars visible toward the dark parts of the nebula.

Fig. 1.6b. An image of Orion taken at 2.2 μm with the Calar Alto 3.5m telescope and an IR camera with a NICMOS3 array, the MAGIC camera described by Herbst et al. (1993). The BN (Becklin-Neugebauer) star is the brightest of the "hidden" infrared stars about $1'$ northwest of the Trapezium.

These stars are at the outer edge of a much more extensive molecular cloud residing behind the HII region. The prominent dark bay in Fig. 1.6a on the left hand side of the ionized gas is a part of this cloud which wraps around the front of the visible nebula.

Only 0.5 kpc away, Orion is the prototype of an active star forming region in which both high and low mass stars are being created. The variety of stars and the different phenomena in the molecular cloud encompass those seen in other parts of the Galaxy, most of them more distant than Orion.

One can see in the near infrared image in Fig. 1.6b many more stars spread throughout the molecular cloud. To a large extent, the stars appear because of the greatly reduced effects of extinction in the infrared. But the brightest infrared sources seen to the northwest of the Trapezium are intrinsically very red; the emission is dominated by the dust shells surrounding these stars; the apparent "photospheric" temperatures are only a few hundred K. The BN (Becklin-Neugebauer) object is the brightest at 2.2 μm. Fig. 1.7 shows spectra of several infrared sources very similar to BN; the very cool color temperatures are apparent at once.

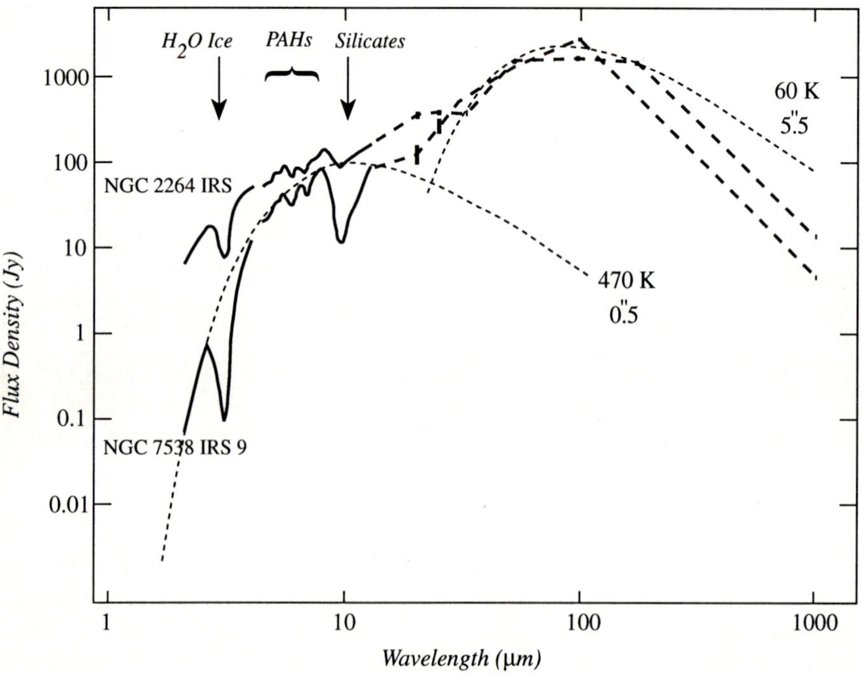

Fig. 1.7. Spectra of two infrared sources in the molecular clouds NGC 2264 (near Orion) and NGC 7538. These spectra are similar to that of the Orion source, BN, but they are more isolated so the far infrared spectral energy distributions may be uniquely attributed to these stars. The figure is from Wynn-Williams (1982) who used data from Allen (1972), Willner et al. (1982), Simon and Dyck (1977), Harvey et al. (1977), and Werner et al. (1979).

The infrared radiation from Orion displays a rich array of phenomena related to star formation. Low resolution spectra of the entire region reveal strong emission from molecular hydrogen excited by shock waves in the molecular cloud. Figure 1.8 is a map of the $v = 1 - 0$ S(1) vibration-rotation line of H_2 (2.122 μm) in the neighborhood of BN. The gas temperature giving rise to this line emission is about 2000 K. There is, in fact, very little hot gas in the molecular cloud. The gas resides in thin layers or filaments behind shock waves driven into the molecular cloud by winds from the stars in the BN region, most probably by an infrared source called IRc 2 (Downes et al. 1981). The H_2 emission shows quite clearly that the young stars generate powerful winds which drive these shock waves. This emission was some of the earliest evidence for these winds (see also Kwan and Scoville 1976; Zuckerman, Kuiper, & Rodriguez-Kuiper 1976), now thought to be ubiquitous among young stars and fundamentally responsible for allowing a star to get rid of angular momentum as it collapses from a large cloud core (Lada 1985; Shu et al. 1992).

Very high resolution spectra of BN show a combination of emission lines and absorption lines from various parts of the near circumstellar environment and the more extended molecular cloud. Many of these lines may be seen in Fig. 1.9. This figure concentrates on the absorption components in the CO lines, especially those originating in two different velocity components within the cold cloud. The emission is even more interesting, however, although somewhat more difficult to interpret. The excitation of CO vibrational emission lines requires very dense, hot gas, and a combination of factors indicate that the gas must be confined to a thin plane close to the star. Scoville et al. (1983) suppose that the emission by CO must come from a hot accretion disk. This was one of the first indications of circumstellar accretion disks around young stars.

The spectrum in Fig. 1.9 is only a small section of the complete spectrum which shows more complex phenomena near BN. The Brα and Brγ lines of atomic hydrogen originate in ionized gas immediately surrounding the star. Velocity profiles of these lines are seen in Fig. 1.10. The lines are broad, with FWHM of more than 50 $\mathrm{km\,s^{-1}}$, much more than the 10 $\mathrm{km\,s^{-1}}$ sound speed in a typical HII region. The ionized gas is, in fact, expanding in a wind coming from BN. The velocity of the wind must fall as the $-\frac{2}{3}$ power of the distance from the star to match the observed profile, indicating that some deceleration takes place in the expanding gas.

Observations such as the ones shown here used to require considerable effort, and only the brightest infrared sources could be detected (Wynn-Williams 1982). Nevertheless, the amount of information available from these infrared data is enormous, tracing properties of the stars, their very near circumstellar environments, and the more extensive molecular clouds. The relative placement of the infrared sources is interesting from the standpoint of star formation and the birth of stellar clusters. The relationship of the stellar and near stellar properties to those of the more extended cloud are likely to yield clues to the initial mass function and the reasons that clouds are supported but ultimately collapse.

The capacity for gathering information in the infrared has increased by many orders of magnitude in only a few years. BN was discovered with a single-detector photometer in the late 1960's by scanning the telescope across the sky. Today, arrays of detectors exist with more than 65,000 elements, each one about 10,000 times more sensitive than the one used to discover BN. Overall, the information rates are a *billion* times greater than they were 25 years ago. The present rate

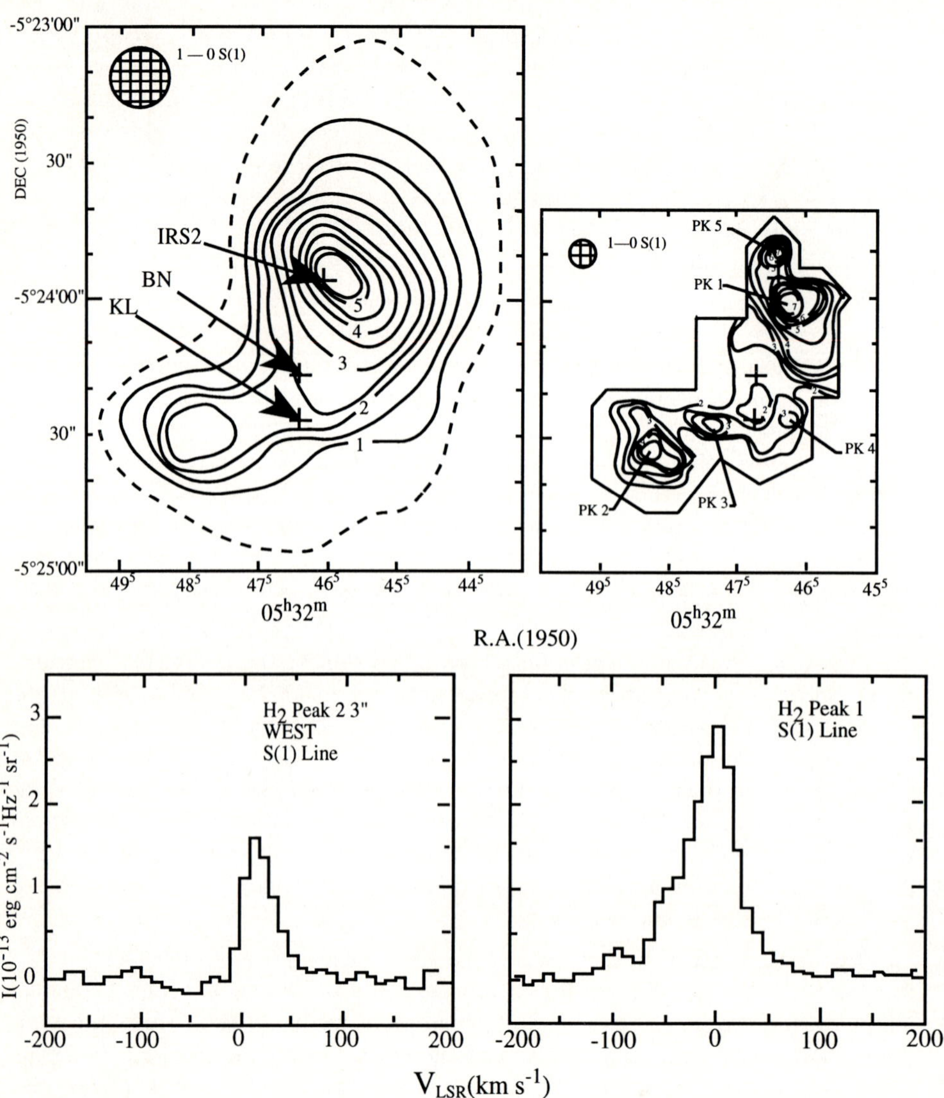

Fig. 1.8. Maps of the $v = 1 - 0$ S(1) line of H_2 emission in Orion with $10''$ (top left) and $5''$ (top right) resolution (Beckwith et al. 1978). The bottom two figures are high resolution spectra of the lines at two of the strongest positions (Scoville et al. 1982) demonstrating the very high velocities in the winds from the IRc 2/BN complex.

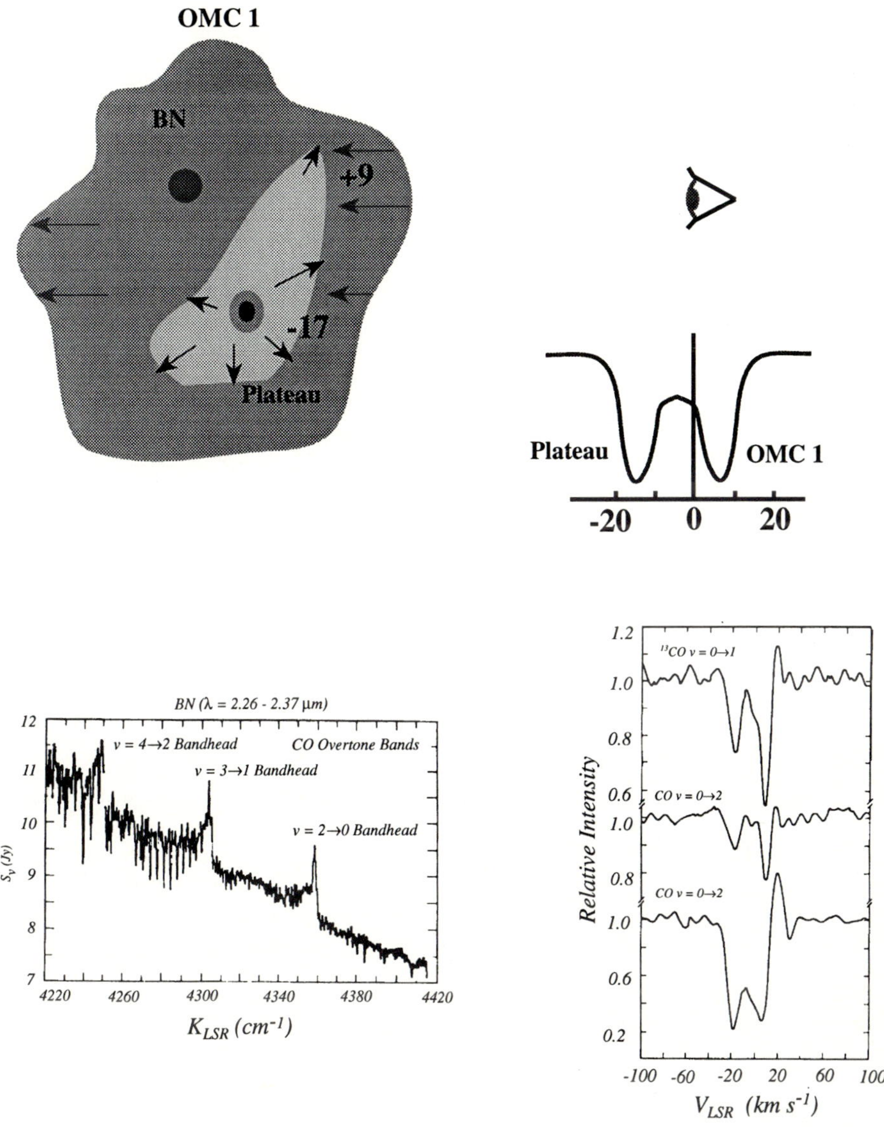

Fig. 1.9. The lower part of the figure shows parts of a spectrum taken by Scoville et al. (1983) of the BN object. The left hand side shows the 2.26 – 2.37 μm section of the overtone vibrational lines of CO; these lines have both emission *and* absorption components, each of which originates from a different part of the circumstellar environment. The right hand side absorption line profiles are sums of many different rotational lines within each vibrational band. The two separate components come from different parts of the cold cloud in front of BN. The upper part of the figure shows a simplified picture of the various components responsible for the absorption lines.

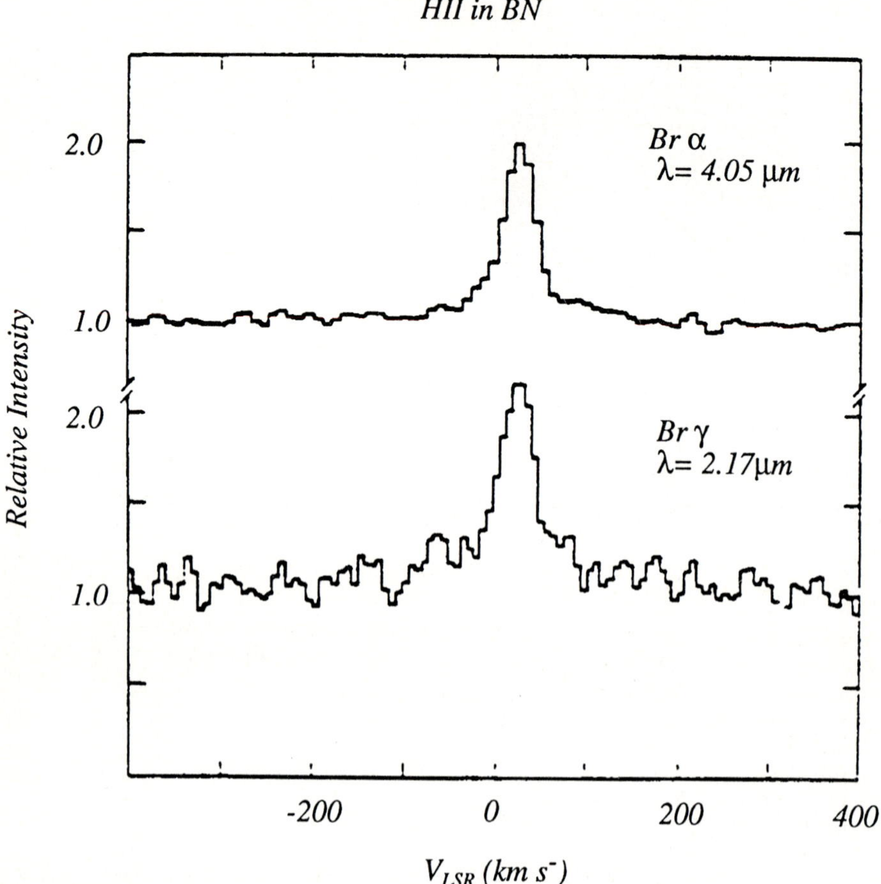

Fig. 1.10. Line profiles of the HI recombination lines from the BN object taken by Scoville et al. 1983.

of improvement is roughly a factor of 16 in area and a factor of 3 in sensitivity every 3 years.

It is, therefore, a very special time for infrared observations of all kinds, but particularly for studies of star formation. In the next chapters, we shall examine the limits set by nature and current technology on observations at infrared wavelengths.

2 Natural Limits to Observation

2.1 Telluric Absorption

Between 1 and about 34 μm, the Earth's atmosphere transmits in a number of narrow *windows*. Figure 2.1 shows the windows between 1 and 5.5 μm, and Figure 2.2 shows the windows at 10 and 20 μm. At longer wavelengths, the troposphere is effectively opaque, opening up again at the submillimeter wavelength of 350 μm. At shorter wavelengths, the visible window is relatively transparent until the ultraviolet at about 0.36 μm. Ground-based observations of the infrared celestial sky are limited to these windows. Airborne and space telescopes overcome this limitation, at great expense and with the limitation of relatively small collecting areas.

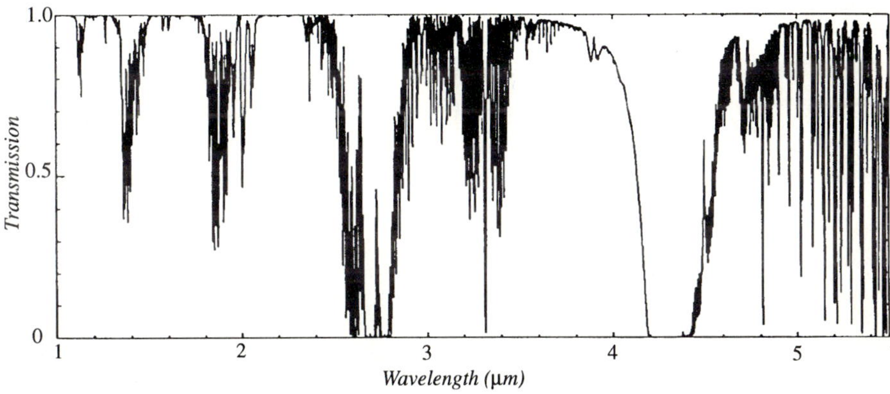

Fig. 2.1. Model transmission of the Earth's atmosphere from 1 to 5.5 μm at 4.2 km (Mauna Kea) with 0.3 mm precipitable water vapor (from McCaughrean 1988).

The atmospheric opacity that defines the windows comes from many closely spaced molecular lines, principally water vapor. Between the windows, the overlap of lines is essentially complete at sea level. As altitude increases, the weaker lines disappear, so that by 10 km there is almost continuous transmission from 1 μm to the submillimeter broken by a few discrete absorption lines. At low altitudes, there is absorption by discrete lines even within the windows. This absorption complicates photometric calibration.

The windows themselves have different characteristic transmissions depending on the principal molecular species responsible for absorption in the neighboring wavelengths. Methane absorption in at the short end of the 3.8 μm window, carbon monoxide and carbon dioxide between 3 and 5 μm, and OH between 1 and 2 μm are especially troubling. The 20 μm window is extremely poor, and the "window" at 34 μm is seen only under conditions of extremely low water vapor;

it is difficult to do accurate photometry at these long wavelengths except at very dry sites such as exist in Hawaii and Chile.

The windows shortward of about 2 μm constitute the near infrared, and those longward of 2 μm are in the *thermal* infrared. The 2.2 μm and 10 μm windows are generally the most important; they are relatively free of atmospheric opacity and are spaced just far enough apart so that the V (0.55 μm), K (2.2 μm), and N (10 μm) photometric bands are sensitive to the widest range of temperatures normally seen in compact objects such as young stars.

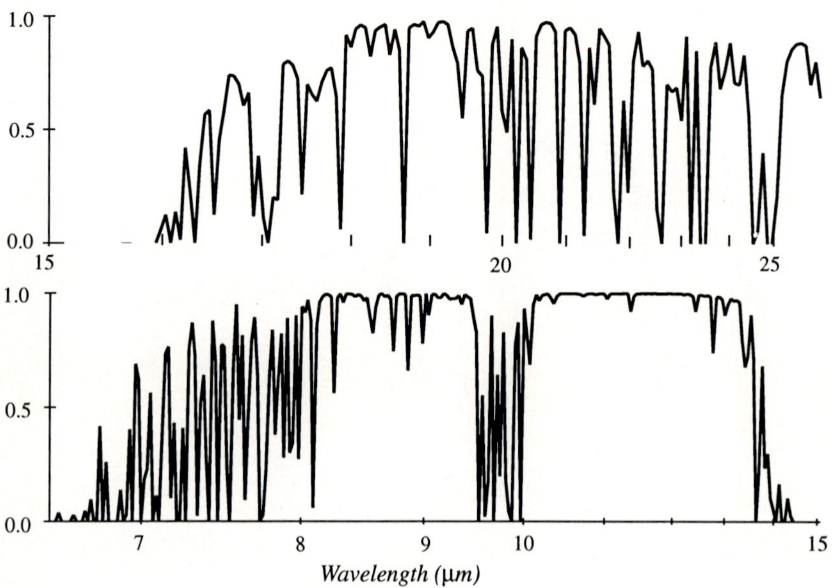

Fig. 2.2. Transparency in the 10, and 20 μm windows calculated for 1 mm precipitable water.

The average transmission through the windows varies with the water vapor content along the line of sight of the telescope, because H_2O is the main source of opacity. It has a low scale height in the troposphere. The average window opacity is not simply a linear function of the column density or airpath. The centers of the "clean" windows at 2.2 and 10 μm are relatively free of absorbing lines. They narrow somewhat as the lines at the edges of the windows become deeper and saturate. The "dirty" windows, such as the M band, suffer badly from the effects of variable water vapor. The photometric quality of a night, indeed, of a site, depends as much on the fluctuations in the humidity as on the cloud cover.

Table 2.1. Standard (Johnson) infrared photometric bands

Band	λ (μm)	$\Delta\lambda$ (μm)	0^m F_ν(Jy)
J	1.25	0.30	1570
H	1.65	0.35	1020
K	2.20	0.40	636
L	3.5	0.55	281
L′	3.8	0.60	235
M	4.8	0.30	154
N	10	5	37
Q	21	8	10

Zero magnitude flux densities from Bessell and Brett (1988) and Beckwith et al. (1976).

Standard photometric filters encompass these windows. The J, H, K, L (or L′), M, and N system is the most common. The bandpasses are listed in Tab. 2.1.

2.2 Thermal Background Emission

The ambient temperature, T, is approximately 270 K. At this temperature, the Earth's atmosphere, the telescope, and the warm optical elements outside the cryostat emit thermal radiation as greybodies. The total emissivity, ϵ, of this radiation is simply 1 minus the overall transmission, so the power onto a detector is:

$$P = \eta \epsilon A \Omega \Delta\nu B_\nu(T), \tag{2.1}$$

where η is the transmission of all cooled optics times the detector quantum efficiency, $A\Omega$ is the area-solid angle product of the instrument (the étendue), $\Delta\nu$ is the frequency bandpass of the filter, and $B_\nu(T)$ is the Planck function. For $T = 270$ K, $B_\nu(T)$ has a maximum near 15 μm. The shorter wavelength emission drops exponentially and becomes unimportant below about 2 μm. At 10 μm, a 4 m diameter telescope with a 1″ square field of view ($A\Omega = 3 \times 10^{-10}\,\mathrm{m^2\,sr}$) and a filter bandpass of 5 μm ($\Delta\nu = 1.5 \times 10^{13}$ Hz) produces power $P = 9 \times 10^{-9}\epsilon$ W on the detector, or $4 \times 10^{11}\epsilon$ photons s.$^{-1}$ If the average transmission of the atmosphere is 0.9, the telescope transmission is 0.9, the instrument window transmission is 0.95, and the instrument is otherwise perfectly efficient ($\eta = 1$), then $\epsilon = 0.23$, the power on the detector is 2×10^{-9} W, and the photon rate is about 10^{11} s.$^{-1}$ At this bandwidth and solid angle, the background, much of which comes from the telescope and instrument surfaces, is equivalent to 1000 Jy arcsec,$^{-2}$ where 1 Jy $= 10^{-26}\,\mathrm{W\,m^{-2}\,Hz^{-1}}$. The sky and telescope are incredibly bright sources of radiation, equivalent to $-3.6^m/\square''$.

Photons arrive randomly, so the noise inherent in their detection is shot noise. The shot noise is just the square root of the number of photons detected[2], which in one second is 3×10^5. Therefore, a celestial source emitting $3 \times 10^5/(1 - \epsilon)$ photons s^{-1} into this solid angle and bandwidth would produce a signal just

[2] one can safely ignore the Bose statistical nature of photons at these infrared wavelengths

equal to the noise in the background radiation from the atmosphere and instrument. Such a source has a flux density of 4 mJy or 10^m at $10\,\mu$m. The noise-equivalent flux density of an excellent $10\,\mu$m instrument with $1''$ pixels is 4 mJy, and it depends entirely on the (unavoidable) background. This noise limit is called the *background limit* or BLIP for **B**ackground **Li**mited **P**erformance.

Shortward of $10\,\mu$m, the 270 K Planck function drops exponentially with wavelength. The photon rate at $3.8\,\mu$m is down to $10^8\,\mathrm{s}^{-1}$, and BLIP is $200\,\mu$Jy. By $2.2\,\mu$m, the photon rate is only $1.3\times10^4\,\mathrm{s}^{-1}$, corresponding to a sky brightness of about $15^m/\square''$and a BLIP of about 20^m. In fact, there is additional emission from OH airglow at the short end of the window, and the actual sky brightness is somewhat greater (and site dependent). However, thermal emission is only important for wavelengths longer than about $2\,\mu$m, and for this reason these wavelengths are said to be in the *thermal* infrared.

We can write the BLIP noise equivalent flux density (NEFD) for an instrument with (cold) transmission and quantum efficiency, η, explicitly as:

$$\mathrm{NEFD} = \frac{1}{\eta(1-\epsilon)}\frac{h\nu}{A\Delta\nu}\left(\eta\epsilon\frac{A\Omega\Delta\nu}{h\nu}B_\nu(T)\right)^{\frac{1}{2}} \tag{2.2}$$

$$= 4.8\ \mathrm{mJy\ Hz}^{-\frac{1}{2}}\ \eta^{-\frac{1}{2}}\left(\frac{\theta}{1''}\right)\left(\frac{4.8\ \mathrm{m}}{D}\right)\left(\frac{\epsilon^{\frac{1}{2}}}{(1-\epsilon)}\right)\left(\frac{\nu/\Delta\nu}{2}\right)^{\frac{1}{2}}\left(\frac{10\ \mu\mathrm{m}}{\lambda}\right)^{\frac{1}{2}}$$

$$\times\left(\frac{206}{e^{5.33\left(\frac{10\ \mu\mathrm{m}}{\lambda}\right)\left(\frac{270\ \mathrm{K}}{T}\right)}-1}\right)^{\frac{1}{2}}$$

The BLIP varies as the inverse square root of the instrument transmission and quantum efficiency. The photon rate is sufficiently small at about $2\,\mu$m that the thermal background is unimportant, especially since other noise sources dominate for ground-based observations. Using Table 2.1, but assuming perfect detector quantum efficiency, Table 2.2 gives the BLIP NEFD and photon rates for the standard photometric bands under these very optimistic assumptions.

Table 2.2. Thermal BLIP(270 K) for an Ideal Detector

Filter	$\lambda(\mu\mathrm{m})$	NEFD(mJy/$\square''$)	Noise Mag/$\square''$	Sky only Mag/$\square''$
K	2.2	0.0004	23	16*
L′	3.8	0.074	16	5.6
M	4.8	0.64	14	2.7
N	10	3.3	10	-2.7
Q	20	24	7	-4.9

* Ignores OH airglow.

Equation 2.2 shows the high desirability of using small pixels on large telescopes to push the BLIP very low; the image sizes must, of course, be small enough to justify the use of small pixels. At short wavelengths, the exponential factor in temperature means there is a noticeable difference between summer and winter observations, the sensitivity improving in the winter by 0.5 mag, or so, depending on the site.

In practice, the limits given by (2.2) are not achieved. Detection systems (optics and detector) are usually far from ideal, with a total transmission and quantum efficiency of 0.5 being unusually good. In addition, a major problem at many sites is a phenomenon called *sky noise*. Sky noise comes from fluctuations in the background rate – see (2.1) – owing to fluctuations in the atmospheric transmission and, hence, emissivity. In this case, the noise in background rate fluctuations completely overwhelm the shot noise in the photon stream.

Sky noise is most obvious when the sky is relatively clear and dry and very slight changes in the line-of-sight water vapor content (say from 0.5 to 1 mm precipitable water) drastically alter the emissivity. Sky *emission* is most obvious when the sky is partly cloudy; the emissivity varies between a small value (less than 0.2) and unity as clouds pass through the line-of-sight. In both cases, the noise increases from the square root of the number of detected photons to a large fraction of the actual number, a very large factor, indeed. But even when there are no obvious clouds present, the water vapor content along the line-of-sight changes sporadically, essentially due to micro-clouds of water vapor not apparent in the visual, and modifies the emissivity.

Sky noise is observed to vary approximately inversely with the sampling frequency (Käufl et al. 1991). At high water vapor sites, such as Palomar, sampling times as short as 20 ms are needed to suppress sky noise to manageable levels, whereas at dry sites such as in Chile or Hawaii, a few Hz sampling frequencies are often adequate. Sky noise is mainly a problem at 5 μm and longer. It tends to be particularly bad near sunset and sunrise, when the temperature is rapidly changing. Figure 2.3 shows a power spectrum of the sky noise in Chile at 10 μm.

2.3 Airglow

In addition to the thermal background, the sky emits strongly in transitions of excited molecules in the upper atmosphere. The emission is called *airglow* and is a common problem at the red end of the visual window. The J, H, and K windows suffer from airglow emission by OH molecules that are excited by energetic electrons produced by the solar wind. The emission constitutes essentially all of the background radiation at J and H and is important at the short end of the K window.

The OH emission actually comes from a large number of discrete lines that are well-separated at high resolution (Ramsay et al. 1992; Maihara et al. 1993a). In the H band, there are about 60 discrete OH lines whose combined width fills only a few percent of the H window. Unfortunately, the spacing is such that all present-day bandpass filters admit all of the lines, resulting in a large background on the detector. Unlike the thermal emission at longer wavelengths, the airglow could, in principle, be eliminated by careful filtering (Maihara et al. 1993b). The relatively rapid time variability sets an upper limit to integration times with most infrared instruments operating in the airglow regime, since, as we shall see,

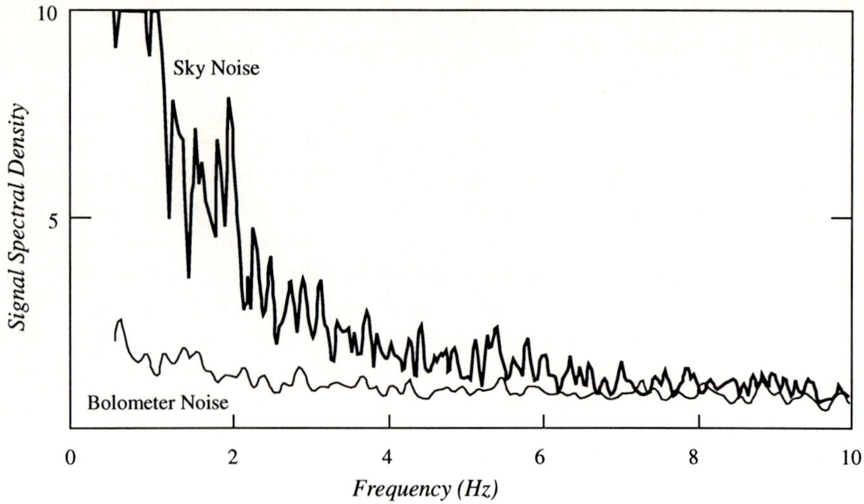

Fig. 2.3. Power spectra of the sky noise vs. frequency and angular separation after Fig. 1 of Käufl *et al.* (1991).

the techniques of sky subtraction rely on having stable sky emission. Figure 2.4 is an example spectrum of the H window at a resolution large enough to separate most of the lines but not actually resolve them.

Airglow varies periodically with time changing on timescales of order 10 minutes. These changes mean that its strength must be measured over much shorter timescales, if it is to be subtracted from observations. The OH has a maximum at about 90 km altitude, so the airglow is not reduced by going to high altitude sites. It is latitude dependent. Some of the best ground-based sites are actually near local maxima in the OH emission blanket. Figure 2.5 shows the strength of airglow over time in a few discrete lines near $1.7\,\mu$m seen from Mauna Kea.

Figure 2.6 plots the total background between 1 and $5\,\mu$m measured by McCaughrean (1988). Notice the enormous increase in background going into the thermal infrared region.

2.4 Extraterrestrial Limits

By going to space, we can overcome the OH airglow and thermal emission of the atmosphere. To overcome the thermal emission of the telescope, the telescope can be cooled to very low temperatures. For the near-infrared, even modest cooling to 100 K or so produces a huge benefit; far infrared telescopes require cooling to a few degrees above absolute zero to achieve significant sensitivity gains, something which is only possible in the near vacuum of space. But even

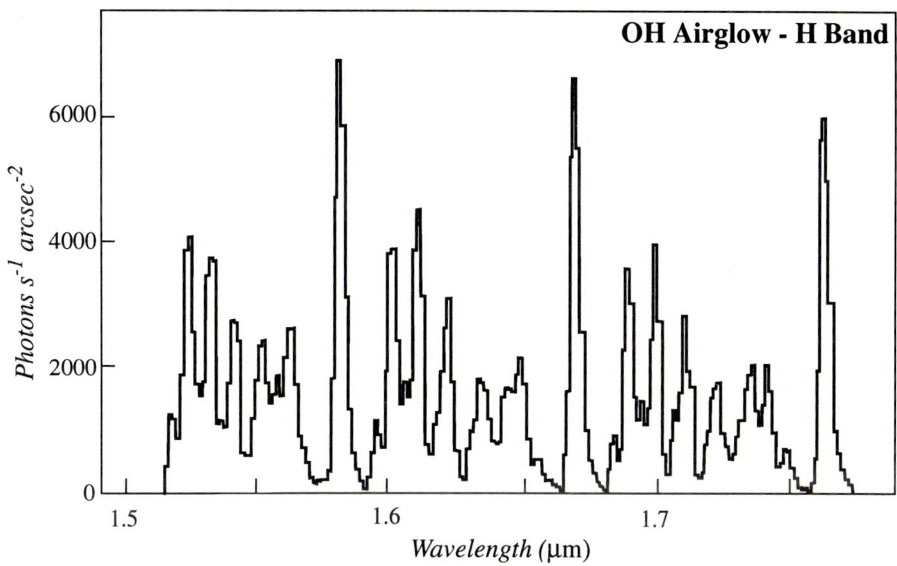

Fig. 2.4. Spectrum of OH airglow over Mauna Kea from Ramsay et al. (1992).

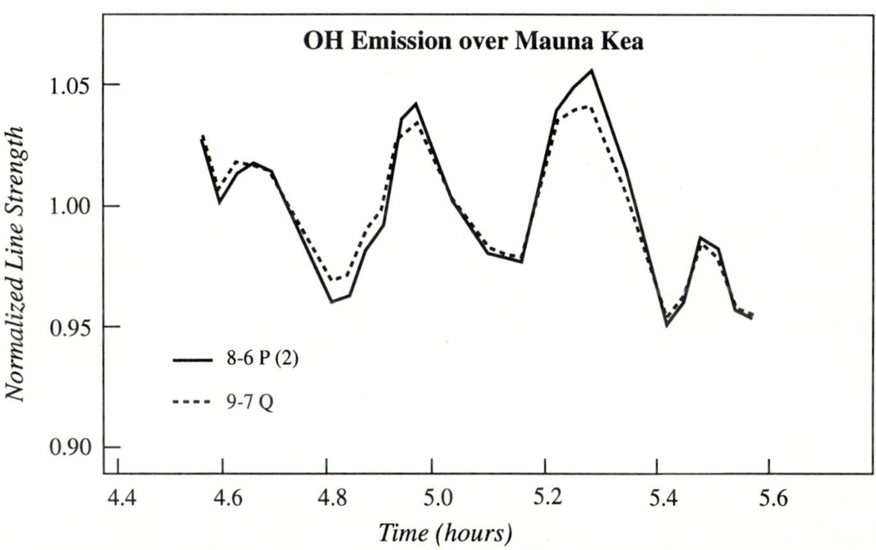

Fig. 2.5. Strength of OH airglow vs. time over Mauna Kea from Ramsay et al. (1992).

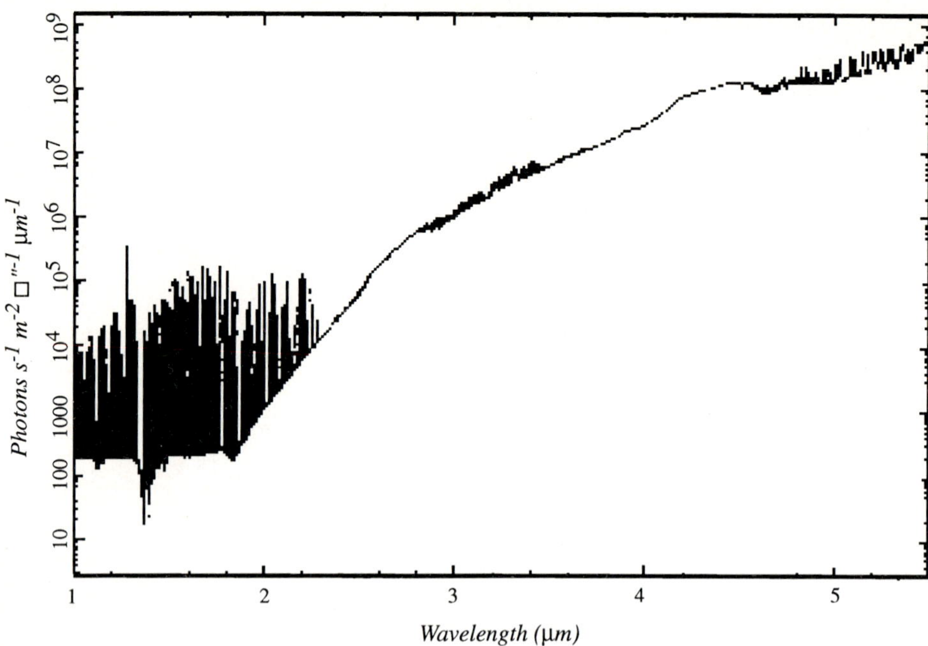

Fig. 2.6. Spectrum of the total background radiation just inside the cryostat window of the UKIRT IRCAM camera from McCaughrean (1988). This figure illustrates the enormous increase in background going from about 2 to 5 μm, between 4 and 5 orders of magnitude.

after cooling the telescope and putting it above the atmosphere, we are plagued by other natural backgrounds.

The most immediate problem in the near infrared is the *zodiacal light.* Zodiacal light comes from very small dust particles orbiting the Sun out to a radius of a few AU. The particles are created by collisions between asteroids and cometary debris. They spiral in towards the Sun in response to the radiation friction of sunlight (the Poynting-Robertson drag). Because they are warmed by sunlight in the same way as is the Earth, they are at about the same temperature of 270 K, and they emit thermal radiation just like the atmosphere. The total opacity of the zodiacal particles is extremely small, however, and rather steady with time although not with direction. Therefore, the background limit from this thermal radiation is relatively small and constant; there is no equivalent of sky noise. It is, nevertheless, measurable and constitutes the primary source of background radiation between about 2 μm and the far infrared for spacecraft observations.

At shorter wavelengths, the zodiacal particles reflect sunlight. At wavelengths longer than about 30 μm, the zodiacal light is less important than emission from interstellar dust particles, the so-called infrared "cirrus" (Low et al. 1984). The emission is also thermal but with a very low temperature typically 15 K or less. It has a patchy structure at all spatial scales thus far observed (minutes of arc and larger).

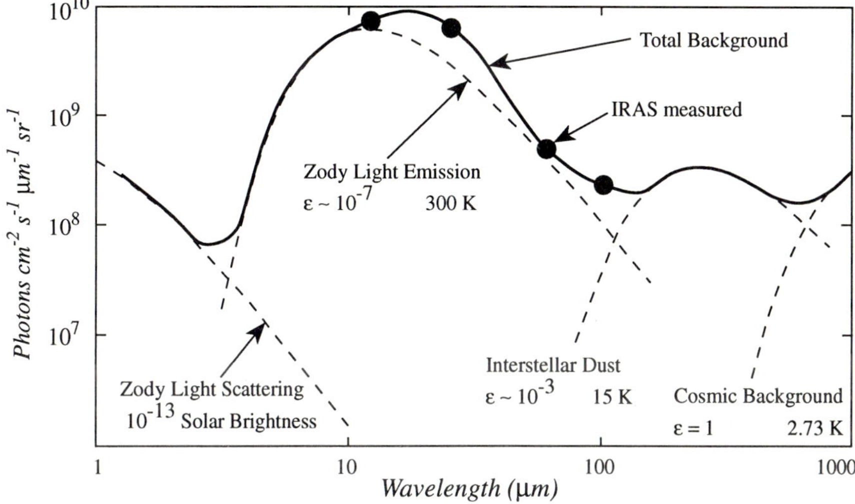

Fig. 2.7. Spectrum of the zodiacal light in the direction of the ecliptic plane and in the direction of the ecliptic pole.

The zodiacal light is a maximum at about 16 μm, falls exponentially until about 3 μm, then begins to rise again rapidly as the reflected sunlight overwhelms the thermal emission. There is a broad minimum in the zodiacal background very near to 3 μm; this wavelength has been proposed as the best wavelength for the most sensitive possible observations of extragalactic backgrounds. Figure 2.7 displays the spectrum of the zodiacal light in the direction of the ecliptic pole.

The background noise from the zodiacal light is calculated in the identical manner as in (2.2) but with $\epsilon \sim 10^{-7}$, and replacing $B_\nu(T)$ with the reflected solar spectrum at the shortest wavelengths. The photon rates are very small, but they are important for the long integration times normally planned for deep, spacecraft observations.

Spacecraft can be launched out of the ecliptic or to distant solar orbit, beyond the 3 AU or so that contains the zodiacal particles. At this point, one might imagine that we would have escaped all but the cosmic background radiation. However, at low enough levels, we then find that the accumulated light of many faint stars in the galaxy and many faint galaxies pose a fundamental limit to observations. There are enough stars and galaxies that for modest sized pixels more than one object is always within a single resolution element. This is called

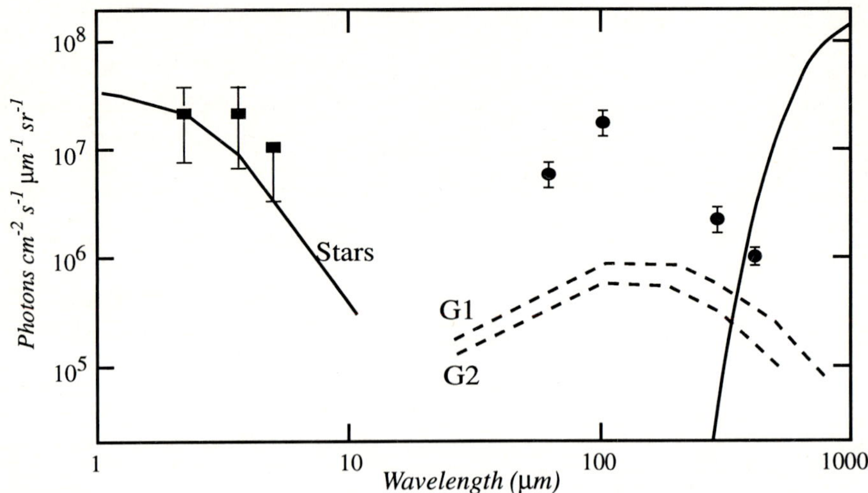

Fig. 2.8. Background radiation due to the sum of many faint stars and galaxies from Franceschini et al. 1991. G1 and G2 are the contributions from normal and active galaxies according to different evolutionary models; Stars is the contribution from starlight in the Galaxy. Filled squares are measurements by Matsumoto et al. (1988); circles are polar emission from the Galaxy by De Zotti et al. (1990).

the *confusion limit*; it depends on where the line-of-sight is with respect to the plane of the Milky Way and, of course, on the pixel size, θ; with larger space telescopes, the confusion limit will be at fainter magnitudes. It is less of a problem for observations toward the Galactic poles, where confusion by distant galaxies will inevitably set the ultimate limit. Figure 2.8 shows the limiting surface brightness set by contributions from faint stars and galaxies in the direction of the Galactic pole (Francheschini et al. 1991).

There are additional practical problems encountered by spacecraft: cosmic rays generate charge pairs in all known detectors creating unwanted background noise. The generation rate depends on the detector size, its sensitivity to high energy radiation, the amount of shielding around the detector, the orbit of the spacecraft, and the solar cycle. No general formulation of this background applies, but it can be an important component of the total noise.

3 Near Infrared Observational Techniques

3.1 Elements of an Infrared Instrument

The radiation from a celestial source can be characterised by a specific intensity, $I_\nu(\theta)$, which is a function of frequency, ν $(= \frac{c}{\lambda})$, and direction on the sky. An instrument ultimately detects only the *power* of the radiation for a specific time interval, usually by converting it to an electrical voltage or current which is then measured with standard techniques. The instrument must define an optical bandwidth or wavelength range, $\Delta\lambda$, a solid angle on the sky, Ω, an area to capture the light, A, and a time over which to make the measurement, t. The quantities $\Delta\lambda$, Ω, and A are fixed by optical elements; the detection time is usually variable and chosen by an observer to meet the sensitivity requirements. The instrument often includes a *modulating element* that periodically switches the field of view from the celestial source of interest to a comparison source, often the sky, so that the radiation measurements are differential. Figure 3.1 is a schematic diagram of a generic infrared instrument.

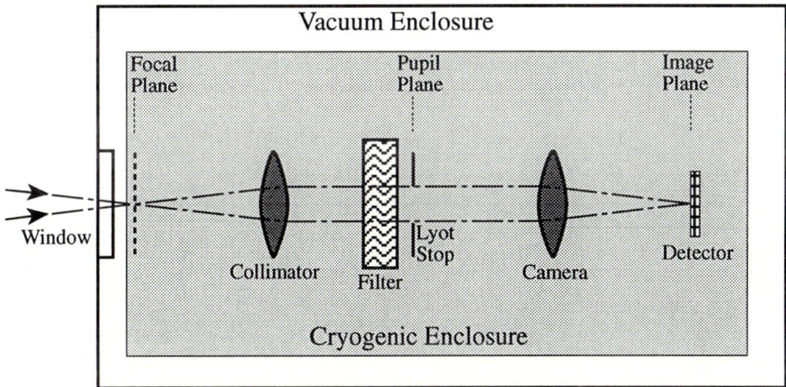

Fig. 3.1. A generic infrared instrument showing the relationship of the important subsystems. The enclosure maintaining the low temperatures and vacuum is called a *cryostat*. The filter could be an interference filter or a complex optical subsystem, such as a spectrometer. The *Lyot stop* rejects unwanted background radiation surrounding the pupil image.

The optical elements consist of a collecting area or pupil, a filter which might be an elaborate optical subsystem itself, and a detector that, together with the pupil, defines a field of view on the sky. An important property of all optical systems is that the area-solid angle product, $A\Omega$ – the throughput or étendue – is a conserved quantity throughout the optical train (Liouville's theorem). Therefore, if an instrument on a 4 m telescope ($A = \frac{\pi}{4} 16\,\mathrm{m}^2$) is sensitive to a solid angle $(1'')^2$, the system étendue is $3 \times 10^{-10}\,\mathrm{m}^2\,\mathrm{sr}$. This is true regardless of whether you use the telescope area and solid angle on the sky or the detector area, which might be only a few microns across, and the solid angle of the light bundle which is focussed onto it, normally of order 1 sr.

Because a detector can accept light from no more (and usually much less) than 2π sr, there is a natural upper limit to the étendue of an optical system with a fixed detector size. As we shall discuss in the next lecture, it is usually necessary to make detectors very small to reduce their intrinsic noise properties, and so it is often difficult to increase the étendue of an instrument by increasing A at fixed Ω, even though this is often desirable. This difficulty is already severe for designers of infrared instruments for the next generation of large telescopes; it is easier to design for high resolution (small $A\Omega$) rather than wide field (large $A\Omega$). For example, the current generation of near-infrared arrays have pixels about 30 μm square. It is difficult to achieve final focal ratios faster than f/1 for detector subsystems, so that $A\Omega \leq 7 \times 10^{-10}\,\mathrm{m^2\,sr}$. On a 10 m telescope, the pixel size is no more than $0\farcs6$ across; in fact, it will usually be much less, since f/1 is difficult to achieve for large arrays. Wide-field instruments will be a challenge to build.

High resolution is limited by diffraction from the optics, usually by the pupil size. Circular pupils of diameter, D, produce the characteristic Airy diffraction pattern, and the full width at half maximum (FWHM) is $\theta_{\mathrm{FWHM}} = 1.22\frac{\lambda}{D}$. Table 3.1 lists diffraction-limited resolutions for various combinations of wavelength and pupil diameter.

Table 3.1. Diffraction-limited Resolution of Large Telescopes

Band	λ (μm)	4 m $\Delta\theta('')$	10 m $\Delta\theta('')$
J	1.25	0.08	0.03
H	1.65	0.10	0.04
K	2.2	0.14	0.05
L′	3.8	0.24	0.10
M	4.8	0.30	0.12
N	10	0.63	0.25
Q	20	1.26	0.50

The optical bandwidth is also limited by instrument size. In general, higher spectral resolution for a fixed étendue requires a larger instrument. A dispersive spectrometer, such as a grating, also has a diffraction limit. The smallest resolvable angular separation of two rays coming from the grating is $\Delta\phi = \frac{\lambda}{L\cos\phi}$, where L is the length of the grating perpendicular to the ruling direction, and ϕ is the angle of the grating relative to the light beam (*cf* Fig. 3.2a). The resolving power of a grating is $\frac{\lambda}{\Delta\lambda} = \frac{\sin\phi}{\cos\phi\Delta\phi}$, so:

$$\frac{\lambda}{\Delta\lambda} = \frac{L\sin\phi}{\lambda}$$

$$\leq 2.3 \times 10^5 \left(\frac{L}{50\ \mathrm{cm}}\right)\left(\frac{2.2\ \mu\mathrm{m}}{\lambda}\right) \tag{3.1}$$

To achieve a resolving power at 2.2 μm of 300,000, equivalent to 1 km s^{-1}, requires a grating at least 66 cm across operating at the diffraction limit. However, the $A\Omega$ for the spectrometer cannot be less than the étendue of the telescope itself (without light loss): $L \cos\phi \Delta\phi \sim \sqrt{A\Omega}$. Thus:

$$\begin{aligned}\frac{\lambda}{\Delta\lambda} &= \frac{L \sin\phi}{(A\Omega)^{\frac{1}{2}}} \\ &\leq 1.2 \times 10^4 \left(\frac{L}{50\ \mathrm{cm}}\right)\left(\frac{10\ \mathrm{m}}{D}\right)\left(\frac{1''}{\theta}\right)\end{aligned} \tag{3.2}$$

In the above example of a 10 m telescope with 0$''$.6 pixels, L must be at least 8 m across to achieve 1 km s^{-1}! A more general description of filtering subsystems is given in Sect. 3.3.

3.2 Background Subtraction

As noted in Section 2, emission from the sky is always present in the infrared, even in space. Modern detectors are capable of reaching the background limit for most observations; as a result, the sky itself is usually the largest source of noise. In fact, the power from the source of interest is often much less than the background power during the observation, and so the background power must be measured accurately and subtracted to yield a meaningful measurement of the source power. The technique for doing so is normally referred to as *sky subtraction*. At thermal infrared wavelengths, it is not uncommon for the background power to exceed the source power by a factor of 10^7. The final accuracy of the source observation depends completely on the effectiveness of canceling out the enormous background emission.

A straightforward method to eliminate the sky power is to measure it separately before and after the source measurement. The measurement is made in a direction offset by a small angle, $\Delta\theta$, from the source direction; $\Delta\theta$ is chosen to be the smallest possible displacement in which there is no emission from the source itself. For extended objects, choice of $\Delta\theta$ is problematic; it is often impossible to get sufficiently far from the source to ensure that its radiation does not contaminate the sky measurement. Measurements are usually made in a series of steps until they reach a large displacement from the object of interest.

Sky emission varies with angle and time. The difference in power between two sky positions grows with $\Delta\theta$ and Δt, the latter being the time difference between the two observations. For effective cancellation, both $\Delta\theta$ and Δt must be as small as possible. An angular separation of a few minutes of arc and Δt of a few seconds are normally adequate for good cancellation in the near infrared under clear sky conditions. At the shortest wavelengths, OH airglow varies significantly on timescales of a few minutes and sets the maximum time between sky measurements. In the thermal infrared, $\Delta\theta$ should be less than about

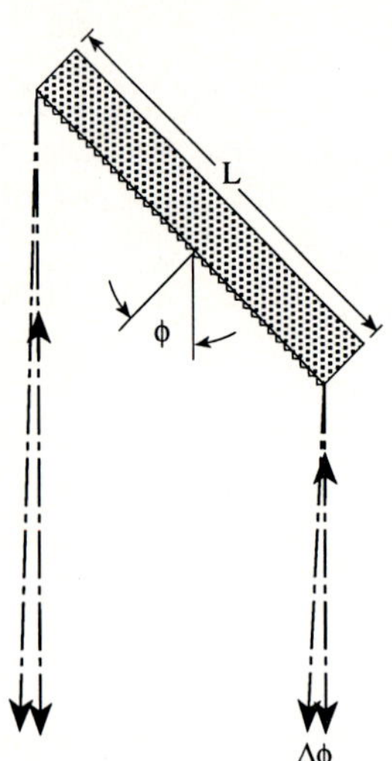

Fig. 3.2a. The quantities used for (3.1) and (3.2) in calculating grating resolution.

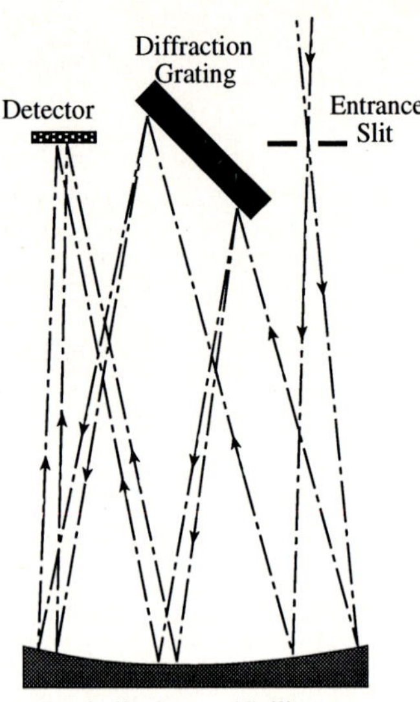

Fig. 3.2b. A diagram of an Ebert-Fastie grating spectrometer.

10″, and Δt less than a few hundreds of milliseconds. The power spectrum of sky noise tends to have an inverse frequency dependence (commonly called $1/f$ noise), which is site dependent.

The demands of small Δt, especially, drive the design of infrared telescopes and the techniques employed for sky subtraction. Infrared photometers, consisting of one or at most a few single detectors with limited fields of view, use a modulating optical element to switch rapidly between two positions on the sky. This form of sky switching is referred to rather prosaically as *chopping*. The modulation is done at a fixed frequency between a few and a few tens of Hz. As discussed below, the best systems articulate the secondary mirror of the telescope, but a variety of optical modulators can be used. The detection electronics usually incorporate lock-in amplifiers or phase-referenced software rectifiers to synchronously demodulate the photometer output which is modulated at the switching frequency.

The recent availability of large format detector arrays at infrared wavelengths has changed the techniques of sky subtraction and the construction of instruments (Joyce 1992). Throughout the near infrared, and even to wavelengths of about 5 μm, sky subtraction is accomplished using *staring* methods, as opposed to rapid chopping. This method is identical to that used with optical CCD's (Tyson & Seitzer 1988), but is generally more effective in the infrared; there the images are background limited in relatively short exposures, so that the sky changes very little between images. The response – number of photoelectrons per photon – of the infrared arrays also appears to be somewhat more stable than for the optical CCDs.

Sky subtraction is carried out by taking several exposures with the images displaced slightly on the detector array, then subsequently taking median averages of the frames without recentering to provide sky measurements for the displaced images. After sky subtraction, the displaced images are aligned and averaged to produce the final picture.

Thermal infrared observations, even with arrays, still require rapid chopping especially in the presence of sky noise. Thermal infrared imagers must be synchronized with the optical choppers, and the relatively high data rates – thousands of full frames per second – drive the electronic requirements. To a large extent, thermal infrared observations with arrays are the same as with single detectors, except there are many more detectors to deal with simultaneously. Sky noise which affects all pixels identically may be eliminated in the final image processing, thereby enabling good observations at otherwise mediocre thermal infrared sites.

3.3 Optical Baffling: Pupil Plane Masks

Because the optical components of an instrument emit infrared radiation at ambient temperature, it is necessary to cool as many of them as possible. Apart from the primary mirror, the window of the cryostat, and usually a secondary telescope mirror, all optical components are cooled. Cooling to 77 K with liquid nitrogen is adequate for $\lambda \lesssim 3\,\mu$m; cooling to 4 K with liquid helium is necessary for longer wavelengths.

To eliminate emission from warm structures, infrared instruments are designed to focus an image of the pupil, either the primary or secondary mirror, on a cold baffle called a *Lyot* stop inside the cryostat. The Lyot stop position is shown schematically in Fig. 3.1. The cold baffle may be slightly undersized, for wavelengths of 2 μm and longer to completely eliminate the possibility of background radiation from surrounding warm structures; it may be exactly sized for the shortest wavelengths. In the J and H bands, where OH airglow is the major source of background, it is often desirable to have warm baffles surrounding the secondary mirror to ensure that the instrument does not see past the secondary to the bright sky.

The need to create first a pupil image on the Lyot stop and then a focal plane image on the detector complicates the optical design of infrared instruments, but also endows them with great versatility. The final image scale can be chosen by the optical designer within broad limits; it depends only slightly on the image scale produced by the telescope itself. Unlike visual wavelengths, where focal changers or focal reducers are specialized instruments, the image scale of almost every infrared instrument can be determined for the convenience of the observer. It is also common for cameras to allow a choice of image scales via cold mechanisms, giving the observer wide latitude to select a scale well matched to the seeing and diffraction limitations of the telescope. Furthermore, because there is a pupil image in the system, it is relatively easy to place filtering elements near the pupil plane. For example, most infrared cameras have mechanisms to put transmission gratings (grisms) near the pupil plane, giving the camera a spectroscopic capability. Special masks to apodize the pupil for coronographic observations or the elimination of diffraction spikes from telescope structure are easily incorporated into standard infrared instruments.

Unfortunately, as we shall see in the next section, the need for cooling greatly exacerbates the difficulty of designing high resolution spectrometers.

3.4 Optical Filtering and Spectroscopy

Imaging of the sky requires two dimensions. The distribution over wavelength of the light from each direction requires a third dimension. Addition of time variation and polarization adds an additional two dimensions, but most instruments are designed primarily to sample the three-dimensional data space of direction and wavelength[3]. Infrared detectors are for the present inherently two dimensional, so a single exposure samples a plane or a rectangular solid in the three-dimensional data space.

Cameras and some low-resolution spectrometers employ interference filters to isolate specific optical bandwidths centered on specific wavelengths. They slice a plane parallel to the two space dimensions and are, therefore, best suited for scientific problems in which the angular distribution of the radiation is of primary interest rather than the spectral distribution. For example, to assess the density of young stars in a star forming region, images at a few widely separated wavelengths should be adequate to reveal the locations of all stars and provide some color information to determine the temperatures and spectral types. Imaging of specific emission lines is often enough to delineate the location of the radiating gas and establish its relationship to the local environment. Imaging devices are most useful for surveys of large areas of the sky, to locate new objects or to study the distribution of many known objects.

Interference filters are readily available from commercial suppliers. They can be designed for virtually all resolving powers, $\frac{\lambda}{\Delta\lambda}$, less than about 100 and all wavelengths less than 20 μm, at which point suitable substrate and coating materials are limited. A nice innovation is the *circular variable filter wheel* (CVF), where a circular segment of substrate (usually glass) is coated with layers whose thickness increases linearly with angle around the circle. The increasing thickness simply increases the central wavelength of the passband, so the CVF has a tunable passband as a function of wheel angle. CVF's have resolving powers between 20 and 100, are available for all wavelengths from the visual to 10 μm, and are small enough to be easily cooled. Perhaps their only disadvantage is that the physical size of the angle over which the central wavelength changes by less than the bandwidth is small, only a few millimeters, so it is difficult to incorporate CVFs into systems with large étendue.

[3] sometimes called a *data cube*, although it is usually rectangular

To reach higher resolving powers than 100 requires a spectrometer. Spectrometers divide broadly into two classes: *dispersive* and *interferometric*, although interference of light is responsible for spectral tuning in both cases. The common dispersive spectrometers use prisms, diffraction gratings, OR a combination in a single transmission grating called a grism. The common interferometric spectrometers are Fabry-Perot interferometers and Fourier Transform Spectrometers (FTS).

Dispersive elements channel light into different angles, the angle varying in proportion to the wavelength of light. The angular variation means that one of the two dimensions in the focal plane is used for wavelength, and the other can be used for spatial resolution. The one-dimensional focal plane resolution is normally limited by a focal plane slit. Dispersive spectrometers usually sample a plane in the data space which is orthogonal to the image plane. Use of image slicers or glass fibres allows instruments which sample the image plane in more general ways, a rectangular solid in the case of an image slicer and a series of separated lines in the case of multiple glass fibres.

Dispersive spectrometers are best suited to problems where spectral distribution is of primary interest. To assess the relative abundance of elements in a nebula or molecular cloud, for example, it is most useful to observe simultaneously a large number of emission or absorption lines in a few directions; imaging adds relatively little additional information. To determine the relative velocities of a number of young stars in a cloud, it is entirely adequate to have a spectrum of a single point from each star; a fibre-fed grating spectrometer with each fibre at the position of a different star in the focal plane is ideal for this problem. For very high spectral resolution, gratings are normally used at high angle of incidence – since $\frac{\lambda}{\Delta\lambda} \propto \tan\phi$. Such gratings are called *echelle* gratings; they are often used with dispersive elements perpendicular to their rulings to spatially separate the different spectral orders which bunch up at high angles of incidence.

Fabry-Perot interferometers and FTS instruments both work to a limited extent as interference filters: they use reflection of light within a resonant cavity to interfere with itself. The constructive interference passes light of specific wavelengths through the cavity, whilst the destructive interference rejects light at other wavelengths. Both devices are inherently imaging, in that they simultaneously pass the two dimensional field of the focal plane (Jenkins and White 1976).

The FTS (Fig. 3.3) passes almost all of the light in a very broad band, but modulates the intensity depending on the relative positions of the two, orthogonally placed mirrors that make up the resonant cavity. By moving one mirror, the output is modulated in proportion to the relative mix of wavelengths in the spectrum. The time variable output is basically the Fourier transform of the spectrum; the spectrum may be recovered by performing a Fourier transform of the output as a function of mirror spacing. The result is a complete sampling of the three dimensional data space, although at the expense of the many required images, one for each mirror spacing. The FTS is a *multiplex* device; at every mirror spacing it samples a modulated piece of the entire data space, multiplexed together on the detector array. Its great strength is the simultaneous measuring of everything the observer could want. It is particularly appropriate when an entire spectrum is necessary, for example a strong continuum source with absorption lines like a star. The beautiful spectra of BN (Scoville et al. 1983) made with the Kitt Peak FTS (Hall et al. 1979) demonstrate the great potential of the FTS.

The multiplex nature of the FTS also proves to be its primary weakness in the infrared; the background radiation from all spectral elements is added to all picture elements at all times. Thus when only a few spectral planes within a data cube are of interest, the noise in these planes is higher than it would be had they been isolated in the first place. FTS instruments are almost always inferior to the other spectrometers under BLIP conditions; they can be useful under the very low background conditions in spacecraft or when detectors are not background limited in conventional high resolution spectrometers. The FTS has the advantage that the spectral resolution may be chosen after the observations by apodizing the Fourier transform – although the resolution of highly dispersive spectrometers may be equally well degraded without penalty in BLIP conditions.

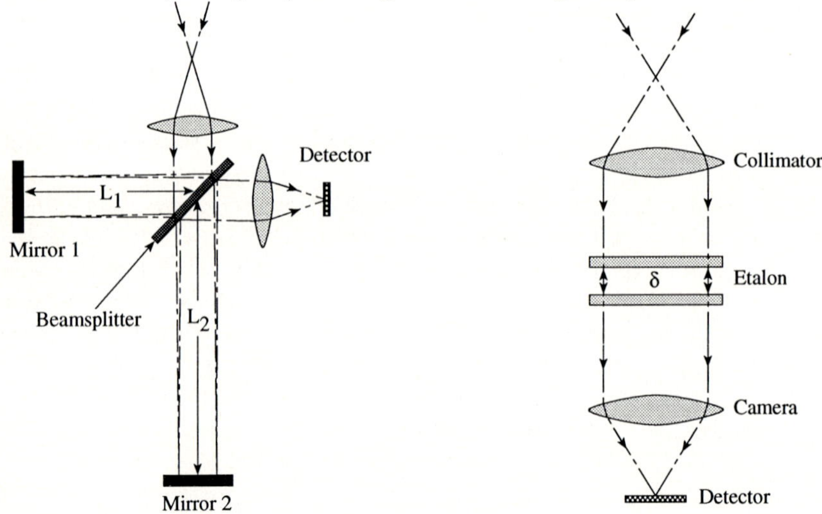

Fig. 3.3. Schematic diagram of a Fourier Transform spectrometer.

Fig. 3.4. Schematic diagram of a Fabry-Perot interferometer.

The Fabry-Perot interferometer (Fig. 3.4) acts very much like a tunable interference filter, albeit usually at much higher spectral resolution. It samples a planar section in data space which encompasses all of the information of interest. Imaging of a narrow emission line over a broad region is accomplished most easily with a Fabry-Perot – witness the map of H_2 emission from the Orion star forming region by Nadeau, Geballe, & Neugebauer (1982). The high spectral resolution rejects almost all the background radiation but none of the line radiation, greatly enhancing the sensitivity of BLIP observations.

The Fabry-Perot is attractive where high resolution is desired from a relatively small instrument. Because the resonant cavity greatly reduces the device size needed to produce interference of the light, the Fabry-Perot provides the largest étendue for a given resolution of all spectrometers. For a number of technical reasons, the best sensitivity is accomplished with cryogenically cooled Fabry-Perots; they are difficult to construct. All these spectrometers are capable of sampling the full three dimensional data space. A grating spectrometer fed by a focal plane slit takes one plane at a time; by moving the slit on the sky, all planes may be sampled for complete coverage. A Fabry-Perot samples planes parallel to the sky plane, but at different wavelengths; varying the tuning allows complete coverage of the data space. An FTS simultaneously samples all data space points, but one must take many samples to interpret the output. In general, the dispersive (e.g. grating) and Fabry-Perot spectrometers are superior to the FTS under BLIP conditions, even when the entire data space must be sampled.

The choice between the dispersive and Fabry-Perot interferometers depends on which variables in the data space are more important for the problem at hand. Generally, the sky background will change slightly between successive exposures, making it difficult to exactly normalize them to the same scale. Relative intensities within a single image are, on the other hand, usually very accurate. If the relative spectral intensities are more important than the intensities at different positions on the sky, stepping a slit spectrometer across the sky is preferable to the Fabry-Perot. If the spatial distribution of intensity (line emission, say) at different wavelengths is more important than the quality of the spectrum, a Fabry-Perot is advantageous. Exactly delineating line emission from a stellar jet, for example, is better done with a Fabry-Perot. Studies of the velocities and elemental abundances of stars in a cluster are best done with a dispersive slit spectrometer.

3.5 Infrared Optimized Telescopes

The mirrors, the spiders, and the surrounding structures of a telescope emit background radiation at all wavelengths longward of 2 μm. An infrared optimized telescope minimizes this background in several ways. For ground-based, ambient temperature telescopes, the mirror coatings must be clean and highly reflective to minimize the emissivity. Warm structures in the optical path – the secondary cage and its supporting spiders – are kept as small as practical, since they provide unwanted background radiation.

The ideal way to rapidly modulate the sky for background subtraction would be to wobble the entire telescope, but this is impossible in the thermal infrared, where frequencies of a few Hz or more are needed. The next best choice is to wobble the secondary mirror. In this way, it is possible to rapidly change the pointing of the telescope without severely altering the optical path. Although one can reimage the pupil onto a mirror near the focal plane and wobble it very rapidly, it then modulates the entire optical path of the telescope and has proven unacceptable as a means of reducing the background.

To minimize the power needed to move the secondary rapidly and to minimize the size of the central obscuration which produces unwanted background radiation, the secondary mirror is designed to be small. The final focal ratio of the telescope is correspondingly slow, typical values ranging from f/35 to f/70. Chopping secondaries are usually only a few tens of centimeters in diameter in marked contrast to conventional Cassegrain designs with large, fast secondary mirrors yielding wide fields in the focal plane.

There are currently only a few infrared-optimized telescopes in the world: the 3.8 m United Kingdom Infrared Telescope (UKIRT), the 3 m Infrared Telescope Facility (IRTF), the 2.3 m Wyoming Infrared Observatory (WIRO), the 1.5 m Telescopio Infrarosso del Gornergrat (TIRGO), and the 1.5 m Mount Lemon infrared telescope. The Gemini telescope project is planned to include an infrared optimized 8 m telescope to be at Mauna Kea in Hawaii.

The most effective way to eliminate telescope background is to cool the entire telescope. On the ground, siting at the South Pole (Bally 1993) gives significant cooling and also very low water vapor, $\lesssim 0.1$ mm precipitable column. For wavelengths near 2 to 3 μm, the background reduction is pronounced, probably enough to render the thermal background negligible. Cryogenically cooled telescopes are possible only in spacecraft. The Infrared Astronomical Satellite (IRAS) was cooled to liquid helium temperatures for its ten month lifetime in low Earth orbit. The 0.6 m Infrared Space Observatory (ISO), scheduled to be launched in 1995, will be cooled to a few degrees K with liquid helium. Eventually, longlife missions will need to evolve designs without cryogens by taking advantage of radiation cooling in outer space.

IRAS was a watershed in the study of young stars, revealing many to have circumstellar disks probably in various stages of planet formation (Aumann et al. 1984; Beichman 1987; Backman & Gillett 1989; Strom et al. 1989), and uncovering scores of young stars embedded in molecular clouds. ISO promises equally profound results through deep infrared photometry of young star clusters and spectroscopy of various objects in young star forming regions. The space infrared telescopes achieve their greatest gains in sensitivity at wavelengths longward of 10 μm, where the ground-based background is very high. At shorter wavelengths, the high backgrounds on the ground are compensated to a large extent by the size of the apertures ($\sim$ 8 m) planned for the next generation.

4 Infrared Detectors

Almost all major advances in infrared instrumentation have come from improvements in detectors. Early thermal infrared detectors were the first to provide BLIP sensitivity owing to the very high backgrounds. Development of the semiconductors PbS, InSb, and HgCdTe paved the way for BLIP detectors at increasingly shorter wavelengths and higher spectral resolutions. There are now detectors which achieve background-limited performance on ground-based telescopes at all infrared wavelengths and resolving powers ($\frac{\lambda}{\Delta\lambda}$) smaller than about 100. Most of these detectors are available in large format arrays; 256 × 256 formats are common; 1024 × 1024 are under development. With one exception, sensitivity is no longer an issue; array format and uniformity are of principle interest.

Detector development has proceeded for many years under the auspices of military research programs. Military development, particularly in the United States, was largely responsible for the start of infrared astronomy. Infrared reconnaissance and surveillance, night vision cameras, and the development of heat-seeking (via thermal infrared emission) weapons are essential components of military missions. Almost all military applications are under high background conditions, however, and in recent years there has been a divergence between military and astronomical requirements. Most of the new astronomical detectors are now funded by space projects (Hubble Space Telescope, SIRTF) and, increasingly, by individual observatories. The result has been better astronomical detectors albeit at much higher cost to astronomical programs. The impact of these new detectors has been enormous, however, with star formation studies, both galactic and extragalactic, being one of the chief beneficiaries of the rapid gains in observational capability.

This chapter discusses basic physical principles for various types of detectors, device limitations, and a listing of important detector types as of early 1993. Two useful references on this subject are Wolf and Zissis (1989) and Vincent (1990).

4.1 Types of Detectors

Detectors may be classified as *incoherent*, those which simply detect total radiative energy, and *coherent*, those which count individual photons. The coherent detectors are preferred when they are available with good sensitivity. The incoherent detectors are much easier to construct; they were among the first to be used and continue to be useful in some wavelength regimes especially the submillimeter and far infrared. The only important type for our purposes are bolometers, mentioned briefly in the next section. Coherent detectors have been developed for almost all near and thermal infrared wavelengths. Their sensitivity is greater than the incoherent detectors for most astronomical applications, and it seems likely that they will gradually displace the use of bolometers for thermal infrared wavelengths.

4.1.1 Bolometers

A bolometer absorbs IR radiation and converts it to heat or lattice energy in a solid. The heat changes the electrical resistivity which is measured as a record of the amount of radiation. The heat is then conducted away so as to maintain the detector at nearly constant temperature. Figure 4.1 shows the components of a bolometer and the principles of operation.

The great virtue of bolometers is that they absorb essentially 100% of the incident light, giving them a quantum efficiency, η, of unity. They are noisy, however, and are useful mainly under very high backgrounds, for example, broad band applications at wavelengths of 10 μm and greater using ambient temperature telescopes. Increasingly, they are being replaced by arrays of photon-counting devices at wavelengths shortward of 30 μm. The reader is referred to Emerson's lectures for a detailed discussion of the operation and use of bolometers.

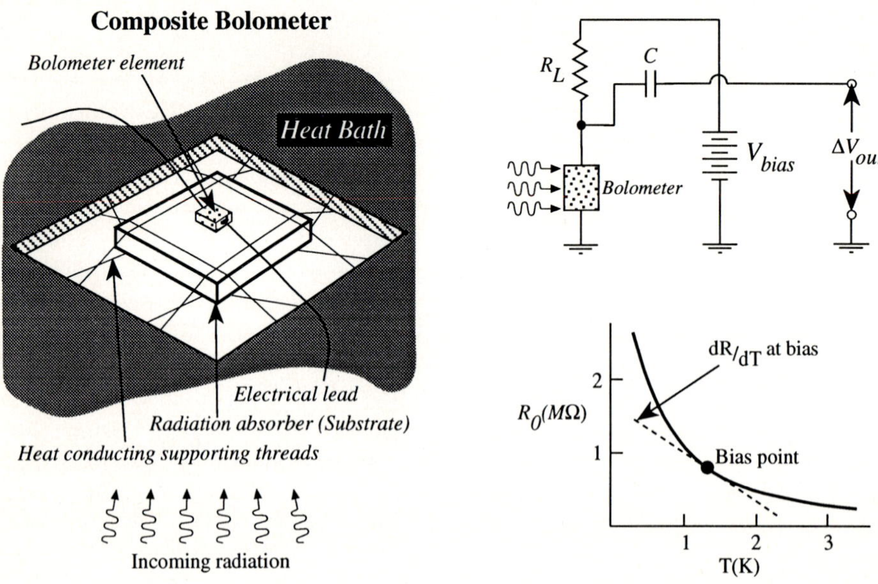

Fig. 4.1a. The principal elements of a *composite bolometer* are shown here. The substrate is a material with small heat capacity and good thermal conductivity which absorbs radiation very well (*e.g.*, materials such as blackened sapphire). The bolometer element, usually germanium or silicon, is a semiconductor whose resistance is a strong function of temperature. The supporting wires (not necessarily metallic) conduct heat away from the substrate to an external heat bath.

Fig. 4.1b. The upper part of the diagram shows the circuit used to measure the bolometer's resistance. The lower graph shows the output response as a function of circuit conditions for several bolometer temperatures (known as *load curves*). Changes in bolometer temperature induce corresponding changes in the output voltage. The voltage source, V_{bias} and load resistor, R_L, hold the bolometer at the *bias point* in absence of radiation.

4.1.2 Photoconductors

Photoconducting detectors absorb light by elevating an electron from a lower to a higher energy band within a semiconducting material. The energy difference between these two bands is called the *bandgap*. The upper band is part of a continuum of energy levels which allow the electron to move freely through the lattice, thus conducting electricity. The electron leaves a *hole* in the lattice, an ion with one fewer electrons than needed to neutralize the positive nuclear charge. The hole may migrate, too, and can carry current just as the electrons: electrons from neighboring ions jump to the site of the hole, leaving another hole and effectively allowing migration from ion to ion in the lattice.

In photoconductors, current carried by either the electrons or the holes dominates the conductivity; the counterpart carrier migrates little or not at all and eventually recombines with a free carrier of opposite charge. Both the *generation* of charge carriers (by photons) and the *recombination* occur in the lattice. As we shall see, the recombination noise makes photoconductors inherently inferior to the photovoltaic detectors discussed in the next section.

Additional energy levels can be created by embedding *impurity* atoms into an otherwise pure semiconducting or insulating crystal in a process called *doping*. Impurity bands reside within the bandgap (*cf*, Fig. 4.2a) and make it an *extrinsic* semiconductor. For example, adding small amounts of antimony (Sb) to a pure silicon crystal creates the antimony-doped silicon (Si:Sb) semiconductor that is important in the thermal infrared. The gap between the ground-state bound level and the first continuous conducting band of levels is about 0.04 eV, allowing photons with wavelength 32 μm ($\frac{hc}{\lambda} = 0.04$ eV) to elevate electrons from the bound level to the conducting band. The holes in this case remain bound, and the electrons are the charge carriers. Fig. 4.2a shows the lattice and energy level structure for Si:Sb.

Table 4.1. Bandgaps of common detector materials

Material	E_B(eV)	λ_0 (μm)	Material	E_B(eV)	λ_0 (μm)
Si:Sb	0.039	32	Diamond	5.33	0.23
Si:As	0.054	23	Si	1.17	1.06
Si:Ga	0.072	17	Ge	0.75	1.65
Ge:Ga	0.011	113	InSb	0.23	5.
Ge:Be	0.025	52	HgCdTe	0.1 – 1	1 – 10

To use photoconductors as radiation detectors, one normally runs a current through the material by means of two electrical leads and a voltage source. Photo-generated carriers increase the current in the circuit which is easily measured. The physics of the photon detection process and the basic circuit for measurement are illustrated in Fig. 4.3a.

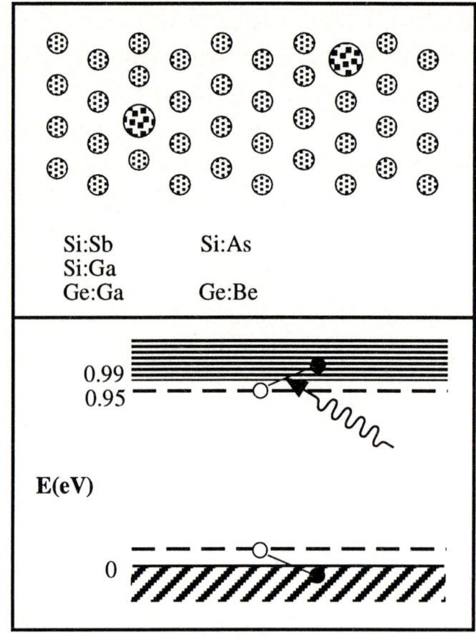

Fig. 4.2a. The upper half shows the lattice structure for a generic extrinsic semiconductor; several examples are listed. The lower half is an energy level diagram for Si:Sb: the intrinsic bandgap in this example is 0.99 eV and the donor levels lie 0.04 eV away from the edges of the band; there is a photo-generated electron-hole pair shown.

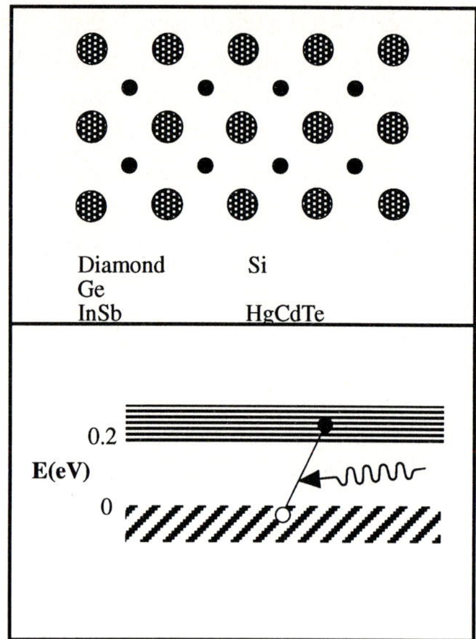

Fig. 4.2b. The upper half shows the lattice structure for an *intrinsic* semiconductor in which there is a single bandgap only. The bottom half shows the energy level diagram and a photo-generated electron-hole pair.

4.1.3 Photovoltaic detectors

A photovoltaic detector is a semiconductor whose doping creates a diode through the matching of two differently doped regions: an *n-type* region, in which there is an excess of free electrons (negative charge), and a *p-type* region, in which there is a slight deficit of free electrons (positively charged). They are normally *intrinsic* semiconductors (Fig. 4.2b). The Fermi energy levels of the electrons must match up, so there is an internal electric field in the semiconductor pointing from the p-type to n-type region.

Photons may be absorbed in the material through the creation of electron-hole charge pairs, just as in a photoconductor. Once the charge pairs are created, however, both the electron and the hole migrate in opposite directions under the influence of the internal electric field. The charge separation changes the electric field between the two regions, charging the detector as if it were a capacitor, and this charge may be measured externally. Photon absorption produces a *voltage* across the detector, hence the name photovoltaic. Fig. 4.3b shows the detection process and internal energy level structure which gives rise to the charge migration.

The principle advantage of the photovoltaic detectors is that only the shot noise associated with charge generation (absorption of photons) enters into the total noise; there are no corresponding recombinations within the semiconductor. Since each noise source is proportional to the square root of the number of carriers created, photoconductors are a factor of $\sqrt{2}$ worse than photovoltaic detectors for the same number of detected photons. Because all other noise sources in the measurement process are at least additive, photovoltaics are always better detectors.

One type of hybrid detector, known as a *blocked-impurity-band* or BIB detector, is made like a photoconductor but with an internal undoped *blocking layer* giving it the noise characteristics of a photovoltaic detector. The BIB is an extrinsic semiconductor with a thin layer that is very heavily doped giving a high concentration of potential photon-generated charges. The rest of the volume is undoped, providing a conduction and valence band but no other bands for generation or recombination of charges. This *blocking* section causes the detector to accumulate charges for subsequent readout without the recombination noise normally associated with photoconductors. It also discriminates against low-energy thermally generated charges (see below). BIB's can have much longer cutoff wavelengths than the corresponding bulk detectors.

4.2 Quantum Efficiency

The relatively low concentration of impurity atoms means that photons must travel some distance before there is a significant chance of absorption. The *absorption length*, l_a, and width of the doped region (called the *depletion zone*) determine the probability of absorption or quantum efficiency. The depletion zone is normally much thinner than the total thickness of the detector to reduce noise and the number of thermally generated carriers (discussed below), and there is a chance that a photon will not be absorbed at all. Because the absorption cross section changes with wavelength, the quantum efficiency is also wavelength dependent. It is almost always the case that the quantum efficiency increases with increasing wavelength until the cutoff wavelength set by the bandgap is reached.

There are various tricks to increase the path length for absorption and thereby increase the quantum efficiency. It is common to coat the surface above the depletion zone with reflecting material which is conducting and serves as one electrical contact to the detector. The detector is then illuminated from the behind (*backside-illuminated*); if possible, the substrate is made thinner by chemical etching or simply by cutting it away with small tools to decrease the chances of photon absorption in the substrate itself. The substrate may also be antireflection-coated with dielectric layers to reduce photon reflection off the first surface; semiconductors almost all have high indices of refraction making the surface reflectivities large. Most infrared detectors require special coating on the exposed surface to prevent chemical reactions with the substrate itself, a so-called *passivating layer*. The quality of the passivation is an important factor in the detector quantum efficiency and lifetime. Non-uniformity in thinning, passivation, and coating across the detector is the main cause of variations in quantum efficiency among different detectors (the *flatness* of the detector).

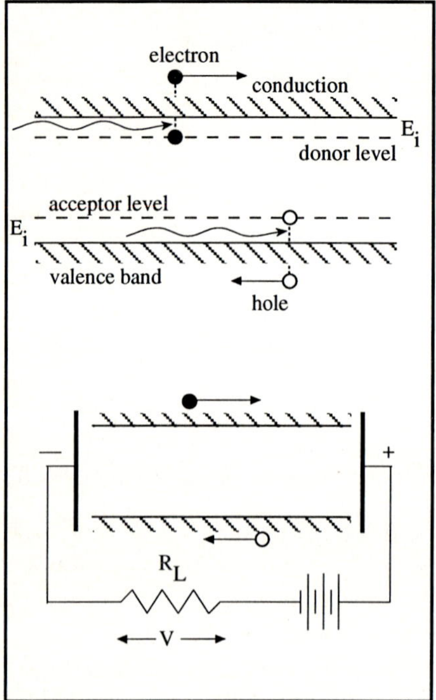

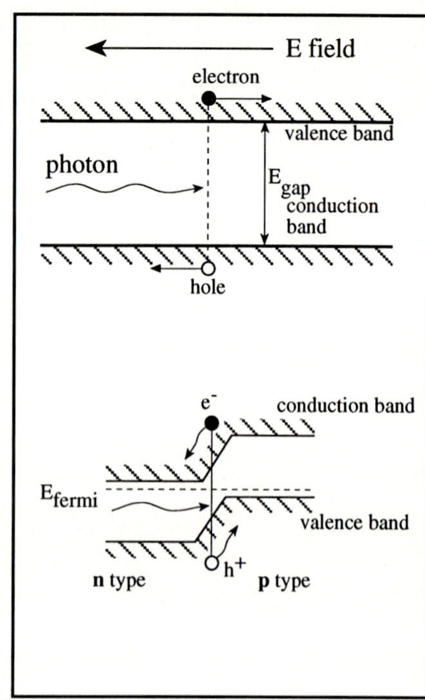

Fig. 4.3a. The upper half shows the energy bands for Si:As as well as the creation of carriers from photon absorption. The lower half indicates the circuit used to measure changes of conductivity: the voltage source and load resistor (R_L) generate a *bias current* through the detector; changes in the current produce changes in the voltage, V, developed across the load resistor.

Fig. 4.3b. The upper half gives the energy band for InSb, a common near-infrared photovoltaic detector. The absolute energy bands, relative to the constant Fermi level, appear in the lower half of the figure. This shows how an electron-hole pair migrates to opposite sides of the detector under the influence of the internal field.

4.3 Charge Storage and Readout

Detector arrays are normally exposed to light for a certain time (the *integration time*) after which the accumulated charge is measured (*read out*) and flushed from the detector (*reset*). In both photovoltaic and photoconductive detectors, the charge is accumulated on a capacitor attached to the detector, normally the combined capacitance of the detector itself and the node capacitance of the gate of a field effect transistor. In photovoltaic detectors, the diode is given a slight reverse bias voltage at the beginning of the integration to enhance the natural electric field at the p-n junction; charge accumulation lowers this bias voltage. By contrast, photoconductors are biased with large voltages to drive a steady current. With this difference, the principles behind the multiplexer operation are similar, so we will examine only a photovoltaic array.

Unlike silicon CCDs, most infrared detectors do not have readout electronics built into the detector substrate. Large arrays of detectors made from one material, HgCdTe, for example, are bonded to a multiplexer chip made from silicon containing a unit cell of electronics for each detector as well as circuitry to address the individual unit cells and transfer the charges to external electronic circuits. The connections between detectors and their unit cells are made with small metal welds called *bump bonds*; the bump bonds provide both electrical connections and mechanical bonds between the detector array and the multiplexer. Differences in the coefficients of thermal expansion between the detector and multiplexer materials set limits to the sizes of arrays, since all these detectors must be cooled to low temperatures for sensitive operation.

Figure 4.4 is a schematic diagram of a hybrid array detector showing the detector array, the multiplexer, and the bump bonds. Figure 4.5 shows a cross section of a NICMOS3 detector unit cell. The NICMOS3 series are HgCdTe devices made by Rockwell Science Center, sensitive from 1 to 2.5 μm. The array is square with 256×256 detectors each 40 μm square. The sapphire wafer serves as a strong mechanical base for a very thin layer of detector substrate; it allows the detector to be very thin for high, uniform quantum efficiency without compromising the mechanical strength needed for long life. The p-n junctions are made by diffusing p-type HgCd into the n-type CdTe substrate, with the band gap set to 2.5 μm.

The unit cell and multiplex circuit for a NICMOS3 detector are also shown in Fig. 4.5. Photoelectrons accumulate on the combined capacitance of the detector and the gate of the FET connected to the output line. The reset FET is normally turned off (open circuit) during integration. When it is turned on by its gate voltage, it acts as a short circuit to ground thus flushing the accumulated charge from the detector. When the output line is enabled, the voltage of the gate (accumulated charge) induces a corresponding voltage on the output which is amplified by the output amplifier and recorded by the off-chip electronics. At any time, the address registers determine which detector is being read out or reset. This unit cell allows the accumulated charge to be read and recorded without altering it – *non-destructive readout*. It is, therefore, possible to read the photo-

Schematic Hybrid Infrared Detector Array

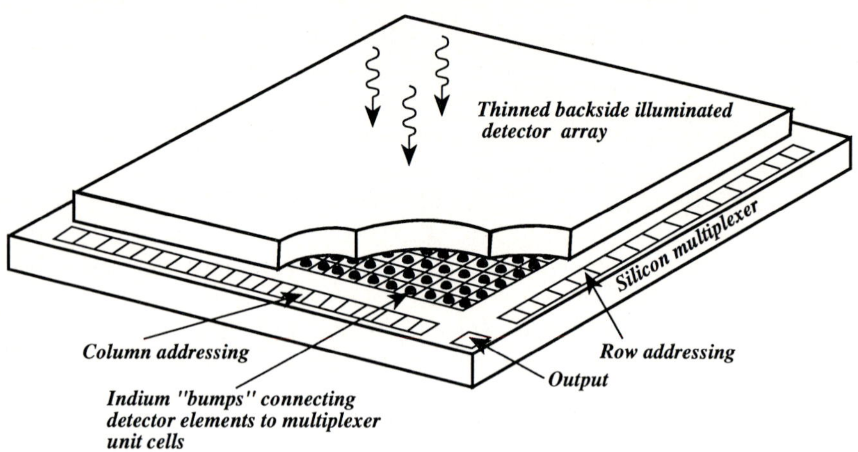

Fig. 4.4. Cross section of a hybrid array detector showing the individual pixels, bump bonds, multiplexer, and detector substrate. The multiplexer consists of unit cells for charge storage and readout, addressing electronics, and output amplifiers. The four quadrants may be read simultaneously.

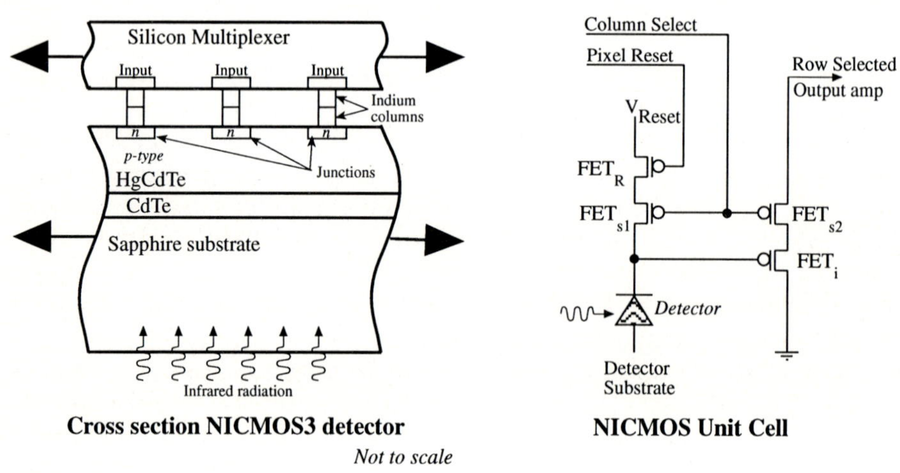

Fig. 4.5a. Cross section of a NICMOS3 unit cell. The NICMOS detector series are Rockwell's HgCdTe arrays using a sapphire base to support the thin detector material. The detectors operate between 1 and 2.5μm The NICMOS3 has a format of 256$\times$256 pixels

Fig. 4.5b. The equivalent circuit for the unit cell electronics. FET_R is closed to reset, open during integration. "Column Select" is enabled (on) when this pixel is being read out. The input FET, FET_i, transfers voltage on the detector to the output amp when selected.

generated charge many times during an integration cycle, a great advantage for some of the applications described below.

Fig. 4.6 plots the voltage on the detector (accumulated charge divided by the capacitance) as a function of time during a typical integration cycle. The detector is reset by enabling the reset FET. There is usually a slight charge from electronic pickup induced on the detector when the reset FET is turned off at time t_1 called the *pedestal* voltage. At time t_2, the detector is exposed to infrared radiation, and the voltage increases as photoelectrons accumulate (we ignore the polarity of the voltage). At time t_3, the integration ends, the detector is reset, and the cycle starts again.

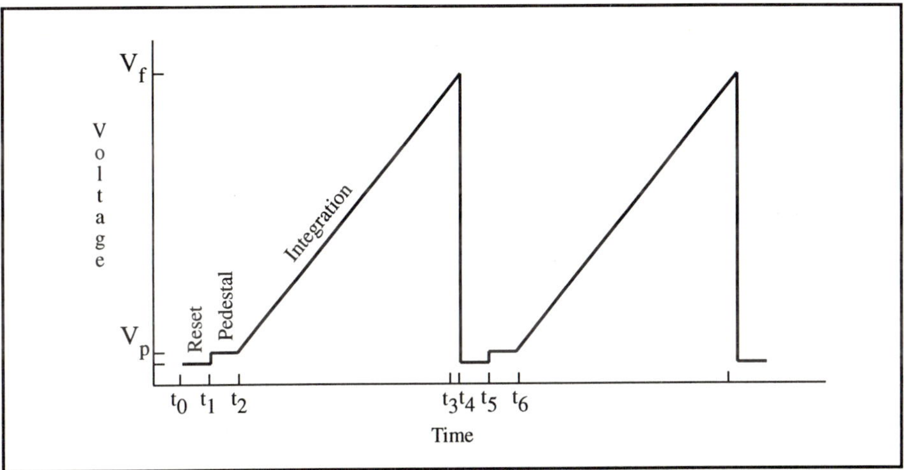

Fig. 4.6. Two integration cycles with voltages and times. The slope of the linearly increasing part is proportional to the photo-generation rate. The voltage is equal to the accumulated charge (photon-generated + dark current) divided by the cell capacitance.

There are many different schemes for reading the detector during this cycle, each with different advantages under different circumstances. The simplest scheme is to record at some time the reset voltage level, V_r, then read the voltage, V_f, at t_3 so that $(V_f - V_r)C$ is the accumulated photocharge which is proportional to the number of detected photons. This method ignores the pedestal level, V_p, which varies with each reset, and is subject to uncertainties in V_r caused by long term drifts in the electronics. Reading V_f and then $V_r(t_4)$ immediatcly afterwards (called *double-correlated sampling*) overcomes electronic drift at the expense of two reads and the still uncertain pedestal level. Reading the detector three times per cycle for several cycles allows one to record the pedestal voltage $(V_p(t_1) - V_r(t_0))$ at the beginning and the final voltage relative to the reset level $(V_f(t_3) - V_r(t_4))$, so the photocharge is: $V_f(t_3) - V_r(t_4) - V_p(t_1) + V_r(t_0)$; the term $V_r(t_0) - V_r(t_4)$ effectively takes out any long term drifts in the electronic levels – this term will be zero if there are no drifts. The three reads per cycle is called *triple-correlated sampling* and is routinely employed to give very low noise performance at the expense of minimum cycle times set by the time to readout the array three times.

Each time a detector is read out, the electronics add *readout noise* to the measurement. The readout noise adds in quadrature to the shot noise of the photocharge. The effect of reading the detector many times and using various combinations of the output to determine the photocharge is to increase this readout noise. For example, the noise associated with triple-correlated sampling is approximately twice the single readout noise owing to the four separate reads needed to calculate the photocharge. There are a number of different readout schemes developed to overcome or reduce the overall readout noise; the most common is to sample the array repeatedly during the integration and use least-squares linear fits to the many different samples to determine the rate at which photocharge is accumulated. Figure 4.7 shows two of the most common schemes: sample up-the-ramp and multiple-endpoint sampling.

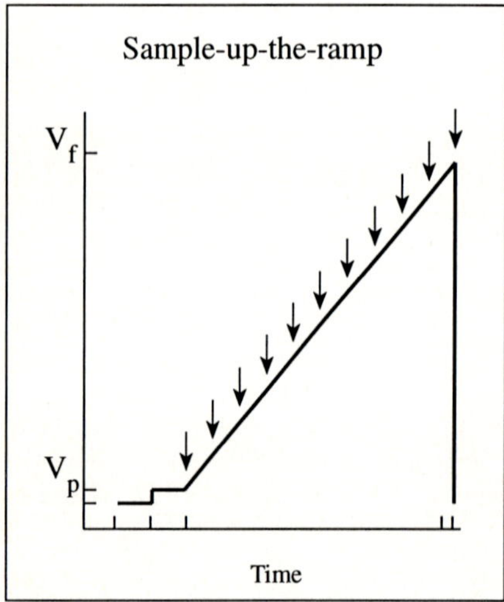

Fig. 4.7a. An integration cycle using the sample up-the-ramp scheme. Each sample is evenly spaced. A straight-line fit to the results gives the photon rate.

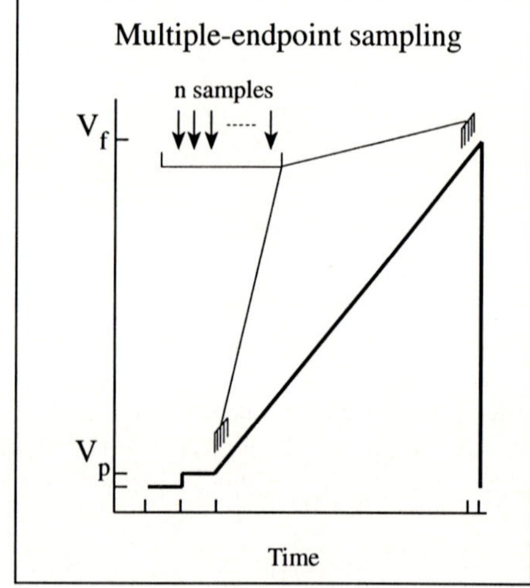

Fig. 4.7b. An integration cycle using the multiple-endpoint sampling pattern; this pattern has been shown to give the lowest readout noises in current detectors (Fowler and Gatley 1990).

Some multiplexers allow random addressing of individual detectors or sections of the array. For very high speed applications, such as lunar occultations or adaptive optics, it is often desirable to read out only a small portion of the array very rapidly ignoring the other sections. Many of the newest multiplexers allow *subaddressing* with subframe rates of a kHz or more.

4.4 Sources of Noise

The most fundamental source of noise is the shot noise in the detection process itself, the generation noise equal to the square root of the number of photogenerated charge pairs. In photoconductors, a recombination noise term increases this fundamental noise by a factor of $\sqrt{2}$. Additional noise may be added by the readout electronics or caused by the addition of charges which are *not* created by photons. Detector manufacturers seek to minimize these other noise sources, of course, but they are almost always present at some level. For many current applications, however, the additional noise sources are smaller than the fundamental noise created by detection of background photons; the detector achieves BLIP for these applications, and the additive noise is inconsequential. BLIP is relatively easy to achieve for wide spectral bandpasses and long integration times. Detector noise is mainly a problem for very high spectral resolution at short wavelengths and at rapid frame rates such as needed for speckle interferometry.

Readout noise, mentioned above, results from a combination of fundamental noise within the readout electronics (for example, Johnson noise in the channel of the output FET) and induced noise by pickup and crosstalk among various components during high speed switching of many transistors. Readout noise is always expressed in units of photocharge, electrons. Constant improvement in the manufacture of multiplexers has steadily decreased this noise from a value of about 2000 in the late 1970's to lower than $30\,e^-$ in the best detectors today. The best silicon CCDs have readout noises as low as $3\,e^-$, so we might expect further improvement in infrared arrays. The best multiple-endpoint sampling techniques record readout noises as low as $10\,e^-$, although these are useful only for integration times of several seconds or more.

Charges generated without light, by thermal excitation, for example, generate noise from *dark current.* Dark current arises most naturally from charge pairs created by heat energy in the lattice of the detector semiconductor. The probability of thermally exciting charges into the conduction band depends exponentially on the band gap and the temperature, so the easiest way to reduce it is to lower the temperature. Figure 4.8 plots the logarithm of thermally generated carriers as a function of inverse temperature for an extrinsic semiconductor showing the steep falloff of carriers created by direct excitation across the bandgap and the slower falloff of charge pairs excited from the donor bands.

Dark current adds additional shot noise through the creation of more charge pairs. If it fluctuates owing to temperature fluctuations during the integration, for example, the added noise can be very large. It is, therefore, important to keep dark current below that generated by background photons in the radiation stream. Dark current is mainly a problem for low background observations requiring long integration times, such as high resolution spectroscopy and spacecraft observations.

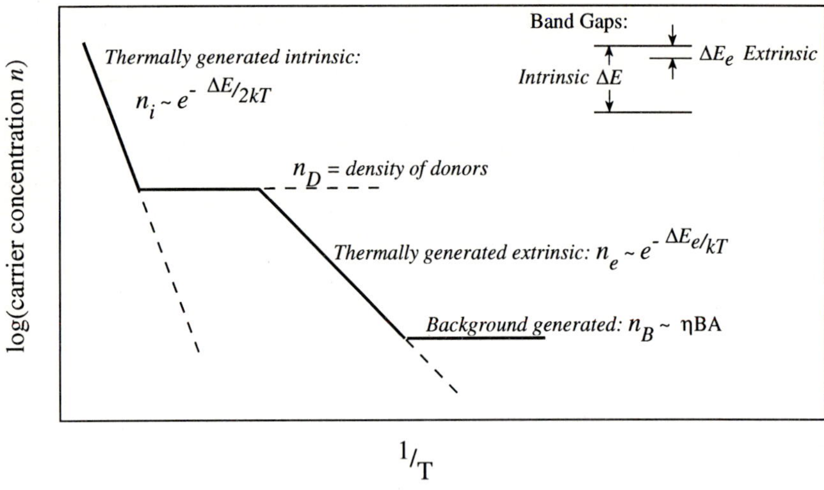

Fig. 4.8. The rate of electron-hole pairs generated thermally (dark current) is plot against inverse temperature for an extrinsic semiconductor. The separate regimes show limiting cases for this detector with equations giving the explicit dependence on detector characteristics (after Fig. 2.7 of Vincent 1990). Detectors should be operated in the region on the right hand side of the figure (low temperature), so that the background generated current is larger than the thermally generated currents.

Finally, there are various noise sources associated with readout electronics and readout schemes. Most modern arrays have made these noise sources sufficiently small that they can be ignored here. One which may continue to be important is added photon noise contributed by residual glow of the on-chip output amplifiers when the arrays are read out many times non-destructively. Added background can be detected in the corners of the arrays where the amplifiers are mounted during multiple endpoint sampling and sample-up-the-ramp reads. A second source of noise is persistence of bright images on the arrays after read and reset. Some arrays continue to show after effects from observation of bright objects for many read cycles. These peculiarities are probably not fundamental to the physics of the arrays, resulting, rather, from idiosyncrasies of the manufacturing techniques. They should be cured in future generation detectors.

4.5 Device Limitations

Table 4.1 outlined the bandgaps and corresponding wavelength sensitivity of common detector materials. The longer wavelength detectors must have smaller energy gaps for photo-assisted charge generation, which means a correspondingly larger dark current at any temperature (*cf,* Fig. 4.8). Long wavelength detectors are more challenging to make while preserving good quantum efficiency and low-noise characteristics. On the other hand, the backgrounds are much higher at longer wavelengths, so the detector performance usually does not have to be as good unless it is part of a spacecraft instrument. And it is normally easier to work with detectors that do not respond to light at longer wavelengths than those of interest.

Pixel sizes are important, because they constrain the subsequent optical subsystem. As noted in the second chapter, $A\Omega$ is a conserved quantity through an optical system. Using large pixels makes the optical design much simpler in general; slower focal ratios generate less distortion, and the total $A\Omega$ is usually limited by the detector size, since Ω can rarely be made larger than 1 sr at the detector. On the other hand, smaller pixels are physically required to make low-noise, large format arrays: smaller pixels mean lower pixel capacitance, which means lower noise, and the total array size is limited by physical strength and thermal mismatch between detector array and multiplexer.

Array formats are very important. Larger formats mean larger possible fields of view and more overall information capacity. The largest format optical CCDs are now approaching 4096^2; the largest low-noise infrared arrays are currently 256^2, but 1024^2 are expected in a few years. Physically these arrays will be a few cm on a side.

Well-depth is the number of charges that can accumulate on a detector before voltage changes induce significant non-linearity, one that cannot be easily corrected through calibration. Well-depth depends mainly on the detector capacitance. Well-depths of $10^4\,e^-$ are low (some InSb detectors), $10^5\,e^-$ are fairly typical (e.g., NICMOS3 detectors), and 10^6 to 10^7 are possible for detectors operating at high background, although there is usually a penalty to be paid in readout noise. The well depth determines how long one can integrate with the combined dark current and background or source-generated photocurrent. Larger well depth is always desired, all other parameters being equal.

The total information capacity of an array is the well-depth times the number of pixels. For an optical CCD, this is currently of order $2048^2 \times 10^5 \approx 4 \times 10^{11}$ or 38 bits of information. Infrared arrays are smaller by a factor of 8 (3 bits) on each side, giving 32 bits of capacity. This capacity already approaches that of photographic plates which have many more effective pixels but only 1 bit capacity per pixel.

The creation of p-n junctions within a substrate to produce many detectors allows the possibility of gaps between detectors. Junctions can be made without gaps, but then there is considerable crosstalk between detectors – charge pairs created over one detector induce voltage on an adjacent detector. On average, the gaps yield a slightly reduced quantum efficiency. The percentage of the total area which is sensitive to photons is called the *filling factor*, and it is often less than 100%. Furthermore, individual pixels are often defective; the array quality depends on the number of good pixels as a percentage of the total (called the *yield*).

For observers, the gaps and bad pixels can be a serious problem, because they mean that the average number of photons detected from a very small object such as a star will be highly sensitive to the position of the source relative to the pixels. Photons can "get lost" between two pixels, greatly reducing photometric accuracy of the arrays, especially on small telescopes where the seeing image can be smaller than the pixels. Careful observers often work with the telescope sufficiently out of focus to spread light from a star over several pixels, thus greatly reducing the position sensitivity of the photometry.

Readout speed is increasingly a limitation as arrays become large and new, high speed applications are desired. Speckle interferometry requires frame rates of 10 Hz or more in the near infrared. For a NICMOS3 detector, this implies pixel readout rates of 650 kHz and data rates of more than 1 Mbyte s^{-1}. The arrays themselves are capable of operating at 20 Hz frame rates or more. The main limitations are external A/D converters and computer recording media – this data rate corresponds to filling a 1 gigabyte DAT tape every six minutes! Lunar occultations and searches for rapid fluctuations (like pulsars) find their main limitation in the array readout speeds. Broad band imaging in the thermal infrared places a heavy demand on speed to flush the detectors before the wells are filled by background light. Table 4.2 gives some examples of high speed applications for these arrays. The typical maximum integration time for the entire array under each application is given as τ in Table 4.2.

Table 4.2. High Speed Applications

Application	τ(ms)	Frame Rate(Hz)
Broad band imaging: 10 μm	< 1	> 1000
Speckle interferometry	< 100	> 10
Lunar occultations	few	few hundred
Pulsar timing	< 1	> 1000
2D Fourier Transform Spectroscopy	~ 1	~ 1000
VLTI beam combining	~ 1	~ 1000
Adaptive optics wavefront sensing	few	few hundred

Table 4.3 lists the common detector materials of importance for near and thermal infrared detectors as of 1993. The first two, the HgCdTe NICMOS series by Rockwell and the InSb detectors from Santa Barbara Research Corporation, are widely available for near-infrared instruments. The Si:As BIBs are available in limited supply within the United States and could become available worldwide, if export restrictions are lifted. There is a fourth type of detector based on silicon technology, the platinum-silicide (PtSi) detectors, which have been used recently

at the Kitt Peak National Observatory. PtSi has very low quantum efficiency, of order a few percent, and is not expected to be important as a detector material of the future.

In principle, the InSb detectors are ideal to cover the entire range between 1 and 5.5 μm. In practice, however, there are still important advantages to the HgCdTe technology when wavelengths shortward of 3 μm are primarily of interest. It is very difficult to baffle detectors in a cryostat to eliminate all stray radiation. The enormous dominance of long wavelengths in stray infrared light puts heavy demands on an instrument using InSb for short wavelength observations especially under low background applications such as spectroscopy. The HgCdTe arrays are easier to handle, since they do not respond to light beyond 2.5 μm; for this reason, they continue to be important for many infrared instruments.

The behaviour of BIB detectors is very good under low background conditions but suffers from effectively low quantum efficiency under high background conditions; the reasons for the low quantum efficiency (of order 5%) are not yet understood. BIBs have not completely replaced bolometers as sensitive detectors in the thermal infrared for broad band photometry. If the sensitivity of BIBs under high background conditions is improved, they will become the standard detector for use in the thermal infrared.

Table 4.3. Near-infrared Detector Arrays (1993)

Material	Type	$\lambda(\mu m)$	Format	Pixel	Q.E.	i_{dark}	R_n
HgCdTe	PV	0.8 – 2.5	256^2	40 μm	60%	0.5 e^-/s	30 e^-
InSb	PV	0.8 – 5	256^2	40 μm	85%	$\leq 1 e^-/s$	50 e^-
Si:As	PC BIB	4 – 25	128^2	75 μm	$\leq 30\%$	$< 10 e^-/s$	$\sim 50 e^-$

5 Future Prospects

The advance of technology will open up many areas of investigation heretofore unreachable through infrared observations. The most exciting are those in which new discoveries could change our understanding of star formation. This author has not the prescience to identify the best areas for new discoveries. However, we can foresee a number of places where technological advances will greatly expand classical lines of research, and avenues of investigation that will be enabled in the infrared through new devices.

This chapter will discuss a few of the places where predictable advances should enable attacks on important problems. The predictable advances are extrapolations from the previous three chapters. Combined with the discussion in the first chapter, this chapter shows why this is a golden age for infrared astronomy.

Most advances over the next few years will be driven by increased detector capability as has historically been the case. We can already see five trends:

- **Larger detector formats.** The next generation (expected by 1995) will be 1024×1024 formats for wavelengths out to 5 μm, in both HgCdTe and InSb. BIB detectors for the thermal infrared are at 128×128; we can expect their sizes to increase as well. There is presently no obvious limit to the sizes of the detectors.
- **Sensitivities limited only by natural backgrounds.** This is already the case for broad band cameras at all wavelengths, for spectrometers of modest resolving power ($\frac{\lambda}{\Delta\lambda} \sim 10^4$) in the near infrared, and for almost all resolutions in the thermal infrared. Quantum efficiency improvements still need to be made for BIB detectors; the others are for all practical purposes nearly perfect.
- **Extension of standard optical techniques to the infrared.** Wide-field and high resolution cameras already exist. Spectrometers of all kinds, including cross-dispersed echelles, fiber-fed spectrographs, imaging spectrometers, as well as more conventional slit spectrographs, will be built. Specialty instruments, such as high speed cameras and spectrometers for interferometric applications, will be built and limited only by natural backgrounds.
- **Reduced backgrounds.** Improvements in filtering, size of telescopes, and reduced backgrounds, especially in space, will produce instruments with sensitivity and resolution near the photon-counting limit in some bands. For example, an OH blocking camera or a camera on a modestly cooled space telescope may be limited by the counting rate of source photons for wavelengths shortward of 3 μm.
- **Order(s) of magnitude improvement in angular resolution.** The use of interferometric techniques and adaptive optics will become common. These techniques are covered in detail by Leinert's chapters in this volume. For our purposes, it is only important to recognize their availability.

A number of problems will derive clear benefit from these advances. A few of these are summarized below.

5.1 Initial Mass Function

The initial mass function (IMF) of star forming clouds, its dependence on the environmental conditions, and its evolution as stellar interactions modify the distribution of masses are still not understood. Surveys of clouds are already underway using large format cameras (*e.g.*, Lada 1991). With increasing format, surveys of large regions for newly born stars become easier. In fact, the current technology makes it somewhat easier to gather the data than to reduce it to a usable form, a problem the next generations of computers and especially software will solve. The new infrared sky surveys currently planned in the United States and Europe will provide complete surveys of the sky down to about 14^m at 2.2 μm. Dedicated surveys of clouds to deeper limits should pick up every young star in each of the clouds, yielding complete *samples* of pre-main sequence stars.

These surveys will probably be inadequate by themselves to determine the IMF of clouds. To determine the mass of a star, its effective temperature and luminosity must be known. Unfortunately, the unknown extinction to stars in a cloud changes the apparent temperature and luminosity, and it is almost impossible to determine the extinction from broad band colors alone. Spectroscopic followup using, for example, multi-object spectrometers will make it possible to determine the IMF on a cloud-by-cloud basis. We may finally be able to identify the most important factors governing the IMF.

5.2 Extragalactic Star Formation

Very deep surveys should uncover the first generations of stars in high redshift galaxies (*e.g.* Thompson, Djorgovski, & Beckwith 1993). It is already possible to measure global star formation in distant starburst galaxies (Soifer et al. 1992). To detect the first generations of stars after the Big Bang will require increases of approximately an order of magnitude in sensitivity and probably one to two orders of magnitude in areal coverage. Large format arrays, reduced background near infrared cameras, and larger telescopes now make this problem tractable.

The study of star formation in nearby galaxies will also benefit from large arrays. Imaging of entire galaxies and clusters of galaxies should penetrate the obscuration which distorts optical photographs. In addition to revealing the old dwarf population, infrared images can uncover the young stars and their relative distribution in galaxies (Telesco 1988).

5.3 Stellar Interaction

Most stars in the sky are members of binary systems or clusters. We have little understanding of the processes leading to star groups, whether fragmentation during cloud collapse or subsequent capture of nearby neighbor stars plays the more important role in creating multiple star systems. Simply assessing the distributions of binary stars as a function of separation has been a difficult task.

High angular resolution coupled with high sensitivity and modest fields of view will make it possible to assess the multiple star distribution in nearby star formation regions. Speckle interferometry on 10 m class telescopes and multi-telescope infrared interferometry – such as planned for the VLTI – should address the "intermediate" separation binaries, those whose separation is too large to be seen as spectroscopic variables but too small to be observed directly in long-exposure images. With the high angular resolution provided by multi-telescope interferometers, one can imagine following the dynamical evolution of a dense star cluster directly. For example, a 100 m baseline at $2.2\,\mu$m gives 2 AU resolution at the distance to Orion. Over a few years, one could easily map the orbits of stars separated by a few AU. Even the effects of neighbors in dense regions might show up at this resolution.

5.4 Circumstellar Disks

Thermal infrared wavelengths are especially important for detection of circumstellar disks: disks tend to dominate emission at 10 to 20 μm relative to other circumstellar processes, but the stellar photospheres can still be detected when disks are not present. Complete surveys for disks will benefit greatly from increases in thermal infrared sensitivity, increases which can come about with the use of array detectors on large telescopes with very small images – from the use of adaptive optics, for example. Spacecraft, such as ISO, will make strong contributions to this problem, as well, and we can hope that complete surveys of disks among stars of different ages will make it possible to study disk evolution for the first time (*e.g.*, Beckwith and Sargent 1992, Beckwith 1993).

The process by which disks accrete and lose mass is poorly understood. Both accreting and outflowing matter can be imaged directly at very high angular resolution. Large baseline interferometers, such as the VLTI, should enable such observations. At a resolution of a few milliseconds of arc, a few AU in the nearby star forming regions, one could detect the time-varying changes in outflowing matter over the course of a year or so; matter accreting directly onto the disk might also be detected, if it has sufficient optical depth. It may also be possible to see gaps in the disks creating by planets in high-angular resolution images.

5.5 Stellar Winds, Large Scale Shocks, and High-temperature Interface Regions

On scales of a few seconds to a few minutes of arc, winds from stars, turbulence within clouds, cloud-cloud collisions, photodissociation regions, and HII regions create high temperature regions within cold molecular material. An example is seen in the H_2 emission from Orion described in the first chapter. These regions result from energy injection by young stars into the surrounding clouds. At the very least, they are a component of the total energy output of a star. But they might well be responsible for the way in which star formation is regulated after the first generation of stars is born: injection of matter and radiation disrupts the clouds and inhibits further star formation.

Imaging spectroscopy of infrared emission lines will be very important for the study of these regions. The winds, shocks, and other interface regions typically extend over several minutes of arc, requiring wide-field imaging. Observations of infrared lines will penetrate much deeper into the clouds than the optical observations do (Edwards, Ray, and Mundt 1992), revealing more of the winds and their interface regions. Large scale shocks cooling through H_2 emission will be studied primarily in the near infrared where the strongest lines are.

5.6 Conclusion

The enormous increase in observing capability brought about by improvements in infrared instruments during the last 25 years has had a major impact on our understanding of star formation. We have identified the sites for creating stars, taken a preliminary census of their content, discovered the energetic processes by which the interstellar medium feeds the stars and the stars spit back some of the residue. We can study star formation not just in the solar neighborhood but also in distant galaxies. All these results were brought about by new observing capabilities, mainly through improvements in detectors.

This improvement shows no signs of abating in the immediate future. We are now at or very close to the sensitivity and resolution limits set by nature over the entire infrared region accessible from the ground. Nevertheless, increasing detector formats will increase the information rate. Larger telescopes with adaptive optics and other interferometric techniques will continue to improve the angular resolution and sensitivity to small sources. Ultimately, space telescopes with modern detectors will bring us to limits that will be exceedingly difficult to overcome: zodiacal light emission, emission from the interstellar medium, and confusion by distant stars and galaxies. At that point, the exploratory phase in the infrared band will cease, and infrared observations will serve primarily as diagnostics. That point appears quite distant now, however. Exploration is likely to continue for some time and should continue to bear fruit in new discoveries about young stars.

References

Allen, D. A. 1972, *Ap. J. Lett.*, **172**, L55.

Abt, H. A. 1983, *Ann. Rev. Astron. Ap.*, **21**, 343.

Abt, H. A. and Levy, S. G. 1976, *Ap. J. Supp.*, **30**, 273.

Aumann, H. H., Gillett, F. C., Beichman, C. A., de Jong, T., Houck, J. R., Low, F. J., Neugebauer, G., Walker, R. G., and Wesselius, P. R. 1984, *Ap. J. Lett.*, **278**, L23.

Backman, D. E., and Gillett, F. C. 1987, in *Cool stars, stellar systems, and the Sun.* ed. J. Linsky and R. E. Stencel (Springer-Verlag:Berlin), p. 340.

Bally, J. 1993, in *The Cold Universe*, ed. T. Montemerle.

Becklin, E. E., Matthews, K., Neugebauer, G., and Willner, S. P. 1978, *Ap. J.*, **220**, 831.

Beckwith, S. V. W. 1993, in *Theory of Accretion Disks II*, NATO Workshop in Garching, eds. F. Meyer, W. Tscharnuter, and W. Duschl.

Beckwith, S., Evans, N. J., II, Becklin, E. E., and Neugebauer, G. 1976, *Ap. J.*, **208**, 390.

Beckwith, S., Persson, S. E., Neugebauer, G., and Becklin, E. E. 1978, *Ap. J.*, **223**, 464.

Beckwith, S. V. W. and Sargent, A. I. 1992, in *Protostars and Planets III*, eds. E. H. Levy and J. I. Lunine, (U. of Arizona Press:Tucson), p. 521.

Beckwith, S. V. W., Sargent, A. I., Chini, R., and Güsten, R. 1990, *Astron. J.*, **99**, 924.

Beichman, C. A. 1987, *Ann. Rev. Astron. Ap.*, **25**, 521.

Bertout, C. 1989, *Ann. Rev. Astron. Ap.*, **27**, 351.

Bessell, M. S. and Brett, J. M. 1988, *Pub. A.S.P.*, **100**, 1134.

Bohren, C. F. and Huffman, D. R. 1983, *Absorption and Scattering of Light by Small Particles* (Wiley: New York).

Cohen, M. and Kuhi, L. V. 1979, *Ap. J. Suppl.*, **41**, 743.

De Zotti, G., Danese, L., Toffolatti, L., and Franceschini, A. 1990, in *Galactic and Extragalactic Background Radiation*, ed. S. Boyer and C. Leinert, (Springer-Verlag:Heidelberg), p. 333.

Edwards, S., Ray, T., and Mundt, R. 1992, in *Protostars and Planets III*, eds. E. H. Levy and J. I. Lunine, (U. of Arizona Press:Tucson), p. 567.

Fowler, A. M. and Gatley, I. 1990, *Ap. J. Lett.*, **353**, L33.

Franceschini, A., Toffolatti, L., Mazzei, P., Danese, L, and De Zotti, G. 1991, *Astron. Ap. Suppl.*, **89**, 285.

Genzel, R. 1992, in *The Galactic Interstellar Medium*, eds. W. B. Burton, B. Elmegreen, and R. Genzel (Springer-Verlag:Berlin).

Ghez, A., Neugebauer, G., and Matthews, K. 1993, *Astron. J.*, in press.

Hall, D. N. B., Ridgway, S. T., Bell, E. A., and Yarborough, J. M. 1979, *Proc. Soc. Photo-Opt. Instrum. Eng.*, **172**, 121.

Harvey, P. M., Campbell, M. F., and Hoffmann, W. F. 1977, *Ap. J.*, **215**, 151.

Herbig, G. H. 1982, in *Symposium on the Orion Nebula*, eds. A. E. Glassgold, P. J. Huggins, and E. L. Schucking, *Ann. N.Y. Acad. Sci.*, **395**, 64.

Herbst, T. M., Beckwith, S. V. W., Birk, C., Hippler, S., McCaughrean, M. J., Mannucci, F., and Wolf, J. 1993, in *Infrared Detectors and Instrumentation*, SPIE Conference Series #1946.

Herbst, T. M., Graham, J. R., Beckwith, S. V. W., Tsutsui, K., Soifer, B. T., and Matthews, K. 1990, *Astron. J.*, **99**, 1773.

Herzberg, G. 1950, *Spectra of Diatomic Molecules*, (Van Nostrand Reinhold:New York).

Jenkins, F. A. and White, H. E. 1976, *Fundamentals of Optics*, (McGraw-Hill:New York).

Joyce, R. R. 1992, in *Astronomical CCD Observing and Reduction Techniques*, ed S. B. Howell, ASP Conference Series, **23**.

Käufl, Bouchet, P., Van Dijsseldonk, A., and Weilenmann, U. 1991, *Experimental Astronomy 2*, p. 115 (Netherlands:Kluwer).

Knacke, R. F. and Young, E. T. 1981, *Ap. J. Lett.*, **249**, L65.

Kwan, J. and Scoville, N. Z. 1976, *Ap. J. Lett.*, **210**, L39.

Lada, C. J. 1985, *Ann. Rev. Astron. Ap.*, **23**, 267.

Lada, E. A., DePoy, D. L., Evans, N. J., II, and Gatley, I. 1991, *Ap. J.*, **371**, 171.

Larson, R. B. 1973, *Ann. Rev. Astron. Ap.*, **11**, 219.

Leighton, R. B. 1959, *Principles of Modern Physics*, (McGraw-Hill:New York).

Leinert, Ch., Zinnecker, H., Weitzel, N., Christou, J., Ridgway, S. T., Jameson, R., Haas, M., and Lenzen, R. 1993, *Astron. Ap.*, in press.

Low, F. J., Beintema, D. A., Gautier, T. N., Gillett, F. C., Beichman,C. A., Neugebauer, G., Young, E., Aumann, H. H., Boggess, N., Emerson, J. P., Habing, H. J., Hauser, M. G., Houck, J. R., Rowan-Robinson, M., Soifer, B. T., Walker, R. G., and Wesselius, P. R. 1984, *Ap. J. Lett.*, **278**, L19.

Maihara, T., Iwamuro, F., Yamashita, T., Hall, D. N. B., Cowie, L. L., Tokunaga, A. T., and Pickles, A. 1993a, *PASP*, September issue.

Maihara, T., Iwamuro, F., Yamashita, T., Hall, D. N. B., Cowie, L. L., Tokunaga, A. T., and Pickles, A. 1993b, in *Infrared Detectors and Instrumentation*, SPIE Conference Series #1946.

Matsumoto, T., Akiba, M., and Murakami, H. 1988, *Ap. J.*, **332**, 575.
McCaughrean, M. J. 1988, *The Astronomical Application of Infrared Array Detectors*, Ph. D. Thesis, University of Edinburgh.
Mundt, R. 1988, in *Formation and Evolution of Low Mass Stars*, eds. A. K. Dupree and M. T. V. T. Lago, (Kluwer:Drodrecht), p. 257.
Nadeau, D., Geballe, T. R., and Neugebauer, G. 1982, *Ap. J.*, **253**, 149.
Palla, F. and Stahler, S. W. 1990, *Ap. J. Lett.*, **360**, L47.
Puget, J. L. and Léger, A. 1989, *Ann. Rev. Astron. Ap.*, **27**, 161.
Ramsay, S. K., Mountain, C. M., and Geballe, T. R. 1992, *MNRAS*, **259**, 751.
Rieke, G. H. and Lebofsky, M. J. 1985, *Ap. J.*, **288**, 618.
Scoville, N. Z., Hall, D. N. B., Kleinmann, S. G., and Ridgway, S. T. 1982, *Ap. J.*, **253**, 136.
Scoville, N. Z., Kleinmann, S. G., Hall, D. N. B., and Ridgway, S. T. 1983, *Ap. J.*, **275**, 201.
Sellgren, K. 1981, *Ap. J.*, **245**, 138.
Sellgren, K. 1984, *Ap. J.*, **277**, 623.
Shu, F. H., Adams, F. C., and Lizano, S. 1987, *Ann. Rev. Astron. Ap.*, **25**, 23.
Shu, F., Najita, J., Galli, D., Ostriker, E., and Lizano, S. 1992, in *Protostars and Planets III*, eds. E. H. Levy and J. I. Lunine, (U. of Arizona Press:Tucson), p. 3.
Simon, T. and Dyck, H. M. 1977, *Astron. J.*, **82**, 725.
Soifer, B. T., Puetter, R. C., Russell, R. W., Willner, S. P., Harvey, P. M., and Gillett, F. C., *Ap. J. Lett.*, **232**, L53.
Soifer, B. T., Neugebauer, G., Matthews, K., Lawrence, C., and Mazzarella, J. 1992, *Ap. J. Lett.*, **399**, L55.
Spitzer, L., Jr. 1978, *Physical Processes in the Interstellar Medium*, (Wiley: New York).
Stahler, S. W., Shu, F. H., and Taam, R. E. 1980, *Ap. J.*, **241**, 637.
Strom, K. M., Strom, S. E. Edwards, S., Cabrit, S., and Skrutskie, M. F. 1989, *Astron. J.*, **97**:1451.
Telesco, C. M. 1988, *Ann. Rev. Astron. Ap.*, **26**, 343.
Thompson, D., Djorgovski, S. J., and Beckwith, S. V. W. 1993, *Astron. J.*, submitted.
Townes, C. H. and Schalow, A. L. 1975, *Microwave Spectroscopy*, (Dover:New York).
Tyson, J. A. and Seitzer, P. 1988, *Ap. J.*, **335**, 552.
Vincent, J. D. 1990, *Fundamentals of Infrared Detector Operation and Testing*, (Wiley: New York).
Werner, M. W., Becklin, E. E., Gatley, I., Matthews, K., Neugebauer, and Wynn-Williams, C. G. 1979, *MNRAS*, **188**, 463.
Willner, S. P., Gillett, F. C., Herter, T. L., Jones, B., Krassner, J., Merrill, K. M., Pipher, J. L., Puetter, R. C., Rudy, R. J., Russell, R. W., and Soifer, B. T. 1982, *Ap. J.*, **253**, 174.
Wolf, W. L. and Zissis, G. J. 1989, *The Infrared Handbook*, (Office of Naval Research:Washington).
Wynn-Williams, C. G. 1982, *Ann. Rev. Astron. Ap.*, **20**, 597.
Zuckerman, B., Kuiper, T. B. H., and Rodriguez-Kuiper, E. N. 1976, *Ap. J. Lett.*, **209**, L137.

High Spatial Resolution Infrared Observations — Principles, Methods, Results

Christoph Leinert

Max-Planck-Institut für Astronomie, Königstuhl 17, D-69117 Heidelberg, Germany

1 Introduction

High spatial resolution in a physical sense means a resolution sufficient to resolve the structures of interest in a given object; studying the zodiacal light or active galactic nuclei thus will result in vastly differing definitions. Intuitively one would require a spatial resolution to go beyond the seeing limit to be considered high. For this survey I follow a formal approach and call high spatial resolution what corresponds to the diffraction limit of a 3-4 m telescope or better. In this context I limit the discussions mostly to the near-infrared range from 1.5 μm to 5 μm. At 10 μm the diffraction limit of a 4 m telescope is no longer smaller than the seeing disk.

The aim of high resolution observations is to detect new structures, like, e.g. an envelope or disk around a star or a hitherto unknown companion, to allow correct interpretation of an object, e.g. as a single star or as a binary, and generally to provide highly resolved pictures for a detailed study of astronomical sources and their environments.

Historically, high spatial resolution infrared observations started on November 20, 1970 with *lunar occultation observations* of IRC+10216, which together with a second observation two months later determined the effective size of the dust shell around this carbon star from 2.2 μm to 10 μm (Toombs et al. 1971). This technique, which is able to give spatial resolutions of up to 1 milliarcsec (mas), continues being used, but it is limited to individual rare occultation events and to sources near the ecliptic. *Near-infrared speckle interferometry* in its typical one-dimensional slit scanning mode was introduced eight years later (Sibille, Chelli and Léna 1979), while Selby et al. (1979) developed a method with a different scanning technique. With its spatial resolution of $0''.1 - 0''.2$, near-infrared speckle interferometry had a considerable number of astrophysical applications from hot spots on Io to the circumstellar environment of pre-main sequence stars to the central regions of nearby Seyfert galaxies. With the availability now of infrared array detectors two-dimensional observing has become standard practice also for near-infrared speckle interferometry. Infrared *Michelson interferometry* developed slowly . The first application by McCarthy and Low (1975) used movable apertures in the exit pupil of the 1.5 m telescope on Catalina Observatory and of the 2.3 m telescope at Kitt Peak. Around the same time Sutton et al. (1977) introduced heterodyne interferometry at 11μm with two closely spaced independent telescopes. True near-infrared long-baseline measurements were first performed with the I2T interferometer at CERGA for the determination of photospheric diameters and effective temperatures of evolved stars, with an accuracy similar to that obtained by lunar occultation observations (Di Benedetto and Conti 1983). A few more small two-telescope interferometers are working in the infrared and plans for future interferometers abound, but still the first interferometer with imaging capabilities, i.e. involving simultaneous measurements with at least three telescopes, only now enters the demonstration phase (Baldwin 1993). This imaging ability will be an essential precondition for the success of the large interferometers associated with the Very Large Telescope and the Keck telescopes which will go into operation around the year 2000. Meanwhile *adaptive optics* has been introduced with the first successful observations on October 12, 1989 on the 1.5 m telescope at the Observatoire de Haute-Provence (Merkle et al. 1989a). This seems to be the first high spatial resolution method which finds wide acceptance in the astronomical community, even right from its beginnings, because it promises to allow direct imaging at high resolution and with better sensitivity than the previous techniques.

In §4–7 of this review I will discuss these four methods in their chronological order, describe their principles, strengths and weaknesses and summarize the main results obtained with them. As a common basis to all techniques §2 deals with imaging from the point of view of high spatial resolution observations and §3 with the characterization and the effects of the turbulent atmosphere. Recent reviews are available for almost every topic covered in this overview, e.g. on imaging (Léna 1989, Roddier 1988), atmosphere (Roddier 1981), speckle interferometry (Haas 1991, Perrier 1988, Roddier 1988, Weigelt 1989), Michelson interferometry (Roddier and Léna 1984 on principles, Ridgway 1992 on results), adaptive optics (Roddier 1992, Beckers 1993). The present review has the purpose to provide an introduction to these various themes. For more detailed information I recommend to start, apart from the mentioned reviews and the references given in the text, with the conference proceedings edited by Merkle (1988a) and Beckers and Merkle (1992a).

2 Imaging

This section is not meant as replacement for the excellent overview on imaging in astronomy by Léna (1989). Rather it emphasizes those relations which help to understand principles and methods of the measurements discussed in the later sections. It turns out that Fourier transforms are an indispensable tool.

The final resolution limit of a telescope is given by diffraction. For a circular aperture of radius D the diffraction limit usually is given as

$$\rho_0 = 1.22\lambda/D \qquad [\mathrm{rad}], \tag{1}$$

which is the angular radius of the first dark ring of the Fraunhofer diffraction pattern. This pattern would appear at infinite distance but is brought to the focal plane by the telescope optics. The diffraction peak enclosed in the dark ring is called the *Airy disk*. It contains 84% of the total energy in a point source image. Imperfect images generally are broader and have a lower peak brightness. One way to measure the quality of an imaging device is by the *Strehl ratio*

$$Z = I/I_0, \tag{2}$$

where I is the actual peak brightness in a point source image and I_0 the corresponding value in the diffraction limited image. The *Rayleigh-Strehl criterion*

$$Z \geq 0.80 \tag{3}$$

defines the quality of an image from the viewpoint of wave optics. If the degradation in the image is due to spherical aberration, this criterion corresponds to maximum allowed wavefront errors of $\pm\lambda/4$. In practice often the quality of a telescope is measured by the smallness of the circle which contains 80% of the light of a point source image. For the New Technology Telescope of ESO the corresponding specification was $\rho = 0.15''$. Below we will see that a strictly physical definition of the diffraction limit leads to

$$\rho_{lim} = \lambda/D \qquad [\text{rad}]. \tag{4}$$

2.1 The Four Planes of the Imaging Process

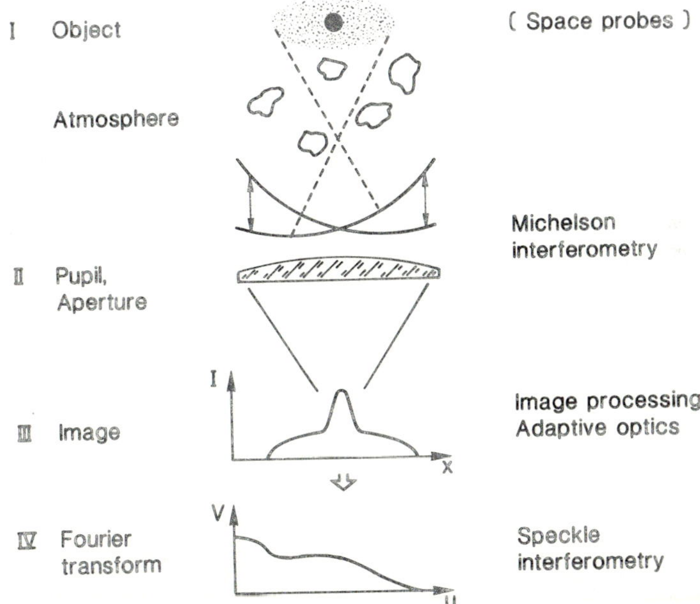

Fig. 1. The four imaging planes

A sketch of the four planes is given in Fig 1.

- The *object* usually is very distant. It is characterised by its brightness distribution $O(\alpha, \delta)$ $[Wm^{-2}sr^{-1}\mu m^{-1}]$. Only a few objects in the planetary system can be visited by spacecraft and be studied in situ.
- The *aperture or pupil plane* contains the information on the object in terms of phase differences as function of the pupil coordinates ξ and η between the wavefronts originating from different parts of the object. These differences can be sensed, e.g. with a Michelson interferometer.
- The light distribution of the *image*, $I(x, y)$ $[Wm^{-2}sr^{-1}\mu m^{-1}]$ is an estimate of the object brightness distribution.
- The *Fourier transform* $\tilde{I}(u, v)$ of the image may be considered as a representation of the image in an independent additional plane. This plane is called the *Fourier plane* or *uv plane.*

In the following I discuss the main relationships between these planes.

2.2 Image and Fourier Plane – the Fourier Transform

The conventional view of an image is that it consists of an ensemble of point sources of different brightnesses. The alternative view is that its intensity distribution is constituted by the superposition of sine waves of different wavelengths and amplitudes, which may be expressed in the two dimensional case as

$$I(x, y) = \int\int_{-\infty}^{+\infty} \tilde{I}(u, v)\, e^{+2\pi i(ux+vy)}\, dudv. \tag{5}$$

Here u and v are the *spatial frequencies* in the x and y direction which measure the number of wavelengths fitting into a unit length or unit angle. This is analogous to wavenumbers in spectroscopy. The unit of spatial frequency used in astronomical measurements is usually taken as [arcsec^{-1}] or [(image size)$^{-1}$]. $\tilde{I}(u, v)$ is called the Fourier transform of I(x,y) and given by

$$\tilde{I}(u, v) = \int\int_{-\infty}^{+\infty} I(x, y)\, e^{-2\pi i(ux+vy)}\, dxdy. \tag{6}$$

The Fourier transform, alternatively written FT(I(x,y)), is a complex quantity, $\tilde{I}(u, v) = |\tilde{I}(u, v)|e^{i\phi(u,v)}$, that is also known as the visibility for reasons which will become clear below. The amplitudes and phases of the sine waves which make up the image are equal to the modulus $|\tilde{I}(u, v)|$ and the phase ϕ of the Fourier transform. Images are real quantities, their Fourier transforms therefore are hermitian:

$$|\tilde{I}(u, v)| = |\tilde{I}(-u, -v)| \quad ; \quad \phi(u, v) = -\phi(-u, -v). \tag{7}$$

In particular $\tilde{I}(0,0)$ is the integral over I(x,y) and $\phi(0,0) = 0.0$. Symmetric images have zero phase at all spatial frequencies. Generally it is the phase, and not the modulus, what dominates the appearance of the image (Figure 2). However,

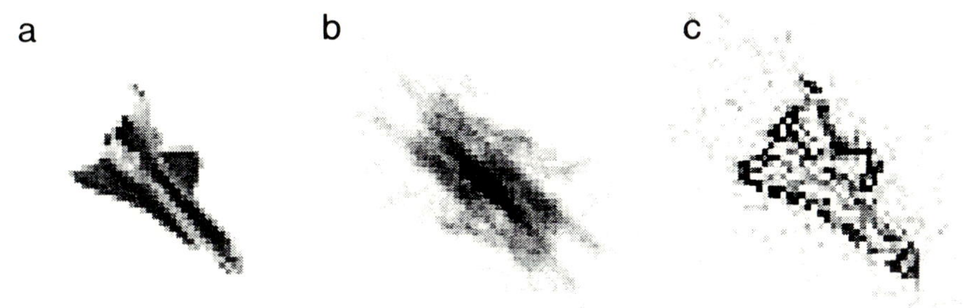

Fig. 2. Example for the importance of phase information for the image. a – digital image of a space shuttle. b – reconstruction from the modulus of the Fourier transform, phases assumed to be zero. c – reconstruction from the phases, modulus set equal to 1.0 for all spatial frequencies. Created by A. Glindemann.

subtraction of a linear term $\lambda u + \mu v$ from the phase simply corresponds to shifting the image by $\lambda/2\pi, \mu/2\pi$.

Often $\tilde{I}(0,0)$ is normalized to 1.0. The Fourier transforms of a few simple functions then read (in one dimension) for

– a point source $\quad \tilde{I}(u) = 1.0$

– a uniform disk of diameter d $\quad \tilde{I}(u) = \frac{sin\pi du}{\pi du}$

– a Gaussian curve $\quad \tilde{I}(u) = e^{-2\pi^2\sigma^2u^2} \qquad (8)$

– a double star with separation d and brightness ratio b/a $\quad |\tilde{I}(u)| = \sqrt{\frac{(a-b)^2}{(a+b)^2} + \frac{4ab}{(a+b)^2}cos^2(\pi du)}$.

("cosine-like curve")

In the case of two-dimensional Fourier transforms we note that

- the 2D transform along any line through the origin is equal to the 1D Fourier transform of the image, after it has been projected onto that line
- in particular for a double star the value of the 2D Fourier transform does not depend on the position along any line perpendicular to the separation vector of the binary components. This gives both the modulus and the phase of the visibility the typical appearance of parallel stripes.

Both relations can most readily be verified by considering the special case of a line along the x axis or a binary with both components on it.

2.3 Image and Object – a Convolution Relation

Imperfections in the optics and inhomogeneities in the atmosphere deteriorate the image of a point source. Instead of the diffraction pattern centered at position x_0, y_0 a light distribution $P(x, y; x_0, y_0)$ around the point x_0, y_0 is observed. This *point spread function* usually has the same shape over a certain field of view, which then results in the functional form $P(x - x_0, y - y_0)$. As far as this condition is fulfilled, we speak of *isoplanaticity*. The field on the sky over which the point spread functions are well correlated is called the *isoplanatic patch*. Within this field the image is obtained by replacing each point of the image with a point spread function weighted according to the brightness of this point, and by incoherent addition of these functions. Mathematically this procedure is described by the convolution

$$I(x,y) = \iint_{-\infty}^{+\infty} O(x',y')P(x-x',y-y')dx'dy', \tag{9}$$

or, in shorter notation

$$I(x,y) = O(x,y) * P(x,y), \tag{10}$$

where O(x,y) is the object as it would appear in the xy plane under diffraction-free perfect imaging. In the Fourier plane, according to the convolution theorem this transforms to the simple relation

$$\tilde{I}(u,v) = \tilde{O}(u,v) \cdot \tilde{P}(u,v). \tag{11}$$

Here, imaging is a process of linear filtering.

2.4 Image and Pupil Plane – a Fourier Transform Relation

According to the Fresnel-Huygens principle elementary waves originate in the pupil with amplitudes equal to the pupil amplitude distribution $G(\xi, \eta)$. Their interference in the focal plane gives rise to the image. Kirchoff's scalar theory of diffraction evaluates this interference explicitly and shows that the amplitude distribution A(x,y) in the image plane is related to the field in the pupil by a Fourier transform (Born and Wolf 1970)

$$A(x,y) = const. \times \iint_{aperture} G(\xi,\eta)e^{-2\pi i(\frac{x}{f}\frac{\xi}{\lambda} + \frac{y}{f}\frac{\eta}{\lambda})}d\xi d\eta \ , \tag{12}$$

if the coordinates ξ,η in the pupil are measured in units of the wavelength and the coordinates in the image as angles (f is the focal length of the telescope). The image is by definition an intensity,

$$I(x,y) = |A(x,y)|^2. \qquad [Wm^{-2}\mu m^{-1}] \tag{13}$$

Changing shape or transmission of the pupil therefore changes the image. Mostly this *apodisation* (Jacquinot and Roizen-Dossier 1964) is done with the aim to sharpen the diffraction peak or to reduce the side lobes, always at the expense of the other quantity. Special apodisation has been proposed for future attempts to detect extrasolar planets by infrared imaging (Angel et al. 1986). In most telescopes the cross of blades carrying the secondary mirror introduces sharp diffraction spikes perpendicular to the long dimension of the blades. Depending on the orientation of the cross, these spikes can interfere with observations of elongated structures as , e.g.,the jet of the quasar 3C273, which runs at a privileged position angle of $\approx 225°$. They can be effectively reduced by adding an irregular sawtooth mask to the arms of the cross (Hardie 1964).

2.5 Pupil Plane and uv Plane – the Michelson Interferometer

The Fourier transform of the Fourier transform of an arbitrary function f(x,y) is equal to f(-x,-y). Equation (12), which relates the Fourier transform of the field distribution in the pupil to the image, therefore suggests a close relation between the amplitude distribution in the aperture and the Fourier transform of the object or image. This relation exists and is the basis for the function of the Michelson interferometer. Figure 3 sketches the interference obtained with two small apertures separated by a distance d. This is the geometry of a Michelson interferometer. In first approximation this is a one-dimensional problem. For a point x in the image the interference between the waves ψ_1 and ψ_2 emanating from the two apertures depends on the pathlength difference $\Delta s(x) = s_1(x) - s_2(x)$, where $s_1(x)$ and $s_2(x)$ are the pathlengths from the apertures to the point x. By the law of Pythagoras we have for d$\ll$f and x$\ll$d

$$\Delta s(x) = \lambda \cdot \frac{x/f}{\lambda/d}. \tag{14}$$

This means that the fringe separation is $\Delta x/f = \lambda/\mathrm{d}$ [sr] and the spatial frequency of the fringe sine wave is d/λ [sr^{-1}]. Since the two apertures may be thought of as parts of a larger aperture with diameter d, the interference fringes represent a partial image, in which we see just one of the manifold of sine waves, the superposition of which makes up the image. It can easily be shown from the relations in Figure 3 that the visibility of the fringes, i.e. the true optical visibility

$$V = (I_{max} - I_{min})/(I_{max} + I_{min}) \tag{15}$$

as a function of aperture separation $\boldsymbol{d}$ and their phase Δ with respect to the zero position is equal to the modulus and phase of the Fourier transform $\tilde{I}(\boldsymbol{d}/\lambda)$ at spatial frequency $\boldsymbol{d}/\lambda$ (here, as in the following, quantities set in ***bold*** type designate vectors, e.g. $\boldsymbol{u}$ stands for the vector (u, v)). This coincidence is the reason that the terms "visibility" and "Fourier transform of the brightness distribution" are usually used interchangeably. However, often by "visibility" one means the modulus only of the complex visibility (6). This use of the word is not perfectly correct, but at least it corresponds to the original physical meaning of (15) that fringes can be seen.

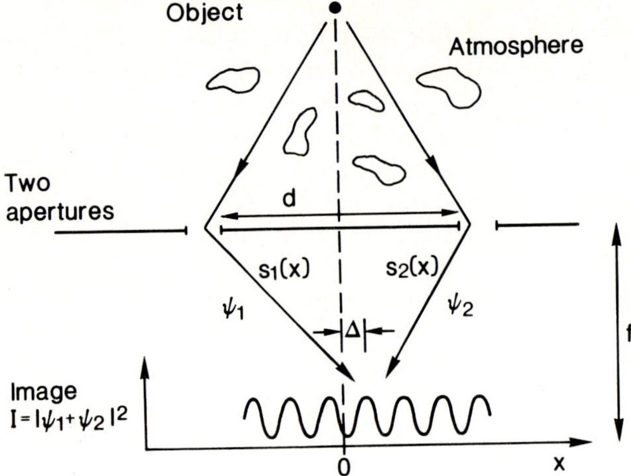

Fig. 3. Principle of the Michelson interferometer: imaging through two spatially separate apertures. The fringe pattern in the image plane results from the superposition of the two light beams ψ_1 and ψ_2. The fringe spacing is $(\lambda/\mathrm{d})\cdot\mathrm{f}$. The offset Δ of the first maximum, expressed e.g. as fraction of the fringe pattern wavelength, is the phase of the fringe pattern. The fringe pattern represents a partial image of the object (see text).

The Michelson interferometer thus acts as a spatial filter in the pupil plane, selecting the particular Fourier component with spatial frequency $\boldsymbol{d}/\lambda$, and the filtered frequency may be changed by changing the separation of the apertures.

2.6 The Telescope as Interferometer – Spatial Resolution and MTF

We may consider a telescope with a circular aperture as a combination of many Michelson interferometers where every subpupil i interferes with every subpupil j (Figure 4). The *Airy pattern* then results as superposition of all possible fringe systems $|\psi_i + \psi_j|^2$, with the interference terms $\psi_i \psi_j^*$, because

$$I(x, y) = |\sum_{k=1}^{N} \psi_k|^2 = \sum_{k=1}^{N} |\psi_k|^2 + \sum_{i=1}^{N} \sum_{j \neq i,=1}^{N} \psi_i \psi_j^* \ . \tag{16}$$

In this concept the limiting *resolution* is given by the maximum spatial frequency of these fringe systems. It is called the *cut-off frequency*,

$$u_c = d/\lambda. \tag{17}$$

This means that structures of angular size λ/d are just resolved and provides the more general definition of resolution mentioned in the beginning.

The efficiency with which a certain spatial frequency is transmitted by the telescope optics is measured by the **M***odulation* **T***ransfer* **F***unction* **MTF($\boldsymbol{u}$)**. This efficiency is given by the number of subapertures for which the second

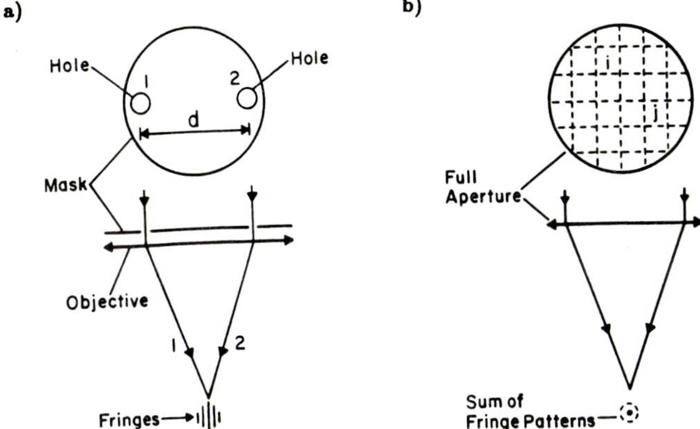

Fig. 4. Imaging – a) through a mask with two holes – b) through a full aperture. From Roddier (1987).

subaperture with separation $\lambda \boldsymbol{u}$ still is contained in the pupil. Figure 5 demonstrates that this number is equal to the common cross section of the aperture

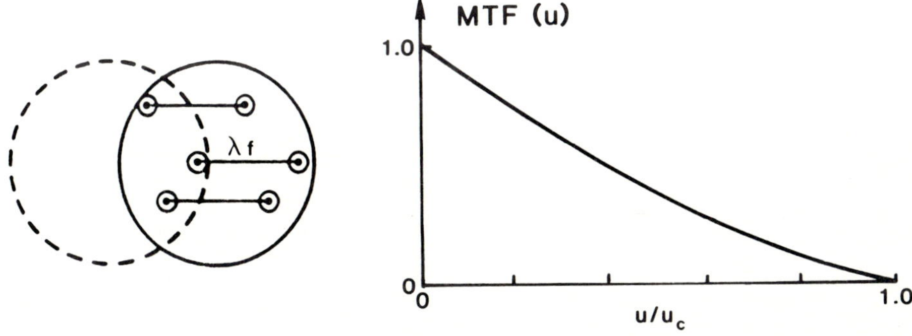

Fig. 5. The modulation transfer function MTF(u), visualized as the number of subaperture pairs with separation λu is given by the overlap area of the shifted pupils. The resulting functional form is shown to the right. Adapted from Roddier (1987).

with its shifted image, or mathematically to the autocorrelation

$$MTF(\boldsymbol{u}) = \int\!\!\int_{pupil} G(\xi, \eta) \cdot G^*(\xi + \lambda u, \eta + \lambda v) d\xi d\eta, \tag{18}$$

or in abbreviated notation

$$MTF(\boldsymbol{u}) = G(\boldsymbol{\xi}) \otimes G^*(\boldsymbol{\xi}) = G \otimes G^*(\lambda \boldsymbol{u}) \tag{19}$$

where again $G(\xi, \eta)$ is the amplitude distribution in the entrance pupil. Normalized to 1.0 for very large structures ($\boldsymbol{u}$=0), the MTF for a circular aperture with diameter D,

$$MTF(\boldsymbol{u}) = \frac{2}{\pi}[\arccos(\frac{\lambda \boldsymbol{u}}{D}) - \frac{\lambda \boldsymbol{u}}{D} \cdot \sqrt{1-(\frac{\lambda \boldsymbol{u}}{D})^2}\,] \tag{20}$$

almost linearly decreases to zero at the cut-off frequency. Thus diffraction limited imaging already contains a considerable degree of apodising, i.e. filtering of high spatial frequencies. In particular, optical images are *band-limited*, i.e. $\tilde{I}(u,v) = 0.0$ for spatial frequencies $|\boldsymbol{u}| > u_c$. A noteworthy additional relation is that the Fourier transform of the MTF (or, more precisely the OTF, see below) as defined in (18) equals the image brightness distribution,

$$FT(MTF(\boldsymbol{u}) = FT(G \otimes G^*(\lambda \boldsymbol{u})) = \tilde{G} \cdot \tilde{G}^* = |A(x,y)|^2 = I(x,y) \ . \tag{21}$$

In verifying (21) we have used the relations $FT^{-1}(f) = \widetilde{f^*}^*$, $\widetilde{f^*}(\boldsymbol{u}) = \tilde{f}^*(\text{-}\boldsymbol{u})$ and $\widetilde{f \otimes g^*} = \tilde{f} \cdot \tilde{g}^*$. In general, the autocorrelation in (18) is a complex quantity, called the *optical transfer function*, which also specifies the relative phase shift as a function of spatial frequency, the MTF($\boldsymbol{u}$) being the modulus of this general transfer function. For easier presentation I have neglected this distinction above.

2.7 Aperture Synthesis – the Interferometer as a Telescope

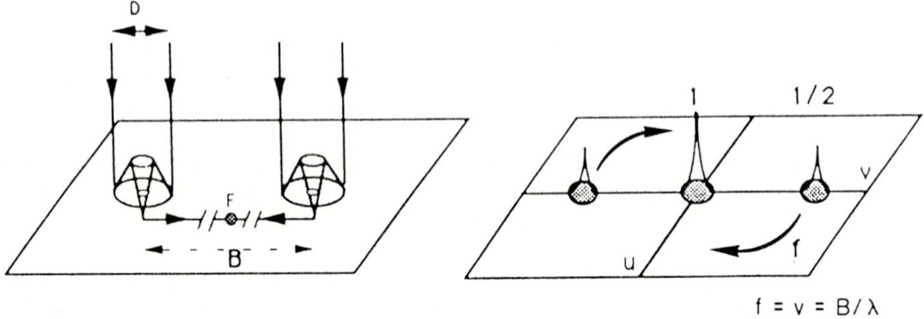

Fig. 6. Geometry of a two-telescope interferometer (left) and the resulting MTF(u) and uv-plane coverage (right). F is the common focus. Adapted from Léna (1989).

Two telescopes of an interferometer, separated by a baseline $\boldsymbol{B}$, result in an MTF($\boldsymbol{u}$) which is different from zero only near the origin – corresponding to coarse structures resolvable with the individual telescopes – and at spatial frequencies $\boldsymbol{u} = \boldsymbol{B}/\lambda$ (Figure 6). For stars out of the zenith the telescopes have a projected separation of $\boldsymbol{d} = \boldsymbol{B} \sin\theta$, where θ is the angle between the baseline and the radius vector to the star. Since θ changes with the diurnal motion of the celestial sphere, each fixed baseline allows one to measure a number of spatial frequencies along a curved track in the uv plane. Figure 7 shows the coverage of the uv plane obtained with four telescopes, which span six baselines, in an observing time of $\approx$ 8 hours. Since the uv plane approximately corresponds to the entrance pupil, as discussed above, this sparse sampling of spatial frequencies, which is not untypical for VLBI radio interferometry, synthesizes a telescope with very incomplete aperture. The resulting image therefore is strongly perturbed. It is called a *dirty map*. The same kind of image for a point source is called

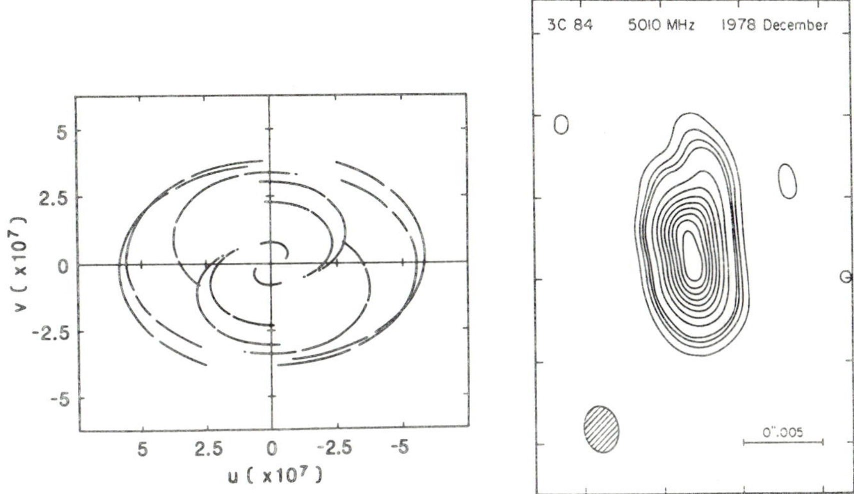

Fig. 7. Left: tracks in the uv-plane obtained from the six baselines of a four-telescope VLBI network during about eight hours of observing. Right: the image of the quasar 3C84 resulting from these observations. From Pearson and Readhead (1981).

a *dirty beam*. Deconvolving the dirty map by the dirty beam with procedures like CLEAN (Högbom 1974) is necessary to obtain acceptable images. As Figure 7 shows, incomplete coverage of the uv plane leads to coarse images. Highly structured images such as those from the VLA radio interferometer are based on a much superior coverage of the uv plane resulting from a large number of telescopes and baselines. If the object is not expected to vary, measurements of different baselines on different days may be combined in the aperture synthesis. The example of Figure 7 may be representative of what optical and near-infrared interferometry is expected to achieve during the coming decade.

2.8 Digital Images and Discrete Fourier Transforms

Near-infrared array detectors of up to 256x256 pixels are now in operation at many telescopes. They provide sampling of the image at an equidistant series of points. Since optical images are band-limited, i.e. $\tilde{I}(\boldsymbol{u}) = 0$ for $|\boldsymbol{u}| > u_c$,

Shannon's sampling theorem applies: it is sufficient to sample the image twice per the shortest spatial wavelength present in it. Vice versa, if an image is sampled at a rate $1/\Delta x$, then the highest spatial frequency which can be resolved, the so-called *Nyquist frequency* f_N, equals $1/2\Delta x$. If the sampling frequency is too low, making $f_N < u_c$, then frequencies higher than f_N can produce spurious signals at the measured frequencies below f_N. This effect, called *aliasing*, is best discussed in the Fourier domain.

The discrete data of digital images require a *discrete Fourier transform*, which is defined in analogy to (6) as

$$\tilde{I}(\frac{n}{NP}) = \sum_{k=0}^{N-1} I(kP) e^{-2\pi i \frac{n}{NP} \cdot kP} \qquad (n = 0 ... N-1), \qquad (22)$$

where P is the pixel size, N the number of pixels in one dimension and therefore NP the size of the image. For convenience and clarity the one-dimensional form is shown. An excellent introduction to the properties of discrete Fourier transforms is given by Brigham (1982). To perform the actual calculation, specialized fast computer codes exist, called Fast Fourier Transforms (FFT). The number of discrete values for moduli and phases needed to specify a discrete Fourier transform is equal to the number of pixels in the image, both in the one- and the two-dimensional case.

The values of a discrete Fourier transform are equal to the corresponding values of the ordinary Fourier transform, if

- the image is periodic
- $\tilde{I}(u, v)$ is band-limited with an upper limit in spatial frequency of u_c
- the spatial sampling frequency is $\geq 2u_c$, i.e. $f_N \geq u_c$

We already know that optical images are band-limited, but the first condition also is fulfilled automatically: it may be shown that the discrete Fourier transform by its formal definition is the transform of the periodically repeated image. Similarly, the discrete Fourier transform also repeats itself periodically at spatial frequency intervals of Δu or Δv of $1/\Delta x = 2f_N$. If $f_N < u_c$, neighbouring periodic repetitions of the digital Fourier transform overlap. The resulting mixing of different frequencies gives rise to the aliasing distortions of the Fourier transform spectrum.

3 The Turbulent Atmosphere

For a thorough, informative description I recommend the excellent review by Roddier (1981).

Atmospheric turbulence follows a Kolmogorov spectrum (given in (27) below). The physical reasoning behind this statement is, that turbulence in the streaming air first starts in very large cells, the size of which is called the *outer scale of turbulence*, and which is of the order of m to km. These motions then decay into smaller and smaller patterns, until they are dissipated. This lower limit of turbulent motion is called the *inner scale of turbulence* and has the size of mm. Qualitatively, if the energy E flows continuously from the largest scale L_0 to smaller scales l, and if the typical lifetime of a turbulence element of size l is taken as equal to the crossing time $t(l) = l/v$, one expects the relation

$$\frac{E(L_0)}{t(L_0)} = \frac{E(l)}{t(l)} = \frac{\frac{1}{2}mv^2}{l/v} = const. \quad \text{or} \quad v^2 \sim l^{2/3}. \tag{23}$$

The turbulent motions give rise to mixing of air of different temperatures, which induces fluctuations of density and hence refractive index. The difference in refractive index n between two points depends on their distance $l = |r_1 - r_2|$ in the same way as the turbulent velocity. Quantitatively this is measured by the *refractive index structure function*

$$D_n(r_1, r_2) = < [n(r_1) - n(r_2)]^2 > = C_N^2 |r_1 - r_2|^{2/3}, \tag{24}$$

where $<>$ designates a time average and C_N^2 is the *refractive index structure constant* describing the optical properties of a layer in the atmosphere at the time of the observations. As a result of the inhomogeneities in the atmosphere, the wavefront arriving on the ground shows fluctuations with a log-normal distribution, which means that $\log Ae^{i\phi}$ and therefore independently the phase ϕ and the logarithm of the amplitude A show a Gaussian distribution. The overwhelming part of the image distortion is due to the phase fluctuations. These are described by the *phase structure function*, obtained by integrating over the density fluctuations along the height of the atmosphere as

$$D_\phi(\xi) = < |\phi(r) - \phi(r + \xi)|^2 > = 2.91 \frac{k^2}{\cos z} \xi^{5/3} \int dh C_N^2(h), \tag{25}$$

where $k = 2\pi/\lambda$ and z is the zenith distance. The relation can be rewritten by introducing the *Fried parameter* r_0 as

$$D_\phi(\xi) = 6.88(\xi/r_0)^{5/3}. \tag{26}$$

The phase fluctuations thus are determined by the integral over the height distribution of the refractive index structure constant. Large contributions come from the lower atmosphere and from a layer at $\approx$ 10 km above ground (Figure 8). During a seeing campaign at La Silla, Chile (ESO VLT report N0.55,1987), 75% of the phase fluctuations were due to air in the height range 30m - 800m.

The two-dimensional power spectrum of the phase fluctuations as function of spatial frequency $\boldsymbol{f}$ is

$$P_\phi(\boldsymbol{f}) \sim f^{-11/3}, \tag{27}$$

while projected to one direction it is

$$P_\phi(\boldsymbol{u}) \sim u^{-8/3}. \tag{28}$$

The latter situation applies if one measures temporal fluctuations resulting from the wind driving the turbulent atmosphere over the telescope.

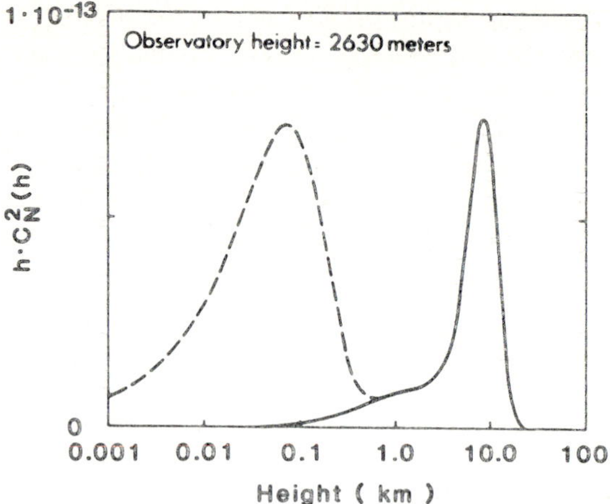

Fig. 8. Average height profile of C_N^2 for an observatory at an altitude of 2630 m (like the future VLT site at Cerro Paranal). Solid line: contribution due to the "free atmosphere", ignoring local seeing. Dashed line: contribution due to the lower atmospheric layers, which are most affected by the diurnal temperature cycle. The given values of C_N^2 correspond to a seeing of 0.63″ at 0.55 μm. In this plot equal ordinate values of the curves correspond to equal contributions per logarithmic height interval. From Beckers (1993).

The image degradation depends on the *coherence function*

$$C(\xi) = < \psi(x)\psi^*(x+\xi) > = e^{-\frac{1}{2}D_\phi(\xi)}, \tag{29}$$

where again $<>$ is the time average and $\psi(x)$ the field distribution at the position of the observer. The MTF is given – according to (18) – as an autocorrelation of $\psi(x)$ over the aperture of the telescope, which results for a long exposure image simply in the product of the telescope modulation transfer function $T(u)$ and the atmospheric attenuation according to (26) and (29),

$$MTF(u) = T(u) \cdot e^{-3.44(\frac{\lambda u}{r_0})^{5/3}}. \tag{30}$$

This nearly is a Gaussian profile (the exponent is 5/3 instead of 2). The deviation – a broadening of the wings of the image – has been confirmed experimentally (see Figure 2 in Roddier 1981).

The significance of the Fried parameter

$$r_0 = [\frac{const.}{\lambda^2}\frac{1}{\cos z}\int C_N^2 dh]^{-3/5} \tag{31}$$

is due to the fact that

- it allows us to describe the influence of the turbulent atmosphere by one single parameter

- it gives the size of coherent patches of the wavefront, in the sense that the phase fluctuation over a distance r_0 is $\sqrt{< |\phi(x) - \phi(x + r_0)|^2 >} \approx 0.8\pi$. For seeing of 1″ $r_0(550nm) \approx 10$cm, $r_0(2.2\mu m) \approx 60$cm.
- to a good approximation the image corresponds to the diffraction limit of a telescope with diameter r_0.
- it allows a simplified estimate of the *correlation time* t_0: if one assumes that frozen-in turbulence is carried over the telescope by a wind with speed v, the resulting estimated time scale is

$$\tau_0 \approx r_0/v, \tag{32}$$

which is of the order of 10 ms at 550 nm and 60 ms at 2.2μm.

Table 1 gives the dependence on Fried parameter r_0, wavelength λ and zenith distance z for several quantities of interest for imaging in the atmosphere. Because the refractive index of the air approaches 1.0 at longer wavelengths, the atmosphere is more well-behaved in the infrared than in the optical wavelength range. As a quality indicator for high resolution observations we take the number of photons which are available in one interferometric exposure (or in one speckle for speckle interferometry). This number is given by the product of coherent patch size, coherence time and allowable bandwidth (see §6.3 below), and it goes with the forth power of r_0 or, for constant seeing, with about the fifth power of the wavelength.

Table 1. Functional dependence of seeing-related quantities

Fried parameter r_0		$\sim \lambda^{6/5}$	$\sim (\cos z)^{3/5}$
Correlation time τ_0	$\approx r_0/v^a$	$\sim \lambda^{6/5}$	$\sim (\cos z)^{3/5}$
Allowable bandwidth $\Delta\lambda/\lambda$	$\approx r_0/D^b$	$\sim \lambda^{6/5}$	$\sim (\cos z)^{3/5}$
Isoplanatic patch size	$\approx r_0/h^c$	$\sim \lambda^{6/5}$	$\sim (\cos z)^{3/5}$
Seeing disk size $\approx \lambda/r_0$	$\sim r_0^{-1}$	$\sim \lambda^{-1/5}$	$\sim (\cos z)^{-3/5}$
Number of speckles^d	$\approx (D/r_0)^2$	$\sim \lambda^{-12/5}$	$\sim (\cos z)^{-6/5}$

^{a}wind speed ^{b} telescope diameter ^{c}height of representative turbulent layer
^{d} see §5

The important aspect of the theory of the turbulent atmosphere is that it is accurate enough in describing the behaviour of the atmosphere to allow predictions for the performance of new instruments. It thus is an indispensable tool for the development of complex new techniques like adaptive optics and truly imaging optical interferometers.

4 Lunar Occultations

A recent comprehensive review of this technique and its applications is not available. The interested reader is referred to Böhme (1978), Stecklum (1985), Warner

(1988) and, for a knowledgeable discussion of present status and future prospects of near-infrared lunar occultation measurements, to the review by Richichi (1993).

4.1 Principles

There is no basic difference between lunar occultations in the visible and the near-infrared, but infrared observations have become more common because the sky around the moon is less bright at longer wavelengths, because at the same time the modulations of the lightcurve are slower (see below), and because many of the interesting objects, such as red giants and young stars, are bright in the infrared. Lunar occultations are singular events which for the planning of observations can be predicted from a lunar ephemeris and the daily rotation of the earth under the assumption of a spherical lunar surface to better than ±5s.

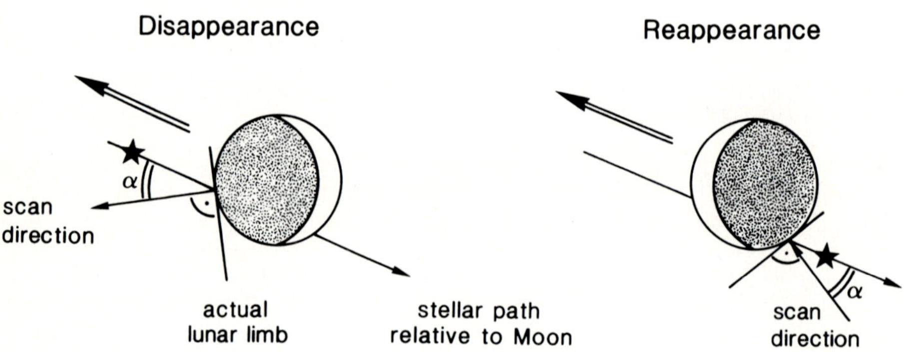

Fig. 9. Geometry of lunar occultation events. The double-lined arrow indicates the direction of lunar motion. For a circular lunar limb profile the scan direction would be equal to the position angle of the contact point (but reversed for a reappearance).

Seen from the observer, the moon moves across the astronomical object and occults it. This leads to two events: a *disappearance* on the forward moving – approximately eastern – limb and a *reappearance* on the western limb (Figure 9). Of these, up to now, only the event happening on the dark side of the moon can be observed. In both cases the occultation of the object occurs in a direction perpendicular to the local lunar limb. The occultation thus corresponds to a one-dimensional scan over the object. The angle between the lunar motion and the direction of this scan (α in Figure 9) is called the *contact angle*. The speed of the occultation then is given by

$$v_{occultation} = v_{moon} \cdot \cos\alpha \qquad [\mathrm{ms}^{-1} \quad \text{or} \quad \mathrm{arcsec\,s}^{-1}]. \tag{33}$$

The motion of the moon is ≈ 1 kms^{-1}, resp. ≈ 0.5 arcsec s^{-1}. Provided the velocity of the moon is known from an ephemeris and the speed of the occultation can be found from the observation, α and hence the local slope of the lunar limb can be determined from equation (33). The slopes are usually less than 10° with respect to the ideal lunar horizon.

The Earth's atmosphere plays no particular rôle in lunar occultation measurements, since the occultation occurs at the lunar limb in the vacuum of space. The telescope simply acts as light collector. Seeing effects therefore are unimportant, except for scintillation which adds unwanted additional brightness fluctuations.

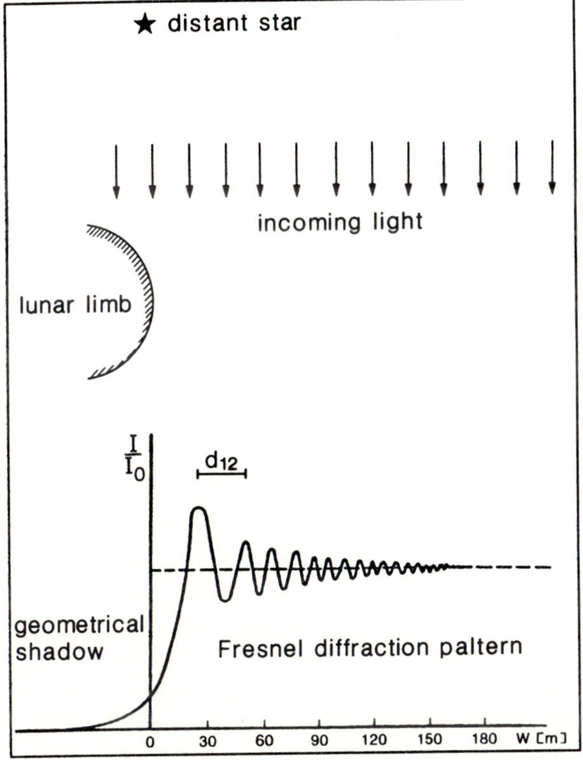

Fig. 10. Lunar occultation lightcurve. The figure shows a snapshot of the diffraction pattern at the edge of the lunar shadow. The linear scale below is given for $\lambda = 2.2$ μm; it scales like $\sqrt{\lambda}$. As the moon drives this pattern across the telescope at a speed of ≈ 1 kms^{-1}, a lightcurve of the same shape results.

Seen from outside, the moon carries its shadow over the telescope. The edge of the shadow is not sharp due to diffraction effects and shows the interference pattern of Fresnel diffraction at a half-plane (Born and Wolf 1970). Each point of the image incoherently adds its own, differently centered diffraction pattern. (See Figure 10). The separation d_{12} of the first, strongest fringes is of the order of $\sqrt{\lambda d_{moon}}$, where d_{moon} is the distance to the lunar limb. At 2.2 μm this corresponds to $\approx$ 30 m or $\approx$ 15 mas or, given the velocity of the moon, to $\approx$ 30 ms with respect to a telescope on ground. Lunar occultation observations therefore require high speed photometry. Obviously, structures of half the fringe separation can easily be recognised by the strong damping and broadening of the fringes they introduce. Narrow structures are unproportionally more difficult to detect, because the source profile is convolved with the diffraction pattern. The smallest stellar diameter which can be resolved for data of good signal-to-noise ratio is about 1 mas while for fainter sources, where the signal gets more noisy, the resolution decreases to 5 mas and less, but still the size of a larger source or the separation of two unresolved sources can be determined with high accuracy. The attractiveness of a method which is capable of a resolution of 1 mas can be visualized by remembering that it takes an interferometer size of 450 m to obtain the same resolution at 2.2 μm.

One may wonder why a ragged lunar landscape should produce the same diffraction pattern as a straight edge. The explanation is that the intensity distribution in the diffraction pattern is essentially determined by integration over a small number of Fresnel zones at the lunar limb. The area over which this integration extends then has a size of a few times $\sqrt{\lambda d_{moon}}$ or 30 m - 100 m, larger than rocks, smaller than mountains, and thus is effectively determined by the local lunar slope.

The presence of an extended object shows primarily in a reduced fringe contrast, but there are at least three instrumental effects which also smear the diffraction pattern: the finite optical bandwidth (leading to a superposition of patterns with different fringe spacings), the diameter of the telescope (averaging over the fringe pattern) and the limited electronic bandwidth (resulting in damping of the oscillations). After these effects have been taken into account by numerical modeling or measurement of a point source, the interpretation of the data is usually done by determining a best fitting parameterised model. Alternatively, deconvolution of the observed occultation trace with the the point source fringe pattern may be performed to get a model-independent estimate of the source brightness distribution. Both, the original iterative process proposed by Lucy (1974) and a version specifically adapted to the problem of occultation observations by Richichi (1989) have been used with success.

For quite extended sources the fringe pattern gets smeared out. In this case, as well as in the observations of binaries, where two independent fringe patterns appear, lunar occultation measurements are based on timing. The size of the occulted object or the separation of its components then is obtained from the actual speed of the occultation by measuring the time it takes to occult the object. This speed, which depends on the unknown local lunar slope, has to be determined by extracting the crossing speed of the fringe pattern from the data.

The quality of lunar occultation observations strongly depends on the sky brightness and sky brightness fluctuations during the measurements. Under favorable conditions full resolution is obtained for objects brighter than K$\approx$4 - 5 mag, while the limiting magnitude for reduced resolution is K$\approx$10 mag for measurements with a single detector and diaphragm of typically 10″. Use of array detectors provides the promise of increasing the sensitivity of lunar occultation measurements by about 2 magnitudes (Richichi 1993) and to allow lunar occultation observations even under conditions of strongly fluctuating sky brightness and perhaps in addition at the bright lunar limb.

4.2 Results

Stellar diameters and effective temperatures This is the classical application for lunar occultation observations. Figure 11 gives an example for the accuracy of angular diameter determinations which may be achieved for data with different noise level (Richichi et al. 1992). If in addition the bolometric flux [Wm^{-2}] of a star can be determined from broad-band photometry covering, e.g.

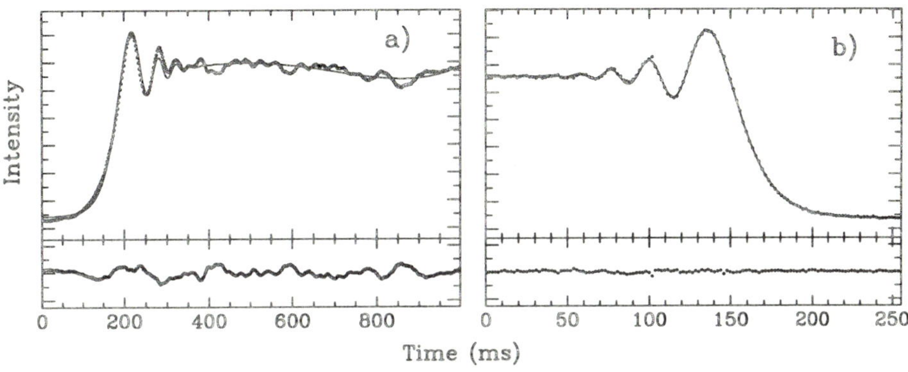

Fig. 11. Repeated diameter determinations of the M7 star DW Gem. a). The lightcurve of a reappearance. The shown best fit gives a uniform disk diameter of $\Phi_{UD} = 6.24 \pm 0.30$ mas. b). A disappearance with better S/N ratio, resulting in $\Phi_{UD} = 5.92 \pm 0.11$ mas. – The lower panels show the residuals of the fits to the same scale as the data. Adapted from Richichi et al. (1992).

the wavelength range 0.4 μm - 100 μm, the measurement of the angular diameter Φ is sufficient to determine the effective temperature, since the unknown distance d and the unknown stellar radius R only enter the relation as ratio R/d $= \Phi/2$:

$$F_{bol} = \frac{4\pi R^2 \cdot \sigma T_{eff}^4}{4\pi d^2} = (\Phi/2)^2 \cdot \sigma T_{eff}^4 \qquad [\mathrm{W/m^2}]. \tag{34}$$

Ridgway et al. (1980) used lunar occultation observations to derive the effective temperature scale for K0 - M6 stars in this way. Their determinations led to effective temperatures higher by up to 500 K and in better agreement with theoretical predictions than earlier estimates based on photometry only. In (34) the real, so-called limb-darkened diameter Φ_{LD} has to be used, which is up to 13% larger than the diameter Φ_{UD} for a uniform disk fitting the same occultation trace. So far the limb darkening correction cannot be derived from the measurements, except marginally for the very brightest stars like Betelgeuse and Antares

(Ridgway et al. 1982, Richichi and Lisi 1990), but has to be adopted from the predictions of model atmospheres. Work on the effective temperature scale has continued both using lunar occultations (Richichi et al. 1992) and optical interferometry (Di Benedetto and Rabbia 1987, see Figure 37). A catalogue of stellar angular diameters measured by lunar occultations has been published by White and Feierman (1987).

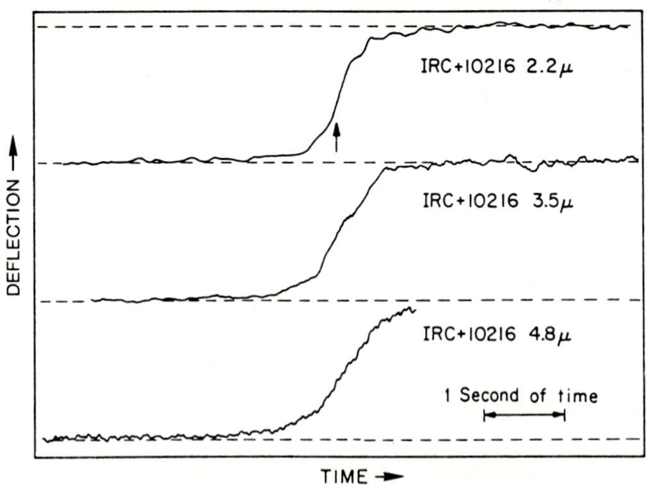

Fig. 12. Near-infrared lunar occultation observations of the circumstellar shell around the carbon star IRC+10216 on 19/20 Nov. 1970. The angular rate of the occultation was 0.29″/s. The increase of shell diameter with wavelength is evident. From Toombs et al. (1971).

Circumstellar envelopes As an example of the measurement of an envelope around a late type star, Figure 12 shows the results of the pioneering observation by Toombs et al. (1971) of the carbon star IRC +10216. The envelope is more extended at longer wavelengths, as expected for thermal re-radiation of dust heated by a central source. A prominent halo has also been found around the Mira star IRC +10011 (Zappala et al. 1974) and a few lesser ones around less extreme stars (Richichi et al. 1988). Speckle interferometry has seemingly the advantage of providing a larger sample of detected halos, however these are often difficult to interpret because of insufficient spatial resolution (see §5.2 below).

The results are similarly sparse for young stellar objects. Here, extended emission has been resolved only around DG Tau (Figure 13, Leinert et al. 1991a, Chen et al. 1992), which carries 25% of the 2.2 μm flux of the system. With help of the specialized deconvolution algorithm mentioned above, the brightness distribution of the extended emission was shown to have approximately a Gaussian shape with a FWHM of 45 $\pm$ 5 mas, corresponding to a radius of only 3.2 AU at the distance DG Tau (140 pc). Probably this emission represents the inner

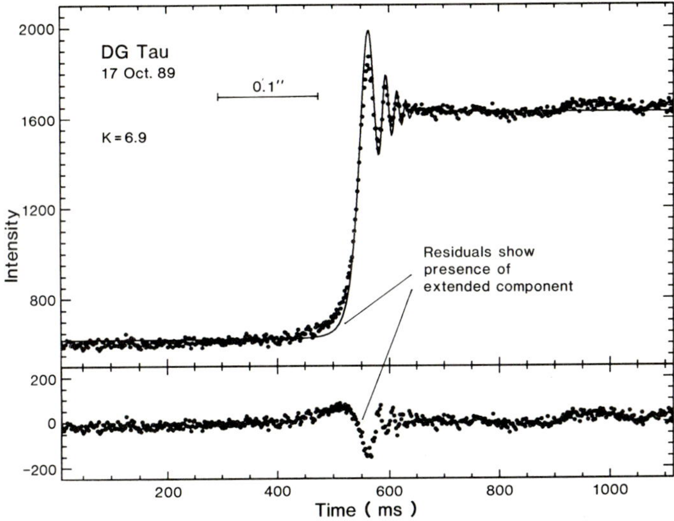

Fig. 13. Detection of extended emission around the young star DG Tau. The data points of the occultation trace are shown together with the model curve for a point source. The systematic residuals both before and after the steep rise in the lightcurve show the presence of extended emission, the FWHM of which was determined to 6.3 AU. From Leinert et al. (1991a).

part of the accretion disk of this active young star, seen in scattered light. Lunar occultations thus provided a first direct view of such a circumstellar disk, the properties of which otherwise have to be inferred indirectly from spectroscopic and photometric evidence.

Binaries Binaries are easily detected by lunar occultations as long as the brightness ratio does not exceed the signal-to-noise ratio because they introduce a characteristic step in the occultation light curve (see Figure 14), and this is true also for rather noisy data. Since the precessing lunar orbit covers two nearby star forming regions in Taurus and Ophiuchus, a study for duplicity of those young stars was started by Simon et al. (1987). These lunar occultations observations have been able to resolve binaries down to a projected separation of 10 mas (DF Tau). At present lunar occultation observations are practically the only method available to resolve in the infrared binaries with separations from 0.01″ - 0.1 ″. We will see below in §5.2 that measurements in just this range are urgently needed to define the period or semi-major axis distribution – one of the basic pieces of information describing a population of binaries – for young stars. So far only 5 binaries falling into this range have been found (Simon et al. 1992a) but the survey is still far from complete.

The Galactic Center Every 19 years there is a two- to three-year period when occultations of the galactic center occur, the last one with several occultation

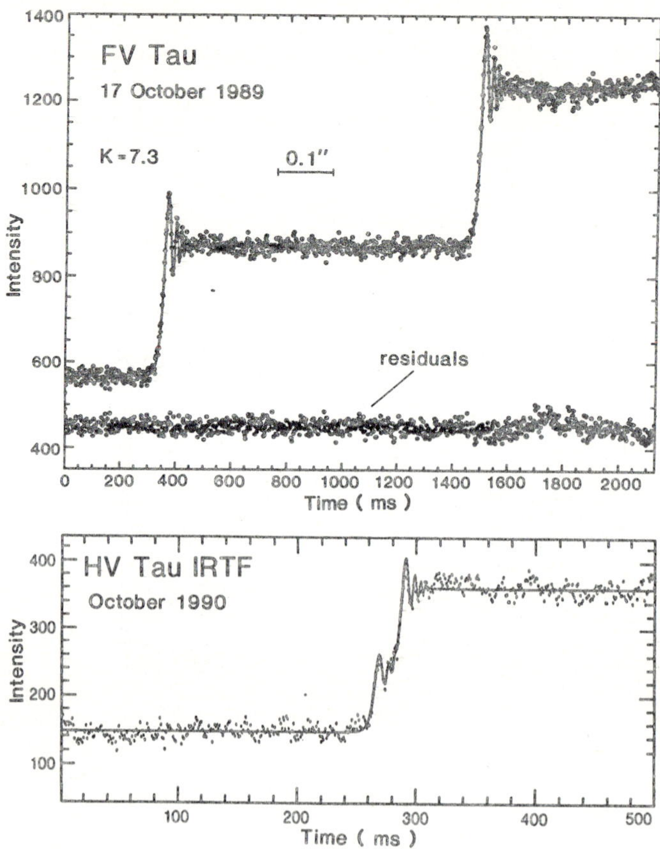

Fig. 14. Lunar occultation traces for a wide and for a close binary. The fits shown with the data result in projected separations of 517 ± 1 mas for FV Tau and 35 ± 2 mas for HV Tau. From Simon and Leinert (1992b).

events between September 1986 and February 1989. The events were observed (in the near-infrared, because the region is highly obscured) from telescopes on La Palma and Mauna Kea (Simon et al. 1990, Simons et al. 1990). The techniques included fast (2 ms) photometry with a single detector as well as observations with infrared arrays, which however were limited in readout rate to 120 to 500 ms. The most informative results were obtained with the fast photometer (Simon et al. 1990). By comparison of the occultation trace with a 2.2 μm image of the region around the galactic center for identification of the individual occultation events, upper limits of the order of 100-200 AU were derived for the size of the known infrared sources in this region. This makes it probable that they simply are bright stars. For the position of the central non-thermal radio source Sag A* only an upper limit for the K brightness could be found.

5 Near-infrared Speckle Interferometry

An exposure of a star, shorter than the coherence time τ_0 breaks into a number of bright points of approximate size λ/d due to the turbulence of the atmosphere (Figure 15). In a simplified picture the occurrence of these *speckles* can be thought of as due to the different slopes in the incoming distorted wavefront. These are distributed irregularly over the aperture, and the more frequent ones give rise to brightness maxima at positions corresponding to the normal to the local wavefront. More correctly, the occurrence of speckles has to be explained as

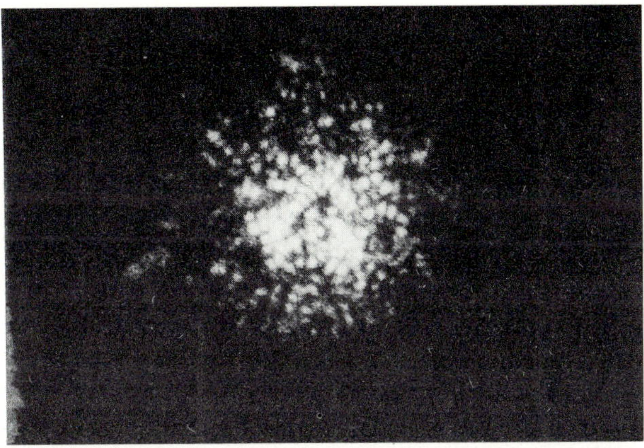

Fig. 15. Speckled image of a point source, taken by G. Weigelt in the visible at the ESO 3.6 m telescope. The size of the speckle cloud is $\approx 1''$. At 2.2 μm, the speckle cloud would be smaller by about 20 %, while the size of the individual speckles would increase by a factor of ≈ 3.5.

interference patterns in the presence of large phase differences in the wavefront. A comparison of these speckles for different sources, including binaries and extended objects, suggests that they carry high spatial frequency information on the object and that to a crude approximation they may be looked at as noisy diffraction-limited images. It was Labeyrie's (1970) idea that this diffraction-limited high resolution information can be extracted by a suitable average over the fast atmospheric variations (a plain average just gives the broad seeing-limited image). Figure 16 shows indeed that this goal can be achieved. The number of speckles is $\approx (D/r_0)^2$, which means that the number of photons per speckle is independent of telescope size and only depends on seeing. Quantitatively, the expression

$$N_{speckle} = 2.3(D/r_0)^2 \tag{35}$$

is used, which leads to a S/N ratio in the power spectra of speckle images which is reduced by exactly the factor $1/N_{speckle}$, which is what one would expect.

Fig. 16. Gain in spatial resolution obtained by speckle interferometry at 2.2 μm on an object with a K brightness of 8.3 mag. Left: Long exposure image, the seeing at K was $\approx$ 2.0″. Right: The image reconstructed from the same data by speckle methods. It is not far from diffraction-limited (0.13″ at the 3.5 m telescope used for these observations). The separation of the binary components is 0.35″. Created by N. Weitzel.

5.1 Methods

Power spectrum analysis ("Labeyrie process") A short exposure speckle pattern of a point source is nothing else than the instantaneous point spread function P($\boldsymbol{x}$) for imaging through the atmosphere. The instantaneous image I($\boldsymbol{x}$) of an object O($\boldsymbol{x}$) is then given by

$$I(\boldsymbol{x}) = O(\boldsymbol{x}) * P(\boldsymbol{x}) \tag{36}$$

or, in Fourier notation

$$\tilde{I}(\boldsymbol{u}) = \tilde{O}(\boldsymbol{u}) \cdot \tilde{P}(\boldsymbol{u}). \tag{37}$$

The simple inversion

$$\tilde{O}(\boldsymbol{u}) = \tilde{I}(\boldsymbol{u})/\tilde{P}(\boldsymbol{u}) \tag{38}$$

cannot usually be used to infer the object, because $\tilde{P}(\boldsymbol{u})$ may have zeros, because $\tilde{I}(\boldsymbol{u})$ is very noisy (indeed it has S/N ≈ 1 for one short exposure even for bright objects), and because $\tilde{P}(\boldsymbol{u})$ is not known unless in the very special case that there is a point source within the isoplanatic patch of the object. Averaging of the Fourier transform (37) again only results in the low resolution of the long exposure image, because for the higher spatial frequencies the average of $\tilde{P}(\boldsymbol{u})$ tends to zero. Labeyrie's (1970) method is based on averaging the power spectra both for the object and a nearby reference point source. This results in

$$\begin{aligned} \text{Object} : < |\tilde{I}(\boldsymbol{u})|^2 >=< |\tilde{O}(\boldsymbol{u})|^2 \cdot |\tilde{P}(\boldsymbol{u})|^2 >= |\tilde{O}(\boldsymbol{u})|^2 \cdot < |\tilde{P}(\boldsymbol{u})|^2 >_1 \\ \text{Reference} : < |\tilde{R}(\boldsymbol{u})|^2 >=< |\tilde{\delta}(\boldsymbol{u})|^2 \cdot |\tilde{P}(\boldsymbol{u})|^2 >= 1 \cdot < |\tilde{P}(\boldsymbol{u})|^2 >_2 . \end{aligned} \tag{39}$$

Because of the low S/N ratio for single exposures, 1000 or more frames have to be averaged to achieve an accuracy of a few percent. In (39), $\tilde{R}(\boldsymbol{u})$ refers to the short exposure image R($\boldsymbol{x}$) of the reference point source, and the subscripts "1" and "2" remind us that the averages do not refer to exactly the same atmospheric conditions but to slightly different viewing directions and observing times. In general, however, these two averages are very similar, so that the power spectrum of the object brightness distribution can be obtained from the observed power spectra by division as

$$|\tilde{O}(\boldsymbol{u})|^2 \approx \frac{< |\tilde{I}(\boldsymbol{u})|^2 >}{< |\tilde{R}(\boldsymbol{u})|^2 >}. \tag{40}$$

Noise terms in the object and reference measurements have to be subtracted before division.

Computation of the MTF($\boldsymbol{u}$) for the object power spectrum from the amplitude distribution in the pupil plane leads to a fourfold integral, in which effects of the telescope and the atmosphere are no longer separable. Figure 17 shows the typical shape of the transfer function: a low-frequency fall-off due to seeing, followed by a plateau up to spatial frequencies not far from the diffraction limit. The gain in high spatial frequency response over conventional imaging is obvious, and a substantial gain is also obtained with respect to the case where

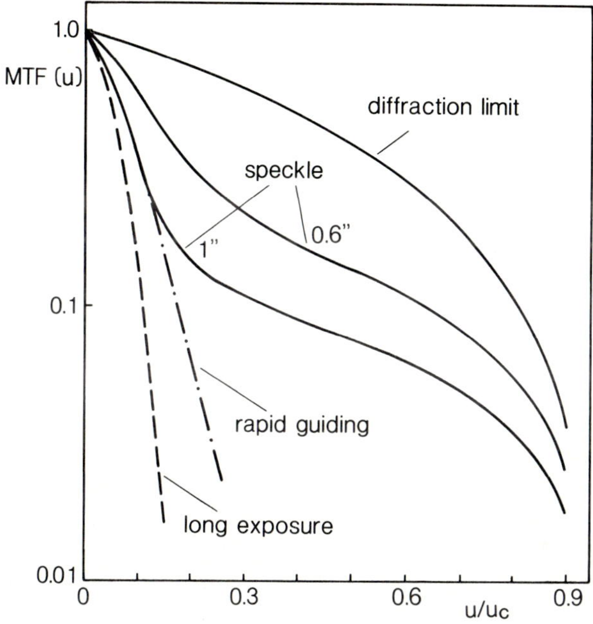

Fig. 17. The modulation transfer function for speckle observations at a 3.5 m telescope with focal ratio f/45 at $\lambda = 2.2\ \mu$m. The cut-off frequency has a value of $u_c^{-1} = 0.13''$. The transfer functions for diffraction-limited imaging, for long exposures and for an exposure obtained with rapid guiding (see Fried (1966) are shown for comparison.

image motion has been removed by rapid guiding. The diffraction limit of the telescope can be reached, although with low sensitivity. However, the object itself is not obtained but only the modulus $|\tilde{O}(\boldsymbol{u})|$ of its Fourier transform. Since the phase remains undetermined, information on asymmetries is lost. A different way to express this is to note that the backwards Fourier transform of the power spectrum

$$FT^{-1}(|\tilde{O}(\boldsymbol{u})|^2) = O(\boldsymbol{x}) \otimes O(\boldsymbol{x}) \tag{41}$$

gives the - symmetric - autocorrelation of the object. In the simple case of a double star there remains a 180° ambiguity on the position angle of the companion. In more complex objects, the structure is lost (compare Figure 2). Therefore, recovery of phase from speckle data is an important part of data evaluation, and in the following section different proposed kinds of averaging of the short exposure will be presented, which achieve this goal.

In the near-infrared, speckle measurements are limited by detector noise (e.g. readout noise of an array or Johnson noise of the feedback resistor in the preamplifier of a single detector). Under these conditions, the sensitivity is essentially the same both for power spectrum analysis and the phase recovery methods to be discussed below. The reachable limiting magnitudes are given in Figure 18 for the case of a slit scanning single detector instrument both for the case of a 3.6 m and an 8 m telescope. The larger telescope not only gives higher spatial

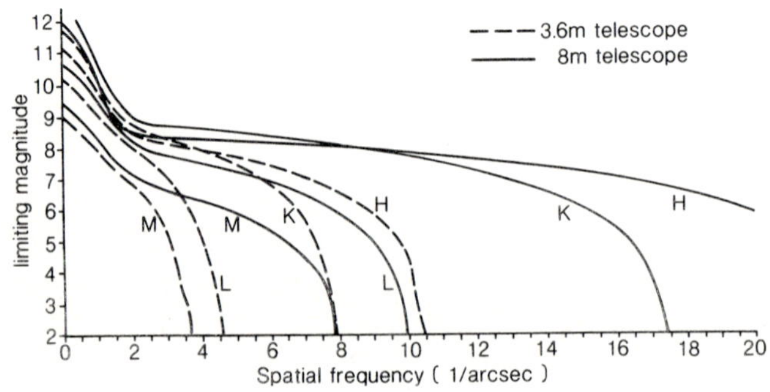

Fig. 18. Limiting magnitude for near-infrared speckle interferometry, for which at a given spatial frequency a S/N ratio of 10 can be obtained in 15 min under seeing of 1″. Adapted from Perrier (1988).

resolution but also provides a gain in sensitivity, which is particularly large ($\approx$ 4 mag) near the cut-off frequency of the smaller telescope. Another gain by $\approx$ 2 mag may be expected by using the now available low read-out noise infrared detector arrays ($\pm$ 30 e^-), but since the estimates of Figure 18 are quite optimistic, it may be safer to use these limits in estimating the performance of detector arrays, too.

Phase recovery In order to be able to reconstruct an *image* from speckle data, also the phase of the complex visibility (visibility = Fourier transform of the object brightness distribution) or equivalent information has to be obtained. Many methods have been proposed, including mathematical schemes relying on the properties of analytical functions (Walker 1981, Bates and Davey 1987). I mention here those used most often in astronomical observations.

Shift and add "Shift and add" is an intuitive method which works with speckle images, but is not a true speckle interferometric method. It assumes that the individual speckles are basically noisy images of the object, in which the noise may be reduced by averaging. As its name suggests, this methods consists of adding the short exposure images after they have been shifted in such a way as to bring the brightest speckle of each frame to the same position. The resulting image profile (Figure 19) then shows a diffraction limited peak (the averaged brightest speckle) sitting atop a broad seeing halo, which is due to smearing of the other speckles which occur at random separations from the brightest one. This is similar to what is achieved with adaptive optics for a good reason (see

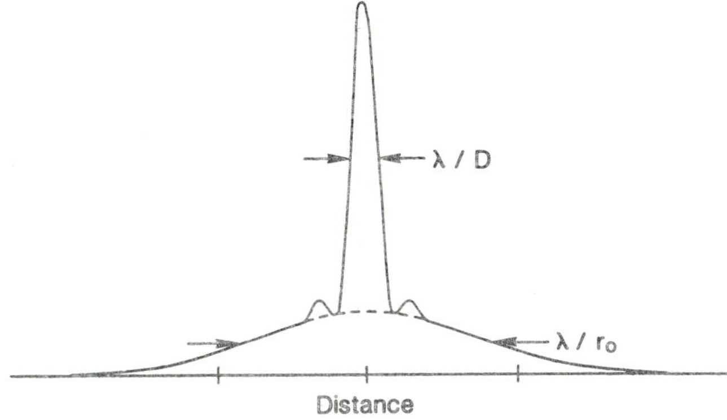

Fig. 19. Typical shape of the point spread function of a "shift and add" image. In §7 we will see that this is also the point spread function typically achieved with partial adaptive optics systems. From Beckers 1993.

§7). We note that by definition of the method, "shift and add" images of point sources do have the highest Strehl ratio achievable without explicit wavefront correction for given seeing conditions .

For objects which are dominated by an unresolved or slightly extended source, "shift and add" provides a convenient and quick look for high spatial resolution features. In many cases, the first diffraction ring can be seen. "Shift and add" has the definite advantage of requiring little computing effort. Another important advantage of the method is that it directly yields, albeit disturbed, images. Provided that a reference point source is measured under approximately the same observing conditions, an additional gain in spatial resolution can be obtained by deconvolving the original "shift and add" image with that of the reference.

However, "shift and add" also has its limitations. It fails for extended sources (because the brightest speckle can occur everywhere in the object) as well as for double stars with similar brightness of the components (because each of the components can contribute the brightest speckle). Christou (1991) found that "shift and add" does not reach the same quality determination of high spatial frequencies as speckle interferometry does, even after deconvolution with the reference source image.

The Knox-Thompson algorithm In 1974 Knox and Thompson proposed to average the cross-spectrum of the short exposure images (as opposed to the power spectrum in the Labeyrie process) in order to determine the phase of the complex visibility. With the notation used in §5.1 this leads to

$$\begin{aligned} &\text{Object} : < \tilde{I}(\boldsymbol{u}_1) \cdot \tilde{I}(\boldsymbol{u}_2) >= \tilde{O}(\boldsymbol{u}_1) \cdot \tilde{O}(\boldsymbol{u}_2) \cdot < \tilde{P}(\boldsymbol{u}_1) \cdot \tilde{P}(\boldsymbol{u}_2) >_1 \\ &\text{Reference} : < \tilde{R}(\boldsymbol{u}_1) \cdot \tilde{R}(\boldsymbol{u}_2 >=< \tilde{P}(\boldsymbol{u}_1)) \cdot \tilde{P}(\boldsymbol{u}_2) >_2, \end{aligned} \tag{42}$$

where the object visibility is $\tilde{O}(\boldsymbol{u}_i) = |\tilde{O}(\boldsymbol{u}_i)| \cdot e^{i\phi(u_i)}$. As in power spectrum analysis the terms due to the instantaneous point spread function $P(\boldsymbol{x})$ cancel to good accuracy , and we obtain after proper normalization of the cross-spectra

$$\frac{< \tilde{I}(\boldsymbol{u}_1) \cdot \tilde{I}(\boldsymbol{u}_2) >}{| < \tilde{I}(\boldsymbol{u}_1) \cdot \tilde{I}(\boldsymbol{u}_2) > |} \cdot \frac{| < \tilde{R}(\boldsymbol{u}_1) \cdot \tilde{R}(\boldsymbol{u}_2) > |}{< \tilde{R}(\boldsymbol{u}_1) \cdot \tilde{R}(\boldsymbol{u}_2) >} \approx e^{i[\phi(u_1) - \phi(u_2)]}. \tag{43}$$

This means that the phase has to be built up recursively, typically in the form

$$\phi(n \cdot u_0) = \phi((n-1) \cdot u_0) + [\phi(n \cdot u_0) - \phi((n-1) \cdot u_0)], \tag{44}$$

where u_0 is the lowest spatial frequency fitting into the image field. In one-dimensional observations, the recursion leads to an accumulation of errors, and the phases for the highest spatial frequencies are not well determined. In two-dimensional observations, there are many paths through the uv plane leading to a desired spatial frequency point. The phases resulting from the different paths then can be averaged, and the problem with the high spatial frequencies essentially disappears.

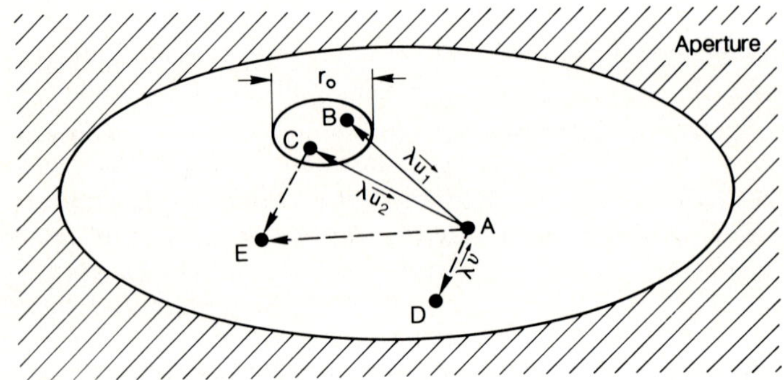

Fig. 20. The phase relations between different spatial frequencies. A given spatial frequency u corresponds to subaperture pairs with separation λu. For demonstration of the principle, pairs with one subaperture at position A are shown. Here, $\phi(\lambda u_1) = \phi(\mathrm{B}) - \phi(\mathrm{A})$, $\phi(\lambda \mathrm{u}_2) = \phi(\mathrm{C}) - \phi(\mathrm{A})$, $\phi(\lambda v) = \phi(\mathrm{D}) - \phi(\mathrm{A})$. Since $\phi(\mathrm{B})$ and $\phi(\mathrm{C})$ are correlated, BC being less than r_0, so are $\phi(u_1)$ and $\phi(u_2)$, while $\phi(v)$ is uncorrelated. Later we will refer to the sum of two spatial frequencies. To illustrate this, we also show point E, separated from A by $\lambda(u_1 + u_2)$.

The Knox-Thompson algorithm is based on the correlation of atmospheric disturbances between the spatial frequencies u_1 and u_2 . Since the correlation in the pupil plane extends only over the range of the size of the Fried parameter, the allowable spatial frequency difference (u_1 - u_2) is limited according to §2.5 to values safely less than the seeing cut-off r_0/λ (Figure 20). Usually, as indicated in (44), the basic frequency u_0 or small multiples thereof are taken as increment in spatial frequency.

An image then may be reconstructed by taking

$$\tilde{O}(\boldsymbol{u}) = \text{Modulus(Labeyrie} \quad \text{process)} \times e^{i\phi(Knox-Thompson)} \tag{45}$$

and

$$O(\boldsymbol{x}) = FT^{-1}[\tilde{O}(\boldsymbol{u})]. \tag{46}$$

Images obtained in this way usually contain considerable high frequency noise, which is suppressed by filtering the high frequencies with a suitable function, e.g. with the MTF($\boldsymbol{u}$) corresponding to diffraction limited imaging (§2.6) or with the *Hanning function*

$$H(u) = \frac{1}{2}(1 + \cos 2\pi \frac{u}{u_{max}}) \ , . \tag{47}$$

which gives a particularly good suppression of high frequency ringing.

Speckle holography We noted above (38) that the full Fourier transform of an object can be recovered from speckle data (in principle even from one frame) if there is a point source within the isoplanatic patch. Obtaining an image with the help of such a close reference is called *speckle holography*. The method works as well in image space. If the composed object, including the close reference at position $\mathbf{x}_R$ is

$$O^+(\boldsymbol{x}) = O(\boldsymbol{x}) + \delta(\boldsymbol{x} - \boldsymbol{x}_R), \tag{48}$$

its autocorrelation (which is what we get from power spectrum analysis)

$$\begin{aligned} O^+(\boldsymbol{x}) \otimes O^+(\boldsymbol{x}) &= \int [O(\boldsymbol{x}') + \delta(\boldsymbol{x}' - \boldsymbol{x}_R)][O(\boldsymbol{x} + \boldsymbol{x}') + \delta(\boldsymbol{x} + \boldsymbol{x}' - \boldsymbol{x}_R)] d\boldsymbol{x}' \\ &= O(\boldsymbol{x}) \otimes O(\boldsymbol{x}) + \delta(\boldsymbol{x}) + O(\boldsymbol{x} + \boldsymbol{x}_R) + O(\boldsymbol{x}_R - \boldsymbol{x}) \end{aligned} \tag{49}$$

contains 2 images of the object, O($\boldsymbol{x} + \boldsymbol{x}_R$) and O($\boldsymbol{x}_R - \boldsymbol{x}$), which are mir ror images with respect to each other. Knowing the position of the reference source, the correct image O($\boldsymbol{x} + \boldsymbol{x}_R$) can be identified. Rarely one finds a bright point source in the isoplanatic patch. Chelli et al. (1984) could use the Becklin-Neugebauer object as reference to obtain an image of IRc2 and to support the conclusion that this source is centered in a cavity of a thick disk.

Speckle masking: triple correlation and bispectrum analysis An authoritative review on this method has been given by Weigelt (1989). The original idea of Weigelt (1977) was to artificially create a reference source image with the same point spread function as the object, in order to be able to apply the method of speckle holography. To this end he produced *masks* by multiplying each speckle image of the object with itself, after having one of the image copies shifted by such an amount as to produce a point source (in the case of a binary this would correspond to shifting by the vectorial separation of the components). While this looked and probably was an academic exercise, it was a fruitful idea, because the generalized version (Lohmann, Weigelt and Wirnitzer 1984) turned out to be perhaps the most important speckle phase recovery method. This generalization is called *triple correlation* because it makes threefold use of each speckle image I($\boldsymbol{x}$). Instead of just one shift to produce a mask it uses all possible shift vectors $\boldsymbol{x}_0$. As was the case for the other speckle methods, the triple correlation expression can be averaged over a large number of exposures to reduce noise:

$$I^{(3)}(\boldsymbol{x}, \boldsymbol{x}_0) \approx < I_n^{(3)}(\boldsymbol{x}, \boldsymbol{x}_0) > = < I_n(\boldsymbol{x}) \cdot I_n(\boldsymbol{x} + \boldsymbol{x}_0) \otimes I_n(\boldsymbol{x}) > \qquad (50)$$

Here n refers to the nth speckle image of the series over which the average $<>$ is performed. On the right-hand side the product of the first two terms corresponds to the creation of a mask, while the correlation of this mask with the unshifted image is similar to the process of speckle holography. Lohmann et al. (1984) show that indeed the object O($\boldsymbol{x}$) can be recovered from the averaged triple correlation.

While correlations often are used in the visible spectral region, where photons are fewer and photon counting cameras exist, for the infrared the Fourier transform of (50)

$$\tilde{I}^{(3)}(\boldsymbol{u}, \boldsymbol{v}) \approx < \tilde{I}_n^{(3)}(\boldsymbol{u}, \boldsymbol{v}) > = < \tilde{I}_n(\boldsymbol{u}) \cdot \tilde{I}_n(\boldsymbol{v}) \cdot \tilde{I}_n(-\boldsymbol{u} - \boldsymbol{v}) > \qquad (51)$$

is more useful. This *bispectrum* ("bi" because it uses two spectral frequencies $\boldsymbol{u}$ and $\boldsymbol{v}$) contains the object phase , because according to (51) we have

$$Phase(\tilde{I}^{(3)}(\boldsymbol{u}, \boldsymbol{v}) \equiv \beta(\boldsymbol{u}, \boldsymbol{v}) = \phi(\boldsymbol{u}) + \phi(\boldsymbol{v}) - \phi(\boldsymbol{u} + \boldsymbol{v}). \qquad (52)$$

The object phases are usually found from a recursive scheme in which the phase of the lowest nonzero spatial frequency is arbitrarily set to $\phi\ (\boldsymbol{u}_0) = 0$. This corresponds - for a symmetrical image - to shifting its center of gravity to exactly the center of the frame. Phases for higher spatial frequencies then are calculated from recursive relations, the first of which is given here as example,

$$\beta(\boldsymbol{u}_0, \boldsymbol{u}_0) = \phi(\boldsymbol{u}_0) + \phi(\boldsymbol{u}_0) - \phi(2\boldsymbol{u}_0), \qquad (53)$$

where $\beta(\boldsymbol{u}_0, \boldsymbol{u}_0)$ has been measured, and the first two terms on the right-hand side already have been determined earlier in the recursion.

An obvious drawback of this method of phase recovery is the extensive need for computing, because for two-dimensional images the bispectrum is a four-dimensional quantity. However, the rapidly increasing performances of available computers greatly reduce this disadvantage. Otherwise, the bispectrum analysis has several strong points:

- it gives a reliable reconstruction of phase, probably slightly superior to the Knox-Thompson reconstructions.
- it allows one to determine phases for disjunct apertures, and therefore it is applicable to multi-telescope interferometry. This follows from the fact that a spatial frequency $\boldsymbol{u}$ corresponds to a baseline vector in the aperture of $\lambda\boldsymbol{u}$, and that - in contrast to the Knox-Thompson algorithm there is no limitation on the relative values of $\boldsymbol{u}$ and $\boldsymbol{v}$.
- the S/N ratio obtained for phases at high spatial frequencies is at least as good as for determination of the power spectrum, in spite of the fact that the bispectrum involves the product of three noisy numbers. Referring to Figure (20) we see that the spatial frequencies involved in computing the bispectrum $\tilde{I}^{(3)}(\boldsymbol{u}, \boldsymbol{v})$ correspond to a closed triangle in the aperture plane. Bispectrum analysis thus contains a strong element of phase closure (see §6.2), which reduces atmospheric disturbances and is considered as the cause for the already mentioned good performance.

5.2 Results

To date, both one-dimensional and two-dimensional speckle interferometry have contributed to noteworthy results in the near infrared. For the future, two-dimensional infrared arrays with low readout noise are expected to be almost exclusively used.

The Galactic Center A recent study at the ESO NTT telescope (Eckhart et al. 1992), based on "short" exposure images under very good seeing conditions, has led to two new results: First, at the position of the galactic center (as defined by the nonthermal radio source Sgr A*), a faint near-infrared source with brightness K=13.7 ± 0.6 mag could be detected; second, the complexes IRS 16, IRS 13 and IRS 29 break up into a much larger number of individual sources than thought earlier. Probably we are seeing a star cluster in the immediate neighborhood of the galactic center.

Different methods were used to get the high spatial resolution image shown in Figure 21. The very good seeing allowed exposure times of up to 1s, which is several times larger than the coherence time, but still gives satisfactory high spatial frequency resolution. For the first one of two overlapping pictures 2000 exposures were subject to the "shift and add" method. For the second one 5000 exposures were evaluated with power spectrum analysis and the Knox-Thompson algorithm. Both parts were then sharpened by deconvolution (Lucy 1974) to $0.''15$ and finally smoothed by a Gaussian beam to the resulting resolution of $0.''25$.

Additional observations at 1.6 μm (Eckart et al. 1993) showed that the fainter stars are probably M giants, while the more luminous sources have the colours of hot stars. For the whole cluster of stars near the center a core radius of 0.15 ± 0.05 pc was derived. The source at the position of the galactic center is also blue (H-K $\approx$ 0.0), the colour of which is compatible with radiation from a hot

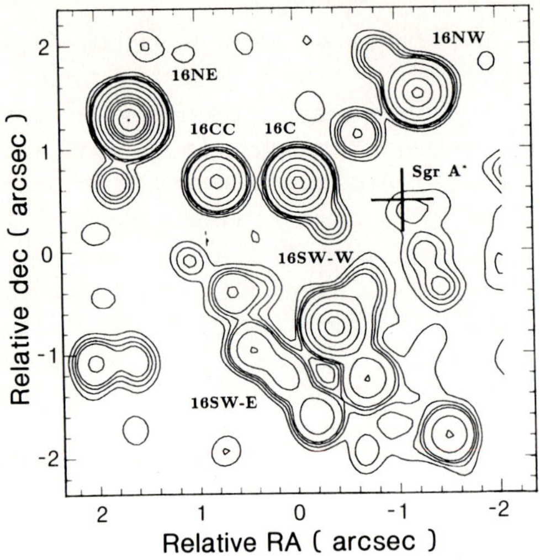

Fig. 21. K band image of the region around the galactic center, as defined by the radio source Sgr A*. The image was obtained by applying speckle methods to moderately long exposures (500ms - 1s). A source with a K brightness of ≈ 13 mag is found at the position of the galactic center. The other sources appear to be stars of a central star cluster From Eckart et al. 1992.

accretion disk. Similar studies are being performed by several other groups with direct imaging under excellent seeing conditions.

Nearby Seyfert galaxies Infrared speckle observations at 3.5 μm (Chelli et al. 1987) showed three components in the central part of the Seyfert 2 galaxy NGC 1068, which appear to be correlated to the maxima of a VLA map obtained at a wavelength of 2 cm. The unresolved central component ($\leq 0.''1$), probably the central core of the galaxy, has an upper limit in linear size of less than 10 pc at the distance of NGC 1068. Measurements of the Seyfert 1 galaxy NGC 4151 (Ayers et al. 1990) showed less new detail. Improved results can only be expected from improved sensitivity, i.e. from the use of the new low-noise infrared detector arrays.

Evolved stars Most red giants emit a strong ($\approx 10^{-5}$ $M_\odot$/yr), slow ($\approx$ 10-30 km s^{-1}) stellar wind, in which dust is formed at some distance r_i from the star. The hope was to be able to determine the extent, and the inner boundary r_i, of the dusty envelope created by this stellar wind and perhaps to measure the diameter of the star. Unfortunately the spatial resolution of 3-4 m telescopes turns out to be inadequate for such a study. Definite photospheric diameters could not be determined, because the objects were not fully resolved (Mariotti et al. 1983). The same is true also for most of the numerous attempts to measure the sizes of dusty envelopes (Dyck et al. 1984, Dyck 1987). Only for the most conspicuous dusty envelopes could detailed results be obtained. For the extreme carbon star IRC +10216 Ridgway and Keady (1988) showed that the envelope, with a north-south extension, is polarised by light scattered at 2.2 μm, while a long-term study by Dyck et al. (1991) showed changes in the structure of the envelope with a timescale of $\approx$ 10 years. The peculiar source known as the "Red Rectangle" (Cohen et al. 1975) was resolved by Leinert and Haas (1989a) into a point-like source (probably coinciding with the visible A star) and a narrow (0.″24 FWHM) and a broader (FWHM $\approx$0.″9) extended envelope, both shifted north by 0.″13 from the unresolved source and possibly containing a second, embedded star, which could then be the source of this beautifully symmetric bipolar nebula.

Low-mass companions around nearby stars Speckle interferometry in the near-infrared is very well suited to looking for low-mass companions to nearby dwarfs, because low-mass objects are expected to be cool, with the maximum of their emission between 1μm and 3μm, and because speckle interferometry is particularly sensitive to the detection of companions, since a typical pattern is produced in the modulus and phase of the visibility by the presence of a double star (see Figure 22).

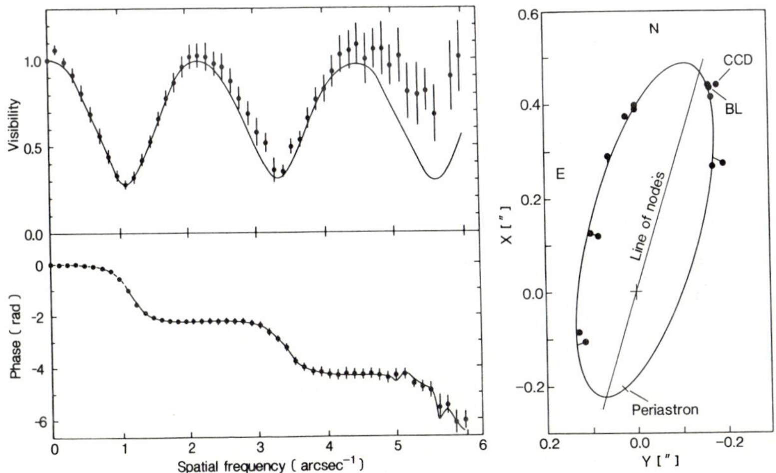

Fig. 22. Low mass companion to the nearby M5 dwarf Gliese 866. Left: visibility and phase show the pattern typical for a double star. The fit curve is for a brightness ratio at 2.2 μm of $I_2/I_1 = 0.54$ and an actual projected separation of 0.44″. Right: the relative orbit of the components of Gliese 866, from which a component mass of $0.38 M_{\odot}$ can be derived. From Leinert et al. 1990.

Don McCarthy at Tucson had a successful program of resolving the unseen companions of astrometric or spectroscopic binaries. The mass determinations showed that the companions to GJ 1245 A and Gliese 234 belong to the lowest mass stars known so far, with masses of 0.074 ± 0.013 $M_{\odot}$ and $0.083 \pm 0.023 M_{\odot}$

(Henry and McCarthy 1993). It should be noted that speckle observations allow us to determine visual orbits for the resolvable binaries and hence the determination of semi-major axis, inclination and combined mass of the system (if the parallax is known). These are exactly the parameters which are needed to extract component masses if in addition a photocentric orbit or the mass function determined from observations of a single-lined spectroscopic binary are known.

A systematic search amongst the nearest stars turned out a couple of hitherto unknown companions, including a companion to the M dwarf Gliese 866, the 11th nearest neighbor of the sun (Leinert et al. 1990, Figure 22). The combined mass of the system is 0.38 $\pm$0.03 $M_\odot$. Radial velocity observations (D. Latham, private communication) showed that it is even a triple system, in which case all three components are indeed of "very low" mass ($< 0.30\ M_\odot$). A systematic search has been completed of the sky north of declination δ = -30° for distances out to 7.5 pc by Henry and McCarthy (1990). They use the data to construct a mass – infrared brightness relation at 2.2μm which is expressed by three straight lines in a log-log diagram (Henry and McCarthy 1993):

$$\begin{aligned}
1.00M_\odot > M > 0.50M_\odot &: \log(M/M_\odot) = -0.105M_K + 0.322 \\
0.50M_\odot > M > 0.18M_\odot &: \log(M/M_\odot) = -0.252M_K + 1.197 \qquad (54)\\
0.18M_\odot > M > 0.08M_\odot &: \log(M/M_\odot) = -0.167M_K + 0.540\ .
\end{aligned}$$

An important result is that a mass at the hydrogen burning limit of 0.08 $M_\odot$ corresponds to an absolute K magnitude of 9.8 mag, or, perhaps easier to remember, $\approx$ 10 mag. Henry and McCarthy also show that the M dwarf mass function, which because of the long main sequence lifetime of low mass stars is at the same time an initial mass function, is rising towards small masses. And they determine the degree of multiplicity (number of double or multiple systems/total number of systems) which for their sample turns out to be 34 $\pm$ 11 %, somewhat lower than the $\approx$ 53% found for solar type main sequence stars (Duquennoy and Mayor 1991).

In principle, speckle interferometry is sensitive enough to detect a brown dwarf companion to a nearby M dwarf. However, the announced first discovery of such a companion to the M dwarf VB8, with K = 11.8 mag and L = 3×10^{-5} $L_\odot$ (McCarthy et al. 1985) could not be confirmed. This, of course, does not invalidate the search method.

Envelopes around young stars "Solar system size halos around young stars" were found by Beckwith et al. (1984), in particular around the low mass T Tauri star HL Tau (see Figure 23) and around the intermediate mass Herbig Ae/Be star R Mon. These halos appear to be seen in scattered light because their brightness relative to the star increases with decreasing wavelength. The inferred masses of 10^{-4} to 10^{-3} $M_\odot$ are much smaller than the masses for the circumstellar disks around young stars as determined by 1.3 mm observations, which are of the order of 0.01 - 0.1 $M_\odot$ (Beckwith et al. 1990). What we see in the near infrared must be light scattered in the outer layers of these disks. This kind of halos is rare among young low-mass stars ($M \leq 1\ M_\odot$), but more common among

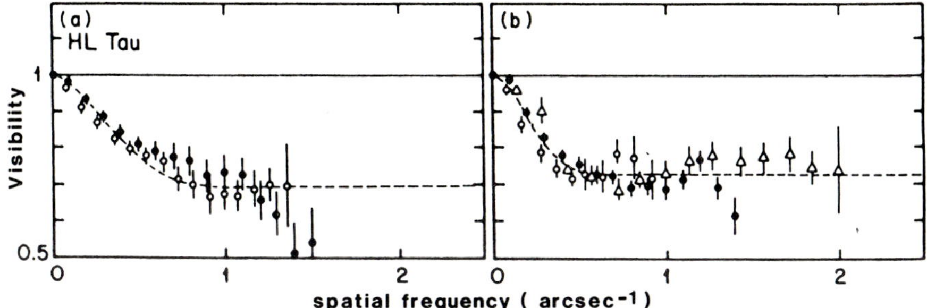

Fig. 23. Halo around the young star HL Tau, measured at 2.2 μm (Beckwith et al. 1984). Both measurements in north-south (a) and east-west direction (b) are shown. The decrease of the visibility curve at low spatial frequencies corresponds to a Gaussian-shaped halo with a FWHM of ≈ 1.6″ (220 AU) contributing 30 % to the total K brightness. The plateau at higher spatial frequencies corresponds to the unresolved star, contributing 70 % of the brightness.

intermediate-mass stars ($M = 2\text{-}5\ M_\odot$). There, infrared speckle interferometry, including measurements in polarized light, has been used to study the inner (d $\leq$ 1000 AU) structure of the Herbig Ae/Be stars LkHα 198, LkHα 233 and V376 Cas (Leinert et al. 1991b, 1993a), and to argue that in the CO outflow source AFGL 490 the near-infrared brightness maximum probably is the driving source (Haas et al. 1992).

Young binaries That young stars also can be binaries was brought to general attention by Dyck et al. (1982), who surprisingly found a companion to the otherwise frequently observed star T Tau. This detection turned out to be an *infrared companion* (brighter in the infrared and more luminous than the optical primary), like several others to be detected in the following years, among which Haro 6-10 (Leinert and Haas 1989b, Figure 24) and Z CMa (Koresko et al.1991) are particularly spectacular cases. In 1989 the companion to T Tau had an outburst by about 2 mag at K (Ghez et al. 1991) which lasted for about one year. The nature of these infrared companions poses an interesting problem for star formation, because more luminous objects are thought to be hotter and to evolve more quickly, in contrast to what is observed here.

Systematic near infrared searches by Ghez et al. (1993) and Leinert et al. (1993b) have found a surprisingly high incidence of binarity among the young stars. In the interval of separations accessible to near-infrared speckle interferometry with 3-5 m telescopes, ≈0.″1 to ≈ 10″, corresponding to 150 AU to 1500 AU at the distance of the nearest star forming regions, approximately twice as many binaries were found than in similar studies for main sequence stars (Figure 25). This means

- either that every or almost every young star is born double
- or that the semi-major axis or period distribution is quite different for young stars and stars on the main sequence.

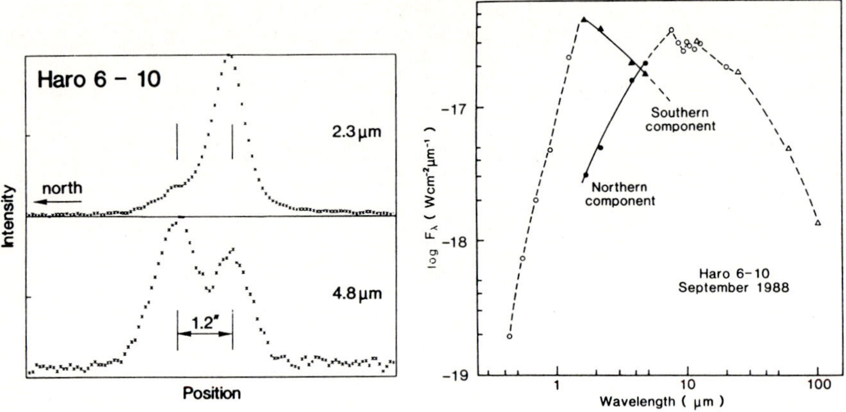

Fig. 24. The infrared companion to the young star Haro 6-10. Left: fast slit scans in north-south direction show the companion 1.2″ north of the point source. Its brightness strongly increases with wavelength and at 4.8 μm already surpasses that of the primary. Right: decomposition of the flux into the contributions of the two components. The companion is the more luminous one. From Leinert and Haas 1989b.

In both cases important changes have to take place during the time of evolution of the young stars to the main sequence. A choice between these two possibilities could be made if continued lunar occultation observations show that the distribution curve of binaries with separation (Figure 25) continues towards smaller separations of 10 - 100 mas in the same way or – alternatively – in a different way than the distribution for main sequence stars does.

Simon et al. (1993) noted an interesting consequence of the high degree of duplicity for the age of the young stars. Since the duplicity of most of these stars had been unknown before, the luminosity for the primaries of these binary systems had been overestimated. This means that after correction for this effect the position of the individual binary components in the Hertzsprung-Russell diagram is lower than assumed before for the system. Because evolutionary tracks lead almost perpendicularly down in this part of the HRD, the age of the young stars, at least in the Taurus star forming region, has to be increased by a factor of two to three from $\approx 10^6$ yr to $\approx 2 \times 10^6$ yr.

Finally, Figure 26 (Koresko et al. 1993) shows that speckle observations (here an observation with a two-dimensional infrared array) are able to provide both spatial details and good dynamic range, a point which often is not appreciated when discussing the merits of this technique.

6 Interferometry

In contrast to speckle interferometry and adaptive optics, which approach the diffraction limit of a single telescope, multi-telescope interferometry can give much higher spatial resolution. For the ESO VLT - interferometer with baselines

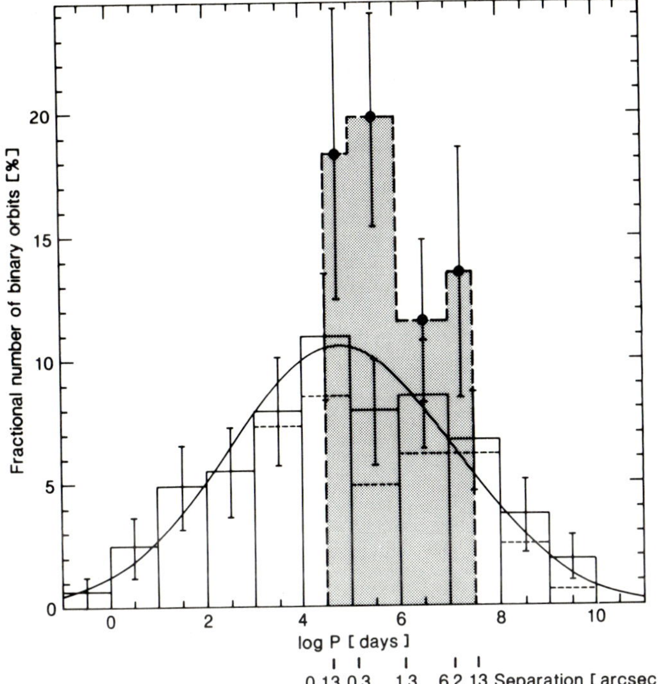

Fig. 25. Comparison of the binary frequency as a function of separation (and log of the orbital period) for the young stars in Taurus (broken line, shaded area) with the observations of nearby solar type field stars by Duquennoy and Mayor (1991) as reproduced from their paper. The ordinate gives in % the number of binary orbits, the periods of which fall into a given interval of orbital period, divided by the total number of stars in the sample. In this representation triples are counted as two pairs. From Leinert et al. 1993b.

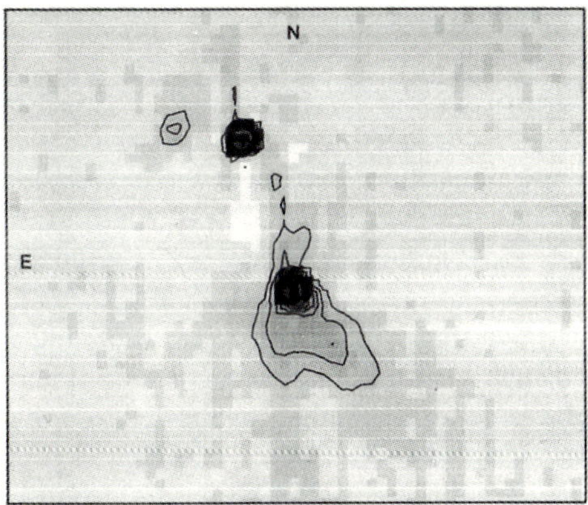

Fig. 26. Reconstructed image of the infrared double source Mon R2 - IRS 3 from speckle observations at 2.2 μm with a two-dimensional detector array. The two main components are separated by 0.80″. The isophotes are in steps of 5 % of the peak brightness. Note the detailed structure at the lowest levels. From Koresko et al. (1993).

of up to 180 m, the resolution limit at 2.2 μm will be 2-3 $\times$ 10^{-3} arcsec. In interferometry, the measured quantities are the visibilities of the fringe patterns associated with the available baselines. Images have to be reconstructed by suitable aperture synthesis. Apart from the professional review by Roddier and Léna (1984), the description of the first actually productive interferometers

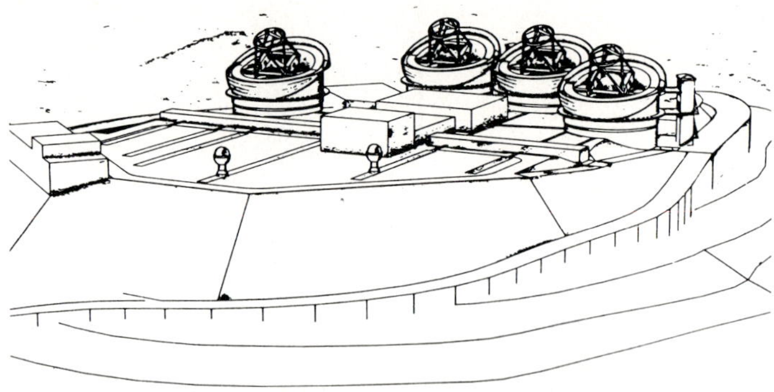

Fig. 27. Sketch of the VLT interferometer on Cerro Paranal as seen from from the southeast. The sketch is not fully up to date. In particular, the buildings near the center have been removed or put underground, and the number of movable small telescopes (here with the appearance of a light bulb) has been increased to three. From Beckers et al. 1992b.

(e.g. Koechlin and Rabbia 1985, Shao et al. 1988b) and a plan for the VLT-interferometer (Mariotti 1992) are very informative on the technique, while Dyck and Kibblewhite (1986) give scientific reasons why it is important to use large baselines.

6.1 Historical development

Michelson predicted the possibility of measuring the angular diameters of stars by interferometry in 1890, and he also performed the first of such measurements on Mount Wilson, with additional flat mirrors extending the baseline of the 2.5 m telescope to 6 m (Michelson and Pease 1921). Attempts to further extend the baseline to 15 m failed, leading to the conclusion that multi-telescope interferometry was impossible, until – again – Labeyrie in 1974 demonstrated its feasibility with measurements on α Lyrae (Labeyrie 1974). Table 2 summarizes physical differences between interferometry in the radio and the optical wavelength range, which explains why the development of optical interferometry has taken so long. Today, the absence of amplifiers for optical waves ($\approx 10^{14}$ Hz) is the most severe limitation on sensitivity and therefore usefulness of optical interferometers. In recent times there has been a renaissance of interest in optical interferometry because

Table 2. Comparison of basic quantities for interferometry at visible and radio wavelengths.

	optical	radio	advantage for radio
λ	0.5 μm	5 cm	easier alignment
ν	6×10^{14} Hz	6 GHz	**amplifiers available**
photons per 10^{-18} Ws	2.5	250000	smoother signal
seeing	$\gg \lambda/D$	$> \lambda/D$	no speckles, simple interference pattern
atmospheric coherence time τ_0	ms	min	higher sensitivity

- faster, more sensitive array detectors have become available,
- lasers allow precise monitoring of instrument dimensions,
- computers increased in capacity and speed, both needed for instrument control and data taking,
- adaptive optics brings the promise to overcome multi-speckle images and to achieve diffraction-limited operation of individual telescopes, which will greatly increase the sensitivity of interferometers,
- the improved theory of the effects of atmospheric turbulence (§3) allows a realistic simulation and improved planning of interferometric instruments.

However, in order to achieve a scientific breakthrough, interferometry should be possible with large telescopes with high sensitivity, and it should be able to provide images, i.e. to measure the phases of the (complex) visibilities of the fringe patterns.

6.2 Closure Phase

In principle, the offset of the fringe pattern from the symmetric position (see Figure 3) measures the phase of the visibility. In practice, the fast atmospheric pathlength fluctuations with a time scale of $\tau_0 \approx 10$ ms and with a rms value of $\Delta s \approx 10\mu$m, slow vibrations (timescale ≈ 1 s) of the telescopes by a few μm, and tracking errors of similar size make a phase measurement for the fringe pattern belonging to an individual telescope pair virtually impossible. However, the *closure phase*, i.e. the sum of the phases measured simultaneously between three independent telescopes, has a well-defined value which only depends on the properties of the object, and in which the telescope based (in particular the atmospheric) pathlength effects cancel (Figure 28). This may be seen by explicitly writing down the sum

$$\text{Simultaneously:} \qquad \begin{aligned} \phi_{12} &= \phi(\boldsymbol{u}_2) + \Delta\phi_2 - \Delta\phi_1 \\ \phi_{23} &= \phi(\boldsymbol{u}_3) + \Delta\phi_3 - \Delta\phi_2 \end{aligned} \tag{55}$$

$$\phi_{31} = \phi(\boldsymbol{u}_1) + \Delta\phi_1 - \Delta\phi_3$$

$$\text{Closure phase} = \text{sum} \ = \phi(\boldsymbol{u}_1) + \phi(\boldsymbol{u}_2) + \phi(\boldsymbol{u}_3)$$

In general ϕ $(\boldsymbol{u})$ is not linear and the closure phase is $\neq 0$. The number of closure phases to be measured in an array of N telescopes is (N-1)(N-2)/2 (Readhead and Wilkinson 1978) and thus smaller than the number N(N-1)/2 of baselines. Some

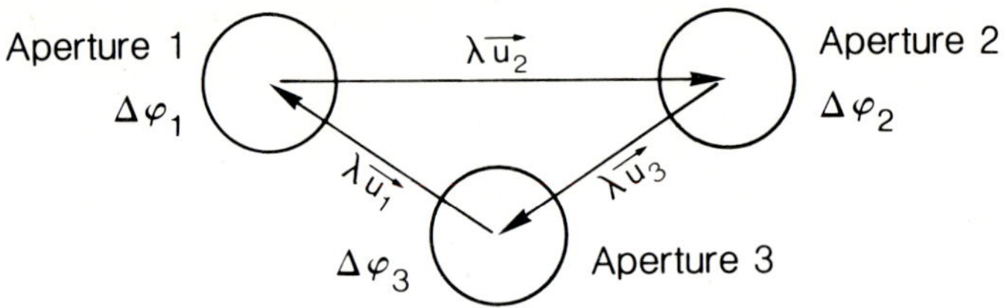

Fig. 28. The geometry of closure phase. Note the condition $u_1 + u_2 + u_3 = 0$.

additional information or assumptions on the object are therefore necessary to reconstruct phase and object. An alternative approach is to fit a model to the data which appears physically reasonable on other grounds.

6.3 Coherence

This is a basic concept needed for understanding interferometry. Light at two positions 1 and 2 separated by $\boldsymbol{r}_1$ - $\boldsymbol{r}_2 = \boldsymbol{r}$ is said to be coherent if superposition of these light beams would lead to interference fringes. More physically, coherence means the correlation of fluctuations of the optical wave field between these two positions. It is quantified by the *degree of coherence*

$$\gamma_{12}(\tau) = \frac{< V_1(t) \cdot V_2^*(t+\tau) >}{< |V_1(t)|^2 >^{1/2} < |V_2(t)|_2 >^{1/2}} \tag{56}$$

where V is the amplitude of the wave field, τ a time difference, and the denominator serves the purpose to normalize $\gamma_{12}(\tau)$ to 1.0 for $\mathbf{r} = 0$ and $\tau = 0$. The degree of coherence thus measures both temporal ($\Delta t = \tau$) and spatial (coordinates $\boldsymbol{r}_1$ and $\boldsymbol{r}_2$) coherence of the wave field.

The degree of coherence vanishes if the light beams at the two positions are not able to produce interference patterns (i.e. have different wavelengths, different directions of polarization or too large a path difference) or have a fixed path difference (e.g. $\lambda/2$) which makes $\gamma_{12}(\tau)$ to zero.

A time difference τ corresponds to a shift $c\tau$ along the incoming light beam. The region where $\gamma_{12}(\tau) \neq 0$ therefore represents a *coherence volume* which limits the relative positions of the telescopes of an interferometer for which fringes can be observed (introducing a delay in the light path is equivalent to shifting the telescope in the direction of the light beam). The *visibility* of the fringes as introduced in §2.5 is equal to the spatial part of the degree of coherence γ_{12}. The degree of coherence again, according to the Van Cittert - Zernicke theorem, equals the Fourier transform of the object brightness distribution,

$$\gamma_{12}(\boldsymbol{d}_{12} = \lambda \boldsymbol{u}) = \text{Visibility}(\boldsymbol{u}) = \tilde{O}(\boldsymbol{u}). \tag{57}$$

As noted above, this is the reason why the Fourier transform of an object image is also called its visibility. Because of the Fourier relation (57), the narrower the source, the larger the (spatial) coherent patch (e.g. for a star of solar size at a distance of 100 pc (angular diameter 9.3 $\times 10^{-5\prime\prime}$) it would be 1100 m.

Coherence length The temporal coherence is related to the spectrum of the source. Assume that interferometry of two light beams is performed with a flat spectrum of spectral bandwidth $\Delta\lambda$ (Figure 29) and that full interference (visibility of fringes = 1.0) occurs at pathlength equalisation. Then introduce a pathlength difference Δs. Fringes will disappear when the length Δs, expressed in wavelengths, differs by 1.0 over the bandpass, i.e. if $(n+1)\lambda_{min} = n\lambda_{max}$, or

$$\Delta s = \pm\lambda \cdot \frac{\lambda}{\Delta\lambda}. \tag{58}$$

(E.g. for measurements at 2.2 μm with $\Delta\lambda = 0.4$ μm, $\Delta s \approx 13$ μm results). Similar expressions are obtained for other spectral shapes.

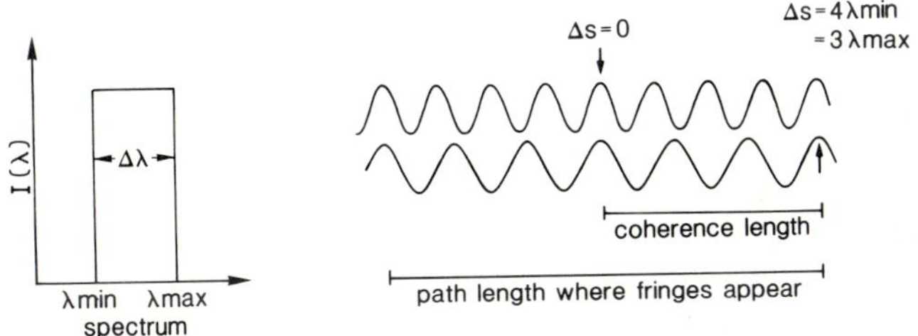

Fig. 29. Coherence length for a source with spectral width $\Delta\lambda = \lambda_{max}/4$.

Double Fourier interferometry For $\boldsymbol{r} = 0$, changing τ resp. the pathlength in one of the beams, is exactly the operation of a Fourier Transform Spectrometer (FTS). The resulting interferometric output signal $I(\tau) \approx 1.0 + \gamma_0(\tau)$ is the Fourier transform of the source spectrum $I(\nu)$, where the spectral resolution is given by the length of the scan in path difference. Again, the inverse relationship (58) between Δs and $\Delta\lambda$ is a consequence of this Fourier relation. For $\boldsymbol{r} = \boldsymbol{d} \neq 0$, changing the pathlength produces a fringe pattern which measures the visibility. However, each wavelength produces a sightly different fringe pattern, and these patterns are superimposed with $I(\nu)$ as weighting function. This is the more complete description of the simple relation (15). The Fourier transform with respect to τ of the measured $I(\tau)$ – at a fixed value of $\boldsymbol{d}$ – then gives the source spectrum $I(\nu)$ multiplied with the visibility corresponding to the separation $\boldsymbol{d}$ (see Mariotti and Ridgway 1988),

$$FT[I(\boldsymbol{d},\tau)] = I(\nu) \cdot Vis(\boldsymbol{d}). \tag{59}$$

If both the dependence on $\boldsymbol{d}$ and τ are investigated in an interferometer, this is called *Double Fourier Interferometry.* To some extent all interferometers are of this type because measuring fringes means changing pathlength (Figure 3). But in the usual setup of an interferometer pathlength changes of only a few wavelengths are used to measure the fringe contrast. The spectral resolution then is so low that in (59) $I(\nu)$ can be considered as constant, and the simple description and determination of visibility as in (15) and (60) applies.

Coherencing and cophasing It cannot be generally expected that the pathlength differences for an individual telescope or in an interferometric telescope array are smaller than the coherence length. *Coherencing* then means enforcing this condition, which allows the occurrence of fringes. A telescope in which for all light beams the optical path differences are limited to a small fraction of a wavelength, is said to be *cophased*, and the same description also applies to an interferometric telescope array.

Limitations by seeing For an undisturbed interferometric exposure, the fringe pattern has to be stable and regular, i.e. the interferometer has to be at least partially and temporarily cophased. Under the conditions of atmospheric turbulence the useful size of the coherent patch is reduced to the size of the Fried parameter r_0. In addition, the allowed exposure time is limited by the atmospheric coherence time τ_0 (§3). The number of photons available for such an interferometric exposure (which is also the number of photons per speckle) then is proportional to the product $r_0^2 \cdot \tau_0 \cdot \Delta\lambda/\lambda$, which with the wavelength dependences summarised in Table 1 is $\sim r_0^4 \sim \lambda^{4.8}$. Clearly there is an advantage of doing interferometry in the near infrared when compared to the optical region.

Pathlength equalisation So far, only optical interferometers with fixed telescope positions have been funded (fixed during one measurement, while transport of smaller telescopes - and even the 1.8 m telescopes of the VLT interferometer - to different stations is foreseen, in order to be able to change the baseline in large steps). The daily motion of a star then leads to large and fast changes in optical pathlength difference (OPD) between individual telescopes. E.g., with a 100 m baseline in an east-west direction the OPD changes for a star on the meridian by 7 cos δ mm/s, where δ is the declination, and for a star which passes the east-west-meridian at an elevation of 30°, the pathlength difference has increased to half the baseline, or 50 m in this example. In fixed-telescope interferometers therefore the delay lines are large structures by themselves. The need for complex delay lines could be avoided by moving the telescopes in order to keep the optical pathlength differences within a range of, say ≤ 1 mm. Labeyrie et al. (1988) have proposed such a concept under the name OVLA (***O**ptical **V**ery **L**arge **A**rray*), in which a large ($\approx$20) number of individual telescopes surround a central beam combining station (Figure 30). This concept would have advantages for a purely

Fig. 30. Artists view of a version of the proposed OVLA interferometer, where the individual telescopes are supposed to move on a platform. Up to now an interferometrist's dream without major funding. From Labeyrie et al. 1992.

interferometric telescope array, e.g. it would allow beam transport by optical fibres of constant length and it would provide a larger interferometric field-of-view, but because of the anticipated complexity of moving telescopes it has not been realised so far. If ever plans for an optical interferometer on the lunar surface should materialize, the OVLA concept of moving telescopes (then on legs) may become more attractive (Labeyrie et al. 1992) because the alternative of maintaining delay lines in this environment would be a very complex task, too.

For interferometric measurements, the optical pathlength differences have to be kept constant to within fractions of a wavelength. This accuracy needs additional control by a device called a "fringe tracker" which uses the interference pattern (typically at a wavelength different from that used for the actual measurements) to stabilize the interference fringes. The Earth's rotation provides a different, simple means to equalize optical path differences in two-telescope interferometers, although only for the short moment which places the baseline perpendicular to the incoming beam. Measurements are performed by setting the interferometer at an OPD of a few mm and waiting until the Earth's rotation has reduced this difference to zero so that the interference fringes appear. In this case phase information is lost.

6.4 Beam Combination – Choices

Several choices have to be taken for the combination of the light beams in an interferometer, and the resulting setups can be used to classify interferometers.

Image plane and pupil plane interferometry Both, beam combination in the image plane and in the pupil plane are possible (Figure 31). In the image plane, given a sufficiently large cophased field-of-view, the interference can be studied separately for different objects contained in the image, while it is difficult to extract this information from pupil plane interferometry.

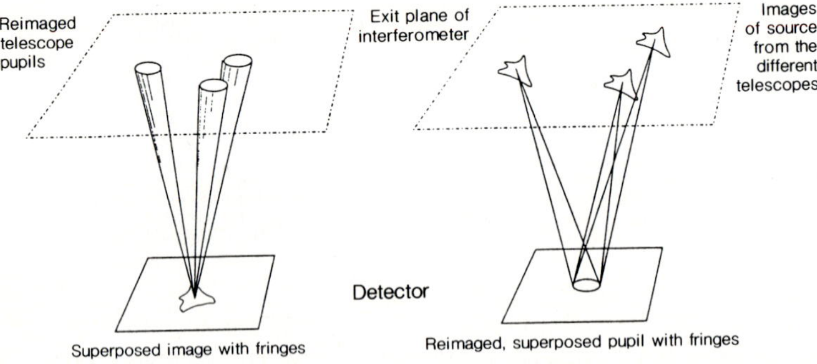

Fig. 31. Geometry for image plane and pupil plane beam combinations from Mariotti (1992).

Multi-mode and single mode interferometry The distinction refers to the quality of the interfering light beams. If they are diffraction-limited, we speak of single mode interferometry. Otherwise multi-speckled images occur and we speak of multi-mode interferometry. The latter, unwanted situation usually cannot be avoided for short wavelengths or for large telescopes. At the expense of complex data analysis, multi-mode interferometry also allows us to make use of all of the incoming light. A 1.8 m telescope is diffraction-limited under excellent seeing conditions ($\approx 0.5''$) for wavelengths larger than 4.2 μm.

Michelson- and Fizeau- configuration The distinction here refers to the relationship between entrance pupil and exit pupil. If the exit pupil is a true image of the entrance pupil , i.e. if the individual telescopes of the interferometer act approximately as subapertures of a very large optical surface, we speak of a Fizeau configuration. In this case a large cophased field-of-view is obtained (see §6.6), but the angular fringe separations $\lambda/\boldsymbol{B}$ may be inconveniently small for large baselines $\boldsymbol{B}$. By reconfiguring the exit pupil with respect to the entrance pupil (Michelson configuration), direction and spacing of the fringes may be accommodated to ones needs, but the usable field-of-view gets considerably smaller, because the interference fringes tend to move out of off-axis images (Tallon and Tallon-Bosc 1992).

Coaxial and multi-axial beam combination The distinction here refers to the geometry of beam combination. Again there are two options (Figure 32). Oblique interference of the light beams (multi-axial beam combination) leads to spatial fringes separated by $\lambda\cdot$ (f/d), where f/d = $1/(2 \tan\alpha/2)$ is the focal ratio corresponding to the angle α between the light beams. By means of a beamsplitter the interfering beams can also be made parallel, with $\alpha = 0$. In this *coaxial* or *flat field* mode the spatial fringes disappear. Instead, by introducing a pathlength difference Δs(t) in one arm of the interferometer, now temporal modulation ("temporal fringes") is obtained in the two outputs I_1 and I_2:

$$I_1(t) = I[1 + V \cdot \cos(2\pi\, \Delta s(t)/\lambda + \phi)]$$
$$I_2(t) = I[1 - V \cdot \cos(2\pi\, \Delta s(t)/\lambda + \phi)], \qquad (60)$$

where I is the flux contained in each of the incoming light beams. From (60) in principle the modulus V and the phase ϕ of the visibility can be determined. If the fluxes F_1 and F_2 in the two interferometer arms are not equal, e.g. due to seeing effects or due to differences in transmission, the measured

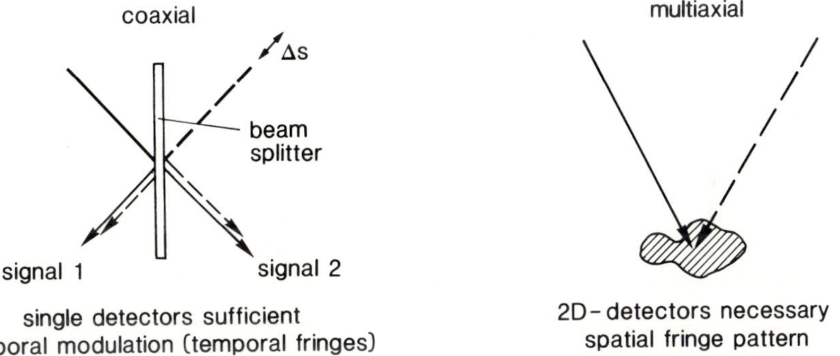

Fig. 32. Geometry for coaxial and multi-axial beam combinations.

modulus of the visibility will be gradually reduced, the reduction factor being $2\sqrt{F_1}\sqrt{F_2}/(F_1+F_2)$. It is an advantage of coaxial beam combination that single element detectors are sufficient for the measurement. However, as a rule this mode of operation can only be used for diffraction-limited beams.

6.5 Beam Combination – Realizations

Beam combining telescope For the VLT interferometer the exit pupil of the interferometer will be imaged to the entrance pupil of a 2 m diameter telescope, which then combines the beams in the image in its focal plane. This setup works with diffraction-limited as well as with multi-speckle images. The simulation of a multi-speckle image for interferometry with the four 8 m telescopes of the VLT is shown in Figure 33. Different fringe systems corresponding to the different baselines can be seen.

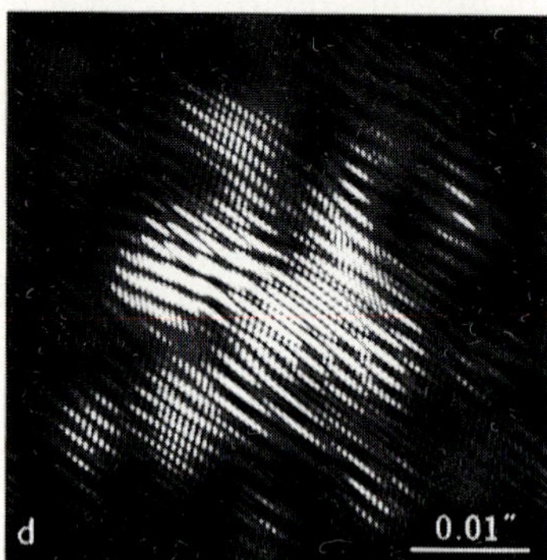

Fig. 33. Simulation of a speckle pattern as it would be produced by combining the light beams from the four 8 m telescopes of the VLT in interferometric mode. $D/r_0 = 4$ has been assumed. The differently oriented and differently spaced fringe patterns correspond to the different telescope-telescope baselines. From Reinheimer et al. 1992.

Dispersed fringes (Figure 34). This method applies to diffraction limited images but also to several speckles, if these are by some means aligned along the entrance slit of a spectrograph. By dispersing the light perpendicular to the orientation of the fringes, the coherence length now corresponds to the spectral resolution $\delta\lambda$ of the spectrograph, and fringes can be observed also if the optical path difference is not close to zero . This method is applied for optical interferometry with the G2T interferometer (two 1.5 m telescopes) on the plateau of Calern. It was this interferometer which supplied the important demonstration to the scientific community that indeed optical interferometry with large telescopes and multi-speckle images is possible (Labeyrie et al. 1986).

Optical fibres . The properties of optical fibres have been described, e.g. by Jeunhomme (1990), while the usefulness of fibres for interferometry has been stressed by Froehly (1981) and Connes et al. (1987) and has been more quantitatively discussed by Shaklan and Roddier (1987) and Rohloff and Leinert (1991). In multi-mode fibres large travel time, i.e. phase differences occur between different parts of a light beam. For interferometry only single-mode fibres are useful. Since standard single-mode fibres actually carry two – interacting – modes with different directions of polarization, polarization maintaining single-mode fibres are to be preferred. In these the interaction between the two polarization directions is largely suppressed, but the difference in propagation between the two

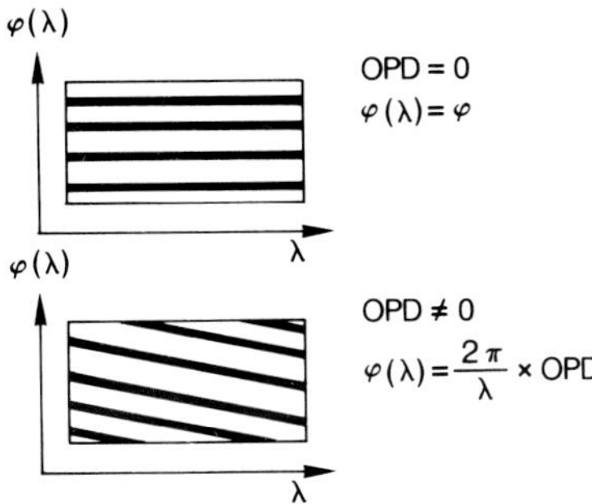

Fig. 34. Spectrum of a fringed speckle for optical path difference zero (above) and for finite optical path difference. This kind of setup allows a direct determination of the path difference from the slope of the fringes. From Koechlin and Rabbia 1985.

modes is enlarged, so that interference has to be performed separately for the two directions of polarization. A single-mode fibre essentially accepts one Airy disk and therefore is well suited for single-mode interferometry only. With optimum adaptation to the acceptance angle α of the fibre ($\sin\alpha$ is called the *numerical aperture*, usually with values of $\approx$ 0.1-0.2), and neglecting reflection losses at the fibre entrance, 78 % of the light in the Airy pattern are accepted by the fibre (Shaklan and Roddier 1988). In a fibre the distinction between image and pupil plane is meaningless and only one well defined type of beam exists. This is an advantage for beam combination. For optical fibres there exists the equivalent of a beam splitter.The necessary coupling between light in the two fibres is achieved by keeping the cores of the fibres closely parallel (within wavelengths)

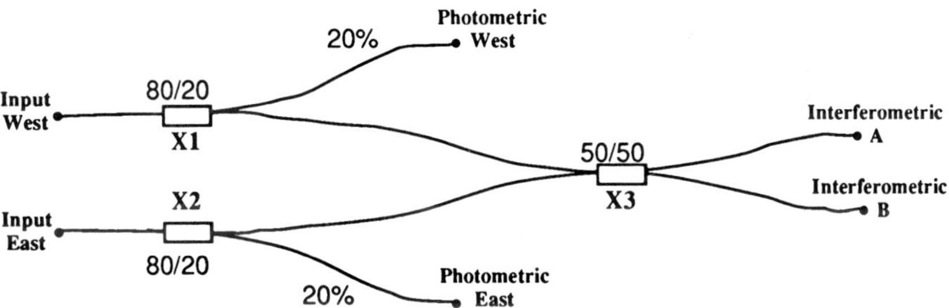

Fig. 35. Principle of the infrared fibre-optics interferometer FLUOR. The input is from two separate 0.8 m telescopes. The X-coupler X3 performs the interferometric coaxial beam combination. Couplers X1 and X2 (with each one input leg cut) separate a small fraction of the incoming light to allow measurements of intensity fluctuations. From Coudé de Foresto and Ridgway (1992).

over several cm. The device is called "X coupler" for obvious reasons (see Figure 35). Single-mode fibres thus naturally provide coaxial beam combination.

The main advantage in using fibres for beam combination is, that once the light is in the fibre, no further optical elements and no further adjustments are necessary, and the light stays well concentrated. The main disadvantage is a strong dependence of the optical length of the fibres on temperature, but to a lesser extent also on pressure and bending. Single-mode fibres for the near infrared, made from zirconiumfluoride glass have been developed during the past years. Coudé de Foresto and Ridgway (1992) demonstrated the operation of a near-infrared fibre-based interferometer (Figure 35) with measurements of the diameter of α Ori. They used the two auxiliary mirrors of the Kitt Peak solar telescope, which define a baseline of 5 m.

Error budget Many factors affect the accuracy with which an interferometer can measure the visibility of an object (Beckers 1990a). In particular, the requirements on guiding needed to obtain accurate (i.e. $\pm$ a few percent) interferometric measurements are quite stringent. The allowed tracking error on the 8 m telescopes of the future VLT interferometer will be as small as $0.018''$.

6.6 Limitations of the Field-of-View

In Figure 6 wavefronts arriving perpendicular to the baseline $\boldsymbol{d}$ have zero optical path difference while for light coming at an angle α to the baseline a path difference of $\Delta s(\alpha) = d \cdot \sin\alpha$ results. For a general viewing direction α now the variation of pathlength difference with a small angle β away from the the axis of the field-of-view similarly is $\Delta s(\beta) \approx d \cos\alpha \cdot \beta$. This strong angular dependence of Δs, unless compensated, leads to a serious limitation of the interferometric field-of-view, down to its natural minimum value, which for nearly monochromatic radiation is the Airy diffraction disk (Tallon and Tallon-Bosc 1992). In the case of interferometry at 2.2 μm at an 8 m telescope, this corresponds to $0.057''$, which is too small for many observing programmes. The field-of-view may be further reduced if the bandwidth is no longer negligible.

Bandwidth and field-of-view. As is the case of speckle interferometry with individual telescopes, most interferometers do not compensate for the path difference associated with the size of the field-of-view. Then Δs has to be kept smaller than the coherence length (58) in order to obtain observable fringes. This limits the allowable bandwidth for a desired field-of-view and vice versa

$$\frac{\Delta\lambda}{\lambda} \leq \frac{\lambda}{\beta d}. \tag{61}$$

Even if the field-of-view is limited to the Airy disk size of an individual telescope of λ/D, still a strong restriction on the allowable bandwidth results: $\Delta\lambda/\lambda \leq D/d$. Examples of the relationship between field-of-view and allowable bandwidth are given in Table 3. For a Michelson interferometer, because of the reconfiguration of the output pupil, the field-of-view is further reduced (Mariotti 1992).

Table 3. Allowable bandwidth for different baselines and field-of-view sizes[a]

d(m)	$\pm\beta('')$	$\Delta\lambda/\lambda^a$	Comment
4	1.0	0.23	Speckle interferometry with single telescope
100	1.0	0.01	VLT interferometer
100	0.057	0.08	Airy disk of 8 m telescope (VLT main telescopes)
100	0.25	0.018	Airy disk of 1.8 m telescope (VLT auxiliary interferometric telescopes)

[a] for $\lambda = 2.2\mu$m

Increasing the field-of-view Beckers (1990b) and Merkle (1988b) have shown that it is possible to compensate for the field-of-view induced path differences $\Delta s(\beta)$ up to the second order in β, if the exit pupil always is a true image of the entrance pupil (as seen from the object) and if the output pupil is approximately coplanar. To achieve this, complex optics has to be introduced i.e. a movable and at the same time deformable mirror. The expected gain in the size of the field, over which $\Delta s(\beta)$ amounts to only a fraction of a wavelength, is considerable: for the 8 m telescopes of the VLT interferometer a field-of-view of 8″, for the auxiliary 1.8 m telescopes 2″ are foreseen (Figure 36).

6.7 Results

Recent results of multi-telescope interferometry have been summarised by Ridgway (1992). The trend of the work is from individual determinations of stellar diameters to treating astrophysical problems, thereby also creating new questions. At the time of writing, closure phase measurements and hence interferometric images are just being demonstrated for the visible wavelength range (Baldwin et al. 1993) and have not yet been performed in the near infrared.

Angular stellar diameters and effective temperatures With individual visibilities measured to ± 1% - 5%, accuracies reaching or surpassing those of lunar occultation measurements have been obtained.

Limb darkening has not yet been measured directly, but the determination of the change of apparent diameter with wavelength allows a comparison with the prediction from model atmospheres (Mozurkewich et al. 1991).

The effective temperature scale for K and M giants has been improved and shows a feature not recognised in earlier lunar occultation measurements (Figure 37, Di Benedetto and Rabbia 1987).

Effective temperatures for supergiants were found to be smaller than for giants of the same spectral type. E.g. 32 Cyg (K5Iab, $R = 215 \pm 33\ R_\odot$) has a measured T_{eff} of 3600 ± 200 K (Di Benedetto and Ferluga 1990a), instead of ≈ 4000 K read from Figure 37. The coolest supergiant found so far is α Her with $T_{eff} = 3260 \pm 40$ K, measured with the simple, earth-rotation scanning two-mirror interferometer IRMA in Wyoming (Benson et al. 1991).

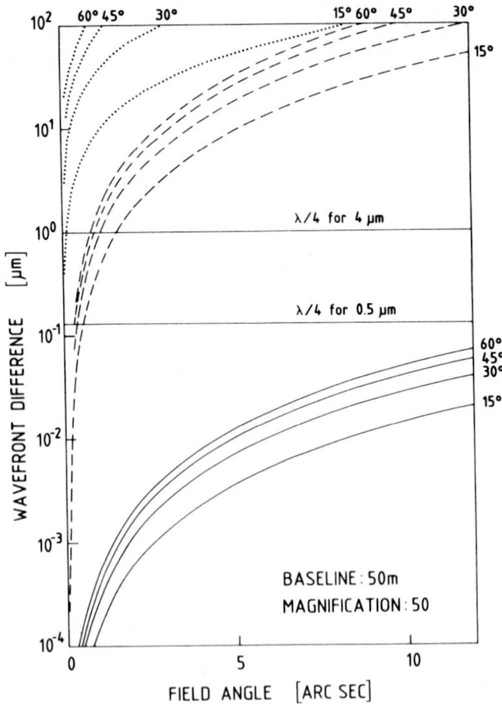

Fig. 36. Dependence of the path difference $\Delta s(\beta)$ on field angle β and zenith angle for different orders of compensation. Dotted lines: on-axis compensation only. Broken lines: projection of the pupil configuration corrected. Solid lines: axial position of the output pupil corrected. The usable field-of-view is taken as the angle where the curves intersect the horizontal line corresponding to a path difference of $\lambda/4$. From Merkle 1988b.

Measurements of surface gravity have been extended to the class of G giants by measuring the angular diameter of the companion to Capella to $6.28 \pm 0.43 \times 10^{-3}{}''$(Di Benedetto and Bonneau 1991).

Small scale structure, possibly due to the presence of a companion, has been inferred for β And, because the measured visibility does not go to zero as expected for a limb-darkened stellar atmosphere (Di Benedetto and Bonneau 1990b).

Circumstellar envelopes The results presented in this section were obtained with heterodyne interferometry at $\lambda \approx 11\mu$m by the group at Berkeley pioneering this technique for astronomical applications. A simple description of the method can be found in in one of their earlier publications (Sutton 1979).

Inner (dust formation) radius for the shells of late giant stars
Measurements at longer baselines showed for the bright carbon star IRC+10216 (Danchi et al. 1990) as well as for the M giant Mira (Bester et al. 1991) that the visibility continues to decrease with increasing baseline (Figure 38). This means that the envelope extends closer to the star than previously thought. In both

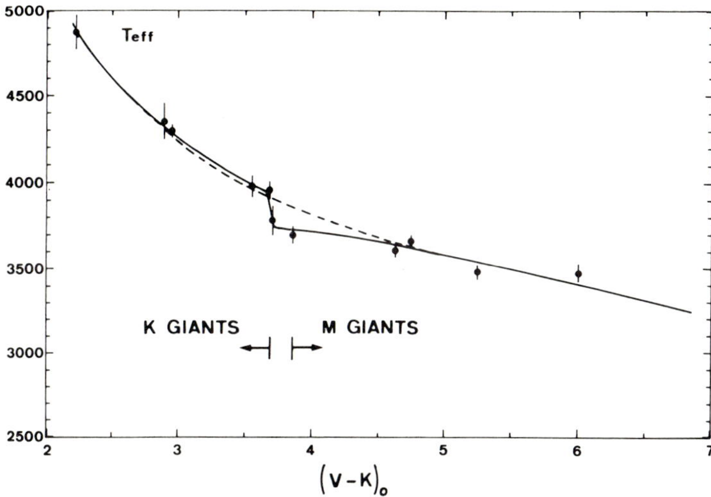

Fig. 37. Measured effective temperature as function of (V-K) colour for K and M giants. From Di Benedetto and Rabbia 1987.

cases the inferred inner radius of the circumstellar shell is at ≈ 3 R_*, at which position temperatures of $\approx$ 1200 - 1300 K are predicted.

Peculiar distribution of dust around α Ori

Signs of a damped oscillation in the decreasing visibility curve, with the position of the second maximum corresponding to a distance of $0.9''$, have been interpreted as due to an optically thin dust shell at about 30 R_* from the star (Bester et al. 1991). The necessary amount of dust and its expected temperature are consistent with earlier observations of CO gas around α Ori by Bernat et al. (1979). This gas expands with a velocity of 11 km/s. Assuming that the dust and the CO gas are coupled, the data could be explained by the ejection of large amounts of mass in an event about 100 years ago.

Astrometry The optical Mark III interferometer on Mount Wilson (Shao et al. 1988a) has been used with success both for wide angle astrometry (internal accuracy $\pm$ 6 mas in δ, $\pm$ 10 mas in α, Shao et al. 1990) and for astrometry of close binary orbits (e.g. α And with a semi-major axis of 24.15 $\pm$ 0.13 mas, Pan et al. 1992). At least the latter type of measurements will have attractive applications in the near infrared. For example, it would allow us to determine masses for young close binaries for an empirical verification of evolutionary calculations.

7 Adaptive Optics

7.1 Overview

A telescope is said to be equipped with *active optics* if its optics can be adjusted any time to minimize the telescope-induced image degradation; e.g., at ESO's

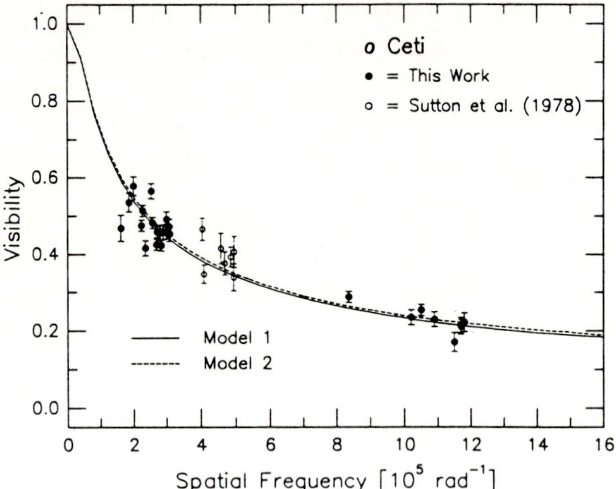

Fig. 38. Extension of the visibility measurements on Mira at 11 μm to higher spatial frequencies, showing that the dust shell extends as close to the star as 3 R_* (2×10^5 rad^{-1} correspond to 1 $arcsec^{-1}$). From Bester et al. (1991).

New Technology Telescope (NTT), the prototype of active optics instruments, this optimization procedure, when observing on a bright star, only takes about 20 minutes. On the other hand, we speak of *adaptive optics* if the image-correcting system is fast enough to correct for the instantaneous image degradation due to the effects of the turbulent atmosphere with their typical timescales of 10 ms (for good correction the adaptive optics system even has to be several times faster than the atmospheric fluctuations). In both cases the correction is based on measurements of aberrations in the shape of the incoming wavefront. With full compensation for the atmospheric degradations, an adaptive optics system will produce diffraction-limited images. If only part of the atmospherically induced wavefront error is corrected, we speak of a *partial adaptive optics* system. Of the imaging algorithms discussed above in §5.1, "shift and add" actually has to be considered as a partial adaptive optics method (Glindemann 1993). As in this particular case, partial adaptive optics systems in general lead to images with a diffraction-limited core on top of a halo the size of which is determined by the seeing (see Figure 19). Generally, improving the wavefront increases the maximum brightness in the image. If the quality of the incoming wavefront is measured by the mean square deviation from a spherical wave $s_\phi^2 = <(\Delta\phi)^2>$, then according to Maréchal (see Born and Wolf 1970) the Strehl ratio can be expressed as

$$Z = I/I_0 = 1 - s_\phi^2(rad) \qquad (62)$$

for small wavefront perturbations. Thus to obtain essentially diffraction-limited images, i.e. $Z \geq 0.80$, the rms wave front error has to be $\leq \lambda/14$.

Babcock (1953) proposed adaptive optics as a means to sharpen astronomical images, useful astronomical adaptive optics systems have only come into operation during the past few years. At present, a full adaptive optics system is only

available for the near infrared; it will shortly be described below. Its application is limited by the wavefront sensor to objects brighter than ≈ 13 mag in the visible (or ≈ 4 mag at 2.2 μm if an infrared sensor is used), respectively to objects with a source of this brightness in the immediate environment.
An obvious gain of adaptive optics lies in the sharpening of the images which allows the resolution of detailed spatial structures. But there are a number of additional advantages (Beckers 1993): it reduces the area of sky contributing to a point source measurement, thereby increasing the contrast to the background by up to a factor of $(D/r_0)^2$ and allowing to detect correspondingly fainter sources. For spectroscopy, it gives both an increase in sensitivity and in possible spectral resolution. For multi-telescope interferometry, by essentially reducing the short exposure image to one speckle, it makes phasing of different telescopes much easier and greatly enhances the sensitivity of the interferometer. Even an adaptation of the point spread function of the telescope to specific purposes (e.g., to searching a star for a nearby faint companion object) seems possible. It is this great versatility which has led to the widespread high interest in this technique.

7.2 Principle

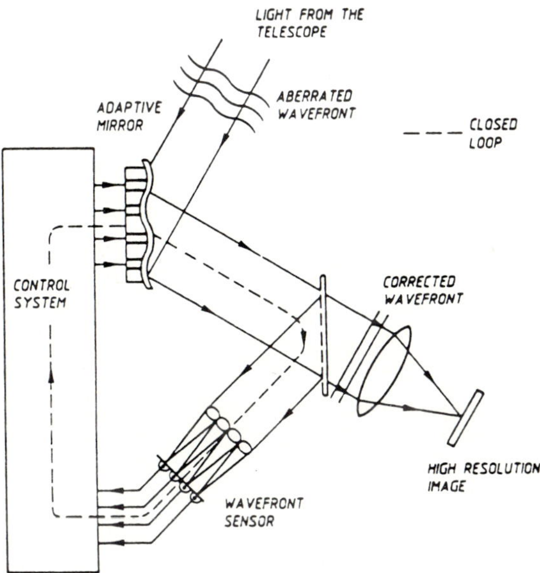

Fig. 39. Principle of an adaptive optics system. From Merkle and Beckers 1989b.

As stated in §3, the image degradation induced by the turbulent atmosphere is almost exclusively due to phase effects (i.e. wavefront errors), while the amplitude fluctuations in the incoming light beam, which give rise to scintillation, play a negligible rôle. Therefore it constitutes no drawback that adaptive optics systems so far only are able to correct the wavefront. The principle of operation is shown in Figure 39. In front of the image plane a *deformable mirror* is introduced into the light path, as well as a beam splitter which extracts part of the light to feed a *wavefront sensor.* The output of this sensor is analysed in terms of wavefront errors, and a feedback signal is computed and sent to the deformable mirror to correct for the measured wavefront errors. If this feedback cycle is sufficiently fast compared to the coherence time τ_0 of the atmosphere so as to allow a correction better than a fraction of a wavelength over the whole aperture, the corrected image will be diffraction-limited (compare (62)). In the visible at a 4 m telescope such a full adaptive optics system would require cycle times several times shorter than 10 ms and a number of sensing and correcting elements of about $(D/r_0)^2 \approx 1600$. Such systems are not yet available. The corresponding numbers for the near infrared at 3.5 μm would be 100 ms and about 15 sensing and correcting elements (compare Table 1). Obviously, adaptive optics can much more easily be implemented in the near infrared than in the visible.

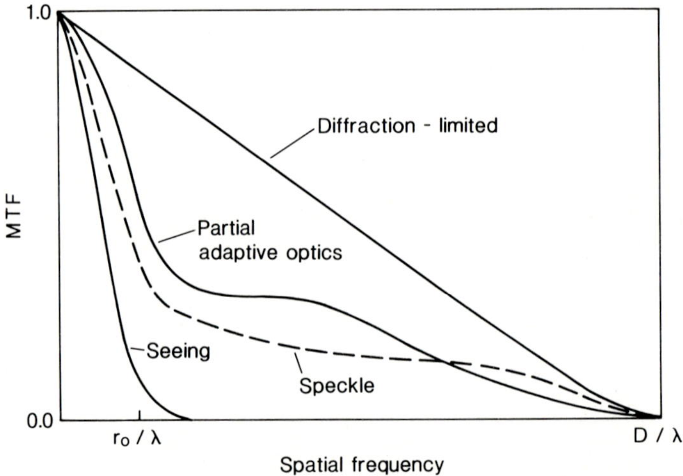

Fig. 40. Qualitative comparison of the modulation transfer functions corresponding to different imaging methods

Figure 40 compares in a qualitative way the sensitivity obtained with adaptive optics at different resolutions to other imaging methods. The general trend is that partial adaptive optics clearly is particularly strong at lower spatial frequencies (low to intermediate spatial resolution), while at the highest spatial frequencies, speckle interferometry still appears to be more effective.

7.3 Measuring and Correcting the Wavefront

The most commonly used wavefront analyzer is the *Hartmann-Shack* sensor which measures the gradient of the wavefront separately for many subapertures, from which the instantaneous shape of the wavefront can be reconstructed and thus the wavefront errors can be determined. The principle of operation is shown in Figure 41. An array of lenslets cuts the aperture into many subapertures and forms separate images for each of these subpupils on a detector array. Any deviation of an image center from the optical axis then is a direct measure of the average tilt of the wavefront over this subaperture. The wavefront sensing for a near-infrared instrument can as well be performed in the visible, since the effective refractive index (n-1) varies by less than 2% from 0.55 μm to 5 μm.

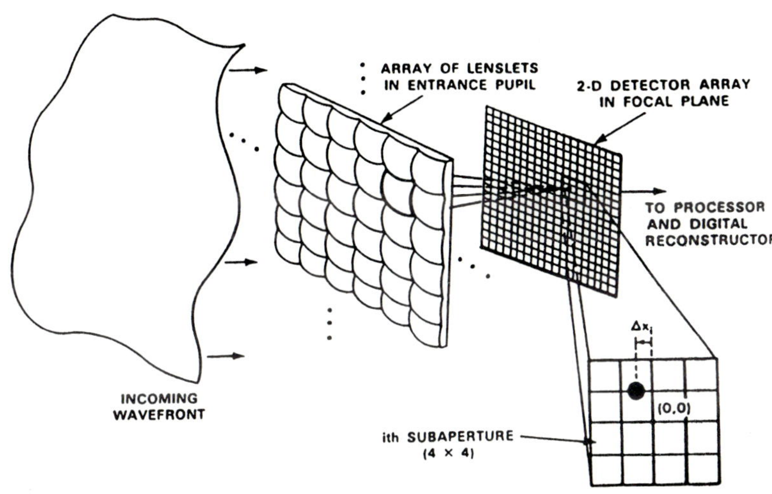

Fig. 41. Principle of operation of a Hartmann-Shack wavefront sensor. Each lenslet forms an image of the object in a different subarray of the detector. Deviations from nominal position reflect the tilt of the wavefront. From Murphy (1992), reprinted with kind permission of Lincoln Laboratory, MIT, Lexington,MA.

It is convenient to describe the measured instantaneous wavefront shape by a complete set of orthogonal polynomial functions, thus discriminating slowly varying low-order modes, which are more easily to correct for, from the highly oscillating high-order modes, which require a more sophisticated correction pattern. Usually one uses the Zernike polynomials (Born and Wolf 1970) which can be separated into the product of a radial polynomial related to the Legendre polynomials and an azimuthal dependence described by sinewaves with an integer number of periods around the aperture (Wang and Markey 1978). (The implicitly defined Karhunen-Loève functions have the additional advantage of having statistically independent coefficients. However, in their lower modes they are almost identical to the Zernike polynomials (Wang and Markey 1978), and we will not discuss them here). The lower order Zernike polynomials correspond to the well known basic optical image distortions, as summarised in Table 4 and are shown for a few typical cases in Figure 42. This helps to get an intuitive feeling for the wavefront corrections to be applied.

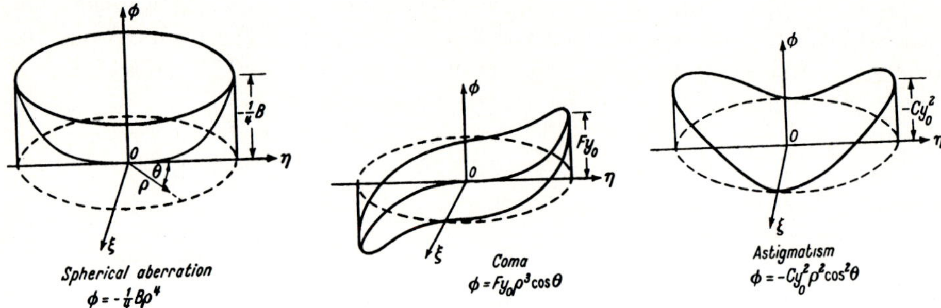

Fig. 42. The wavefront aberrations of a few basic image distortions. Adapted from Born and Wolf 1970.

Table 4. Low order Zernike polynomials and the corresponding optical aberrations

Radial degree (n)	Azimuthal frequency(m) 0	1	2	3	Mean square phase error $< (\Delta\phi)^2 >$ due to each mode
0	Z_1 Piston				–
1		Z_2,Z_3 Tip,tilt			43.5 %
2	Z_4 Defocus		Z_5,Z_6 Astigmatism		2.2 %
3		Z_7,Z_8 Coma		Z_9,Z_{10} Trifoil	0.6 %
4	Z_{11} Spherical aberration		Z_{12},Z_{13}		0.22 %

There is a further advantage in describing the wavefront by orthogonal functions: to determine the low order modes the shape of the wavefront need not be determined exactly, and to correct for them a smaller number of actuators behind the flexible mirror suffices. For example, to determine the tip/tilt correction only requires a single quadrant detector for the image formed by the whole pupil, and to correct for the image motion detected by this device the correcting mirror needs not to be flexible at all but only movable around two axes. In general it can be shown that if the number of actuators of the flexible mirror is insufficient for a full wavefront correction, the best result is obtained by correcting for the low order Zernike modes first (Roddier 1992). The efficiency of an adaptive optics system therefore can be measured by the equivalent number of Zernike modes it corrects for (note that the first one, piston, corresponds to a small shift of the instantaneous wavefront towards or away from the telescope and is irrelevant for the formation of the image). For the actual correction, however, the Zernike modes cannot be used directly; instead one has to use the vibration modes of the mirror plate, which are broadly similar.

The main errors reducing the performance of an adaptive optics system are those related to the fitting of the wavefront, to the time delay between the beginning of wavefront analysis to the end of wavefront correction and to the limited, indeed quite small size of the isoplanatic patch. Errors in the wavefront correction occur because both the number of analyzing subpupils and of actuators behind the flexible mirror is finite or even definitely insufficient for a full correction. The effect of a time delay is obvious: by the time the correction is done, the wavefront may already have changed again. Problems with isoplanaticity do not occur if light from the object itself (e.g. from a different wavelength range than the one to be measured) can be used as reference for the wavefront analysis. However, if a different nearby star is used as reference, small residual wavefront errors always remain, and if the correlation between the wavefront shapes at the two positions is not at least as good as 50%, no improvement at all is obtained. The size of these errors may vary with time and atmospheric conditions in an unpredictable way. Therefore, while in principle adaptive optics should provide fully phase-corrected (i.e. diffraction-limited) pictures, in practice possible phase errors should be calibrated, as in speckle interferometry, by frequent near-simultaneous observations on a nearby unresolved reference star. The term *reference* in adaptive optics is applied both to a nearby object used for wavefront analysis and for a point source used for phase calibration and deconvolution. Instead of inventing a new terminology I will try to be clear about the context and rely on the care of the reader to infer the correct meaning.

7.4 Gain in Resolution

Roddier et al. (1991) and Roddier (1992) have discussed the gain in resolution provided by adaptive optics with respect to conventional imaging. Their computational results are shown in Figure 43 for different levels of wavefront correction as function of the "seeing-effective" telescope size D/r_0. They discuss the Strehl resolution R, which is defined by the average of the $MTF(\boldsymbol{u})$ of the image-forming system, and which measures the maximum brightness of the image. Here it is normalized to the pure seeing-limited resolution ($\approx (r_0/\lambda)^2$,because two-dimensional resolution is considered), which is called R_{max} because it is the

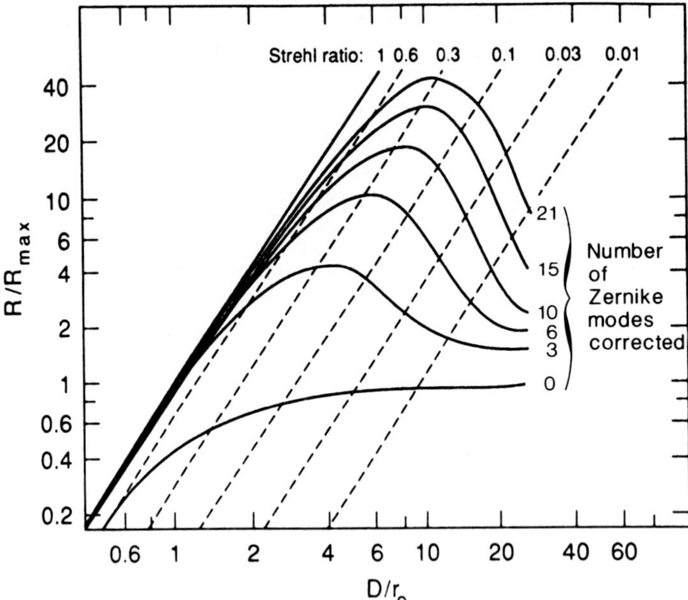

Fig. 43. Normalized angular resolution as function of effective telescope size D/r_0 for different degrees of wavefront correction. The numbers of corrected Zernike modes is given with the individual curves. Zero corresponds to no correction. Adapted from Roddier 1992.

maximum spatial resolution which can be obtained in long exposure images for given seeing conditions (but only can be obtained for large values of D/r_0). In this representation the diffraction limited resolution is a straight line $\sim (D/r_0)^2$ and the lines of constant Strehl ratio (see Equ. 2) are parallel to it. It can be seen that for a given order of compensation the gain in resolution is at maximum with respect to the uncorrected image short of the diffraction limit (which would require a Strehl ratio ≥ 0.80) at a Strehl ratio of about 0.3. This may appear as less than perfect image quality but still is twice as good as the Hubble Space Telescope before optical correction. In the case of a tip/tilt correction (radial order of Zernike polynomial $= 1$) this maximum occurs at $D/r_0 \approx 3$. The gain in the number of necessary actuators with respect to the number $(D/r_0)^2$ needed to approach the diffraction limit is obvious.

Generally, the correction of lower order modes of the wavefront error is easier to perform than the full correction

- because fewer elements are needed for wavefront analysis and wavefront correction
- because for lower order corrections a lower feedback cycle frequency, and hence a lower bandwidth of the system, is required and
- because the isoplanatic patch gets larger the lower the order of correction is, being about $1'$ for simple tip/tilt systems.

The first two of these relations mean, apart from the implied lower complexity of the system, that more photons are available per element of the wavefront sensor per measuring cycle and hence the magnitude limit for the reference stars is appreciably fainter than in the case of full correction. The last relation states that this reference star can be selected from a larger field which increases the chances of finding one. Therefore, Roddier et al. (1991) advocate the use of low-order adaptive optics systems. Near-diffraction-limited resolution then can be obtained for these systems by afterwards deconvolving the partially corrected images of the object with comparable images taken on a reference point source.

The success of lower order correction is related to the fact that a large part of the rms phase fluctuation of the incoming wavefront is in the lower order modes. For example, simple tip/tilt correction will already reduce the rms value of these fluctuation to 36% (see Table 4). However, the above arguments may give too optimistic an impression of what can be achieved and at what cost. The better a correction of an image is achieved, the higher the frequency of low order corrections has to be in order not to distort the improved image, and the smaller the isoplanatic patch size for the low order modes, even for tip/tilt correction (Rigaut and Gendron 1992a).

7.5 The Guide Star Problem

Natural stars

Finding a reference star for atmospheric correction (often called "guide star"), which has to be of sufficient brightness and close enough to the object, turns out to be a problem in many cases. Figure 44 shows on the basis of the Galaxy models by Bahcall and Soneira (1980) that, except for tip/tilt systems, the probability of finding such a star will be small even under optimistic estimates for the necessary limiting magnitude. This appears to be a serious drawback for the application of adaptive optics to the observation of extragalactic objects.

Laser guide stars

Luminous spots, created by lasers high in the atmosphere were proposed as artificial guide stars by Foy and (again!) Labeyrie (1985). This appeared as an elegant and perfect way out of the guide star problem, because these spots can always be placed close to the target. Figure 45 shows that the light from such a spot essentially suffers the same atmospheric distortions as light from a star. The laser beam creating the spot can be sent up through the observing telescope itself. Contamination of the measurements of the object by laser light can be avoided, e.g., by pulsing the laser output and closing the detector during the time of the laser pulse. Such laser systems, originally developed for military purposes, are slowly becoming available for astronomical applications (Primmerman et al 1991, Fugate et al. 1991). Figure 46 shows the image sharpening achieved on a bright double star by this technique (Fugate et al. 1991).

However, difficulties remain. First, in the experiments performed so far, atmospheric correction based on lasers did not reach the same quality as what could be done with natural reference stars. In part this may result from the fact that so far backscattering of the laser light from the troposphere has been

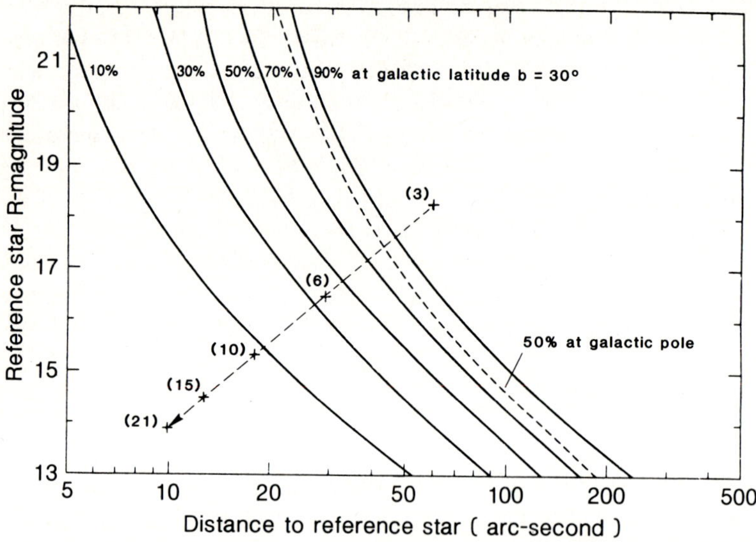

Fig. 44. Probability of finding a reference star for adaptive optics wavefront correction. For a given number of corrected Zernike modes (shown in parenthesis) the crosses indicate the minimum brightness of the reference star needed and the maximum allowable distance from the object. The curved lines give the probability of finding a star of a given brightness at a given distance. Adapted from Roddier (1992).

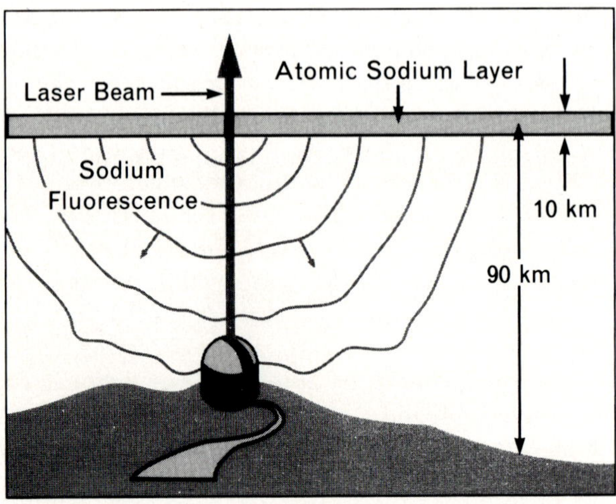

Fig. 45. Schematic drawing showing how an artificial guide star can be created by laser excitation of the mesospheric sodium layer. From Th.H. Jeys 1991, reprinted with kind permission of Lincoln Laboratory, MIT, Lexington, MA.

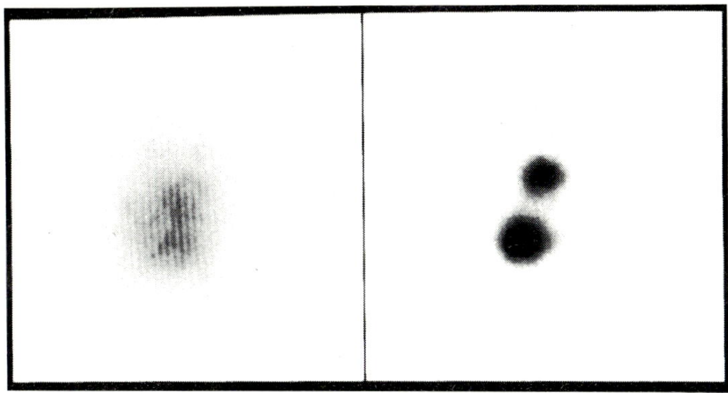

Fig. 46. Demonstration of an adaptive optics system using a laser guide star with a 1s exposure on the binary ξ UMa (separation $\approx 1.3''$). Left: uncorrected image. Right: with adaptive optics correction. A Strehl ratio of 0.1 is obtained. The observations were performed at 0.88 μm with the 1.5 m telescope at the Starfire Optical Range, New Mexico from Fugate (1991).

used to create the artificial stars. The height of these sources then is limited by the extent of the dense lower atmosphere to about 10 km. Light from these sources reaches different parts of the telescope with different angles - in contrast to light coming from a true star. Exciting resonance radiation from the sodium layer at 90 km (see Figure 45) would largely avoid these difficulties. It has been demonstrated as a valid technique in principle by several groups (see Table 6 in Beckers 1993) but needs lasers of very high optical power of the order of several kW. In addition, laser created artificial stars have the shortcoming that they are not sensitive to image motion. Whatever tip and tilt deviation the laser beam wavefront may experience on the way up through the atmosphere, will cancel again when the laser light comes back through the same turbulent atmospheric layers. Thus, still a natural guide star is needed as reference for image motion. In principle one still should gain because of the reduced brightness limit on the star if only image motion (tip/tilt) is to be corrected. But the systems are getting more complicated than thought of by many earlier workers, and truly effective working solutions cannot be expected within the next few years.

7.6 Results

Because of the rapid evolution of the field of adaptive optics, this summary necessarily has an incomplete and preliminary character. Various adaptive optics systems are in different stages of their development (see, e.g., Roddier 1992, Beckers 1993), including adaptive optics using artificial stars created by laser beams (Fugate et al. 1991, Primmerman et al. 1991). Full adaptive optics systems for the visible require too many corrective elements to be feasible at the moment. For the near-infrared so far one full adaptive optics system has come into operation as a user instrument, the COME-ON-PLUS system with its pre-

Table 5. Characterization of the adaptive optics COME-ON systems

	COME-ON[a]	COME-ON PLUS[b]
Type of wavefront sensor	Hartmann-Shack	Hartmann-Shack
Number of subapertures	5×5	7×7
Measured two-dimensional wavefront gradients	25	49
Number of actuators	19	52
Corrected Zernike terms	≈ 10	≈ 20
Wavefront sampling	100 Hz	400 Hz
Bandwidth for correction	9 Hz → 25 Hz	30 Hz
Diffraction-limited for seeing ≈ 1″ at	≥ 2.2 μm	≥ 1.2 μm
Limiting magnitude for		
Strehl ratio ≥60 %	V ≈ 8 mag	V ≈ 9 mag
Strehl ratio ≥10 %	V ≈13 mag	V ≈ 14 mag

[a]Rigaut et al. 1992b [b]Rousset et al. 1992, Hubin et al. 1993b

decessor COME-ON, both on the 3.6 m telescope at La Silla.[1], and its characteristic parameters are given in Table 5. It can be considered as a precursor of the future VLT adaptive optics instrumentation. While the system can provide full correction at infrared wavelengths, partial correction will occur for the shortest wavelengths, for faint sources or for unfavourable seeing. In these cases, COME-ON-PLUS allows one to choose the number of Zernike modes to be corrected in order to take advantage of the better sensitivity of low-order correcting systems mentioned above.

The results obtained with this system are just beginning to be published. It is noteworthy that an appreciable gain can be obtained with this system also for wavelengths shorter than 1 μm, with expected Strehl ratios of up to 0.3 even at 0.5 μm (Hubin et al. 1993a). The astronomical results (those shown were still obtained with the COME-ON system and with a near-infrared wavefront sensor) include a determination of the rotation axis of Ceres (Figure 47) and a new model for the dust distribution around the young star Z CMa (Figure 48). Although the latter results have to be taken with considerable reservation because the pixel scale available at the time of the measurement was of the same size as the object structures to be resolved, they provide a good example of the kind of information to be obtained with fully correcting adaptive optics systems.

[1] The name simply lists the initials of the institutions involved in the development of the system, CGE, Observatoire de Meudon, ESO, ONERA). The system has Been described by Rousset et al. (1990, 1992) and Hubin et al. (1993b

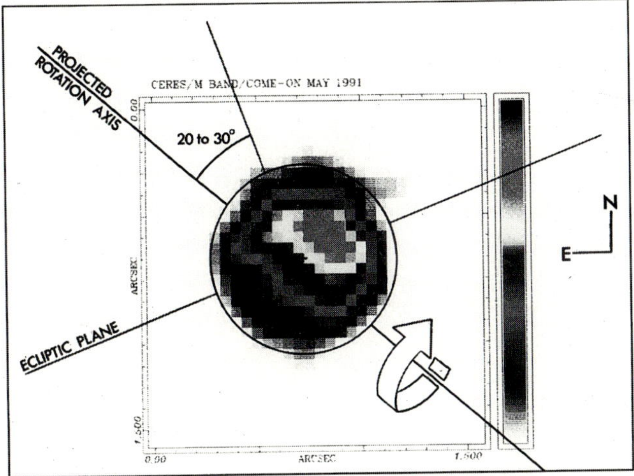

Fig. 47. Determination of the rotation axis of Ceres. This preliminary M band (4.7 μm) image shows the thermal emission of the surface of this asteroid. The offset of the region of highest temperature from the subsolar point is interpreted as due to the asteroid's rotation, which allows the authors to determine the orientation of the spin axis. The diameter of Ceres is $0.9''$. Adapted from Merkle et al. (1991), reprinted with kind permission of O.Saint-Pé and M.Combes (Obs. de Meudon).

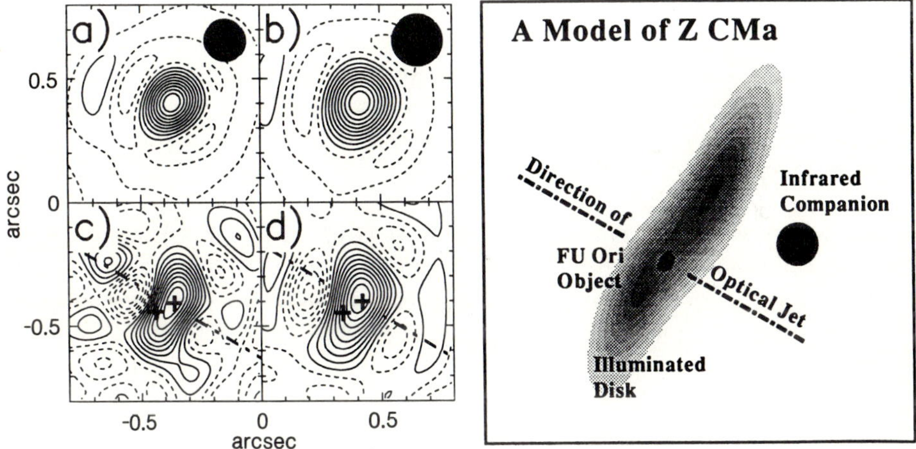

Fig. 48. Possible distribution of circumstellar dust around the young binary and FU Orionis object Z CMa. **Left: A** a reconstructed image at $L'(3.9\mu m)$. **b** A reconstructed image at M (4.8μm). **c** and **d** Remaining elongated structure after a binary with the known separation and with brightnesses of the components as determined from a best fit has been subtracted from the data. The crosses indicate the positions of the binary components, the black dots give the FWHM of a reconstructed point source profile. **Right:** Suggested interpretation of the data from Malbet et al. (1993).

8 Concluding Appraisal

The different near-infrared high spatial resolution techniques discussed in these lectures have different degrees of recognition in the astronomical community:

Lunar occultation observations remain an indispensable tool for reaching very high (few milliarcsec) spatial resolutions. However, their importance is limited because the occultation events are rare and objects cannot be selected at will. Therefore they are only followed by a small number of observers.

Speckle interferometry has been the working horse of near-infrared high spatial resolution observations. Considerable gains are to be expected by the new class of 8 - 10 m telescopes, expected to be available within a few years. It remains to be seen whether and to what extent speckle interferometry will be replaced by the developing adaptive optics systems.

Multi-telescope interferometry, e.g. with the ESO Very Large Telescope VLTI promising to allow routine observations at spatial resolutions of several milliarcsec. However, it is still an open question what the final sensitivity of these instruments will be and how well multi-telescope interferometry can be applied to extragalactic research. Since, e.g. the VLT interferometer will not be operational before around the year 2000, many consider it too large an investment of time to contribute to this demanding project. On the other hand, multi-telescope interferometry may well be "the best hope of astronomy for future dramatic breakthroughs in observational capability" (Ridgway 1992, a member of the team preparing the Bahcall report).

Adaptive optics is considered to be the most promising high spatial resolution technique by the majority of astronomers. Still it has to prove that it is able to combine the high sensitivity of long exposure images with diffraction-limited resolution. Quite possibly, adaptive optics observations followed by processing of the obtained images with speckle interferometric or similar techniques will give the best approximation to this ambitious goal.

Acknowledgments. I want to thank Andreas Glindemann and Andrea Richichi for helpful advice and discussions, the editors for their patience and Doris Anders, Karin Dorn and Martina Weckauf for the careful and professional production of the numerous figures.

References

Angel, J.R.P., Cheng, A.Y.S., Woolf, N.J.: Nature **322** (1986) 341

Ayers, G.R., Benson, J., Carels, K., Dyck, H.M., Spillar, E.: Astrophys. J. **360** (1990) 471

Babcock, H.W.: PASP **65** (1953) 229

Bahcall, J.N., Soneira, R.M.: Astrophys. J. Suppl. **44** (1980) 73

Baldwin, J.E.: Presentation at a meeting on broad band interferometry in Edinburgh, July 13, 1993

Bates, R.H.T., Davey, B.L.K., in: Interferometric imaging in astronomy, ed. J. Goad, Tucson 1987, 219

Beckers, J.M., in: SPIE Proc. Vol. **1236** (1990a) 364
Beckers, J.M., in: SPIE Proc. Vol. **1236** (1990b) 379
Beckers, J.M.: Ann. Rev. Astron. Astroph. **31** (1993) 13
Beckers, J.M., Merkle, F., eds.: High-resolution imaging by interferometry II, ESO Conf. Proc. No. **39** Garching (1992a)
Beckers, J.M., Faucherre, M., Koehler, B., von der Lühe, O., in: High-resolution imaging by interferometry II, eds. J.M. Beckers and F. Merkle, ESO Conf. Proc. No. **39** Garching (1992b)
Beckwith, S., Zuckerman, B., Skrutskie, M.F., Dyck, H.M.: Astrophys. J. **287** (1984) 793
Beckwith, S.V.W., Sargent, A.I., Chini, R.S., Güsten, R.: Astronom. J. **99** (1990) 924
Benson, J.A., Dyck, H.M., Ridgway, S.T., Dixon, D.J., Mason, W.L., Howell, R.R.: Astronom. J. **102** (1991) 2091 and **105** (1993) 736
Bernat, A.P., Hall, D.N.B., Hinkle, K.H., Ridgway, S.T.: Astrophys. J. Letters **233** (1979) L135
Bester, M., Danchi, W.C., Degiacomi, C.G., Townes, C.H., Geballe, T.R.: Astrophys. J. Letters **367** (1991) L27
Born, M., Wolf, E.: "Principles of Optics". Fourth edition. Pergamon Press, Oxford 1970, p.212, p.385, p.433, p.464, p.469
Böhme, D.: Astron. Nachrichten **299** (1978) 243
Brigham, E.O.: "FFT – Schnelle Fourier-Transformation" Oldenbourg Verlag, München 1982 Original English edition by Prentice-Hall Inc., 1974
Chelly, A., Perrier, C., Léna, P.: Astrophys. J. **280** (1984) 163
Chelly, A., Perrier, C., Cruz-González, I., Carrasco, L.: Astronom. Astrophys. **177** (1987) 51
Chen, W.P., Howell, R.R., Simon, M., Benson, J.A.: Astrophys. J. Letters **387** (1992) L43
Christou, J.C.: Experim. Astron. **2** (1991) 27
Cohen, M., and 16 co-authors: Astrophys. J. **196** (1975) 179
Connes, P., Shaklan, S. and Roddier, F., in: Interferometric imaging in astronomy, ed. J. Goad, Tucson (1987) 165
Coudé de Foresto, V., Ridgway, S.T., in: High-resolution imaging by interferometry II, eds. J.M. Beckers and F. Merkle, ESO Conf. Proc. No. **39** (1992) 731 Interferometric imaging in astronomy, ed. J. Goad, Tucson (1987) 165
Danchi, W.C., Bester, M., Degiacomi, C.G., McCullough, P.R., Townes, C.H.: Astrophys. J. Letters **359** (1990) L59
Di Benedetto, G.P., Conti, G.: Astrophys. J. **268** (1983) 309
Di Benedetto, G.P., Rabbia,Y.: Astronom. Astrophys. **188** (1987) 114
Di Benedetto, G.P., Ferluga, S.: Astronom. Astrophys. **236** (1990a) 449
Di Benedetto, G.P., Bonneau, D.: Astrophys. J. **358** (1990b) 617
Di Benedetto, G.P., Bonneau, D.: Astronom. Astrophys. **252** (1991) 645
Duquennoy, A., Mayor, M.: Astronom. Astrophys. **248** (1991) 455
Dyck, H.M., in: Late stages of stellar evolution, S. Kwok and S.R. Pottasch, eds., Reidel, Dordrecht 1987, 19
Dyck, H.M., Simon, Th., Zuckerman, B.: Astrophys. J. Letters **255** (1982) L103
Dyck, H.M., Zuckerman, B., Leinert, Ch., Beckwith,S.: Astrophys. J. **287** (1984) 801
Dyck, H.M., Kibblewhite, J.: PASP **98** (1986) 260
Dyck, H.M., Benson,J.A.,Howell, R.R., Joyce,R.R., Leinert, Ch.: Astronom. J. **102** (1991) 200

Eckart, A., Genzel, R, Krabbe, A., Hofmann, R., van der Werf, P.P., Drapatz, S.: Nature **355** (1992) 526

Eckart, A., Genzel, R, Krabbe, A., Hofmann, R., Sams, B.J., Tacconi-Garman, L.E.: Astrophys. J. Letters **407** (1993) L77

Fried, D.L.: JOSA **56** (1966) 1372

Froehly, C., in: ESO Conf. on "Scientific importance of high angular resolution", eds. M.-H. Ulrich and K. Kjär, Garching (1981) 285

Fugate, R.Q., Fried, D.L., Ameer, G.A.,Boeke, B.R., Browne, S.L., Roberts, P.H., Ruane, R.E., Tyler, G.A.,Wopat, L.M.: Nature **353** (1991) 144

Foy, R., Labeyrie, A.: Astron. Astrophys. **152** (1985) L29

Ghez, A.M., Neugebauer, G., Gorham, D.W., Haniff, C.A., Kulkarni, S.R., Matthews, K., Koresko, C., Beckwith, S.: Astronom. J. **102** (1991) 2066

Ghez, A.M., Neugebauer, G., Matthews, K.: Astronom. J. **106** (1993) November 1, in press

Glindemann, A.: JOSA section A, in review (1993)

Haas, M.: Sterne und Weltraum **30** (1991) 12 and 89

Haas, M., Leinert, Ch., Lenzen, R.: Astron. Astrophys. **261** (1992) 130

Hardie, R.H.: PASP **76** (1964) 173

Henry, T.J., McCarthy Jr., D.W.: Astrophys. J. **350** (1990) 334

Henry, T.J., McCarthy Jr., D.W.: Astronom. J. **106** (1993) 773

Högbom, J.A.: Astron. Astrophys. Suppl. **15** (1974) 417

Hubin, N., Rousset G., Beuzit, J.L., Boyer, C., Richard, J.C.: ESO Messenger No. **71**, (1993a) March, p.50

Hubin, N. et al.: SPIE 1780, (1993b) in press, ESO Technical Preprint No. **48** (1992)

Jacquinot, P, Roizen-Dossier, B.: Progress in Optics Vol.**III** (1964) 29

Jeunhomme, L.B.: Single- mode fibre optics – principles and applications. Dekker, New York (1990)

Jeys, Th.H.: MIT Lincoln Lab. J. Vol. **4** (1991) 133

Knox, K.T., Thompson, B.J.: Astrophys. J. Letters **193** (1974) L45

Koechlin, L., Rabbia, Y.: Astronom. Astrophys. **153** (1985) 91

Koresko, Ch.D., Beckwith, S.V.W., Ghez, A.M., Matthews, K., Neugebauer, G.: Astronom. J. **102** (1991) 2073

Koresko, Ch.D., Beckwith, S.V.W., Ghez, A.M., Matthews, K., Herbst, T.M.: Astronom. J. **105** (1993) 1481

Labeyrie, A.: Astron. Astrophys. **6** (1970) 85

Labeyrie, A.: Astrophys. J. Letters **196** (1974) L71

Labeyrie, A., Schumacher, G., Dugué, M., Thom, C., Bourlon, P, Foy, F., Bonneau, D, Foy, R.: Astron. Astrophys. **162** (1986) 359

Labeyrie, A., Lemaitre, G., Thom, Ch., Vakilii, F., in: High-resolution imaging by interferometry, ed. F. Merkle, ESO Conf. Proc. No. **29** (1988) 669

Labeyrie, A., Cazalé, C., Gong, S., Morand, F., Mourard, D., Kessis, J.J., Rambaut, J.P. Vakili, F., Vernet, D., Arnold, L ., in: High-resolution imaging by interferometry II, eds. J.M. Beckers and F. Merkle, ESO Conf. Proc. No. **29** (1988) 669

Leinert, Ch., Haas, M.: Astron. Astrophys. **221** (1989a) 110

Leinert, Ch., Haas, M.: Astrophys. J. Letters **342** (1989b) L39

Leinert, Ch., Haas, M., Allard,F., Wehrse, R., McCarthy Jr., D.W., Jahreiss, Perrier, Ch.: Astron. Astrophys. **236** (1990) 399

Leinert, Ch., Haas, M., Richichi, A., Zinnecker, H., Mundt, R.: Astron. Astrophys. **250** (1991a) 407

Leinert, Ch., Haas, M., Lenzen,R.: Astron. Astrophys. **246** (1991b) 180

Leinert, Ch., Haas, M., Weitzel, N.: Astron. Astrophys. **271** (1993a) 535
Leinert, Ch., Zinnecker, H., Weitzel, N., Christou, J., Ridgway, S.T., Jameson, R., Haas, M., Lenzen, R.: Astron. Astrophys. (1993b) October 20, in press
Léna, P., in: Evolution of galaxies. Astronomical observations, eds. I. Appenzeller, H.J. Habing, P. Léna, Lecture Notes in Physics No.**333** (1989) 243
Lohmann, A.W., Weigelt, G., Wirnitzer, B.: Applied Opt. **22** (1984) 4028
Lucy, L.B.: Astronom. J. **79** (1974) 745
McCarthy Jr., D.M., Low, F.J.: Astrophys. J. Letters **202** (1975) L37
McCarthy Jr., D.M., Probst, R.G., Low, F.J.: Astrophys. J. Letters **290** (1985) L9
Malbet, F., Rigaut, F., Bertout, CV., Léna, P.: Astron. Astrophys. **271** (1993) L9
Mariotti, J.-M., ed: Coherent combined instrumentation for the VLT interferometer, report by the ESO/VLT Interferometry Panel, annex to ESO/STC-131, Garching April 1992
Mariotti, J.-M., Chelli, A., Foy, R., Léna, P. Sibille, F., Tchountonov, G.: Astron. Astrophys. **120** (1983) 237
Mariotti, J.-M., Ridgway, S.T.: Astron. Astrophys. **195** (1988) 350
Merkle, F., ed.: High-resolution inaging by interferometry, ESO Conf. Proc. No. **29** Garching (1988a)
Merkle, F., in: High-resolution imaging by interferometry, ed. F. Merkle, ESO Conf. Proc. No. **29** (1988b) 921
Merkle, F., Kern, P., Léna, P., Rigaut, F., Fontanella, J.C., Rousset, G., Boyer, C., Gaffard, J.P., Jagourel, P.: ESO Messenger No.**58** (1989a) 1
Merkle, F., Beckers, J.M.: SPIE Vol. **1114** (1989b) Sterne und Weltraum **28** (1989b) 708
Merkle, F., Hubin, N., Gehring, G., Rigaut, F.: ESO Messenger No. **65**, September 1991, 13
Michelson, A.A., Pease, F.G.: Astrophys. J. **53** (1921) 249
Mozurkewich, D., Johnston, K.J., Simon, R.S., Hutter, D.J., Colavita, M., Shao, M., Oan, X.P.: Astronom. J. **101** (1991) 2207
Murphy, D.V.: MIT Lincoln Laboratory J. **5** (1992) 25
Pan, X.P., Shao, M., Colavita, M., Armstrong, J.T., Mozurkewich, D., Vivekanand, M., Denison, C.S., Simon, R.S., Johnston, K.J.: Astrophys. J. **384** (1992) 624
Pearson, T.J., Readhead, A.C.S.: Astrophys. J. **248** (1981) 61
Perrier, C., in: High-resolution imaging by interferometry, ed. F. Merkle, ESO Conderence Proc. No. **29** (1988) 113
Primmerman, Ch.A., Murphy, D.V., Page, D.A., Zollars, B.G.,Barclay, H.T.: Nature **353** (1991) 144
Readhead, A.C.S., Wilkinson,P.N.: Astrophys. J. **223** (1978) 25
Reinheimer, T., Hofmann, K.-H., Weigelt, G. in: High-resolution imaging by interferometry II, eds. J.M. Beckers and F. Merkle, ESO Conf. Proc. No. **39** (1992) 827 Astrophys. J. **223** (1978) 25
Richichi, A.: Astron. Astrophys. **226** (1989) 366
Richichi, A., in: Very high angular resolution imaging, eds. W.J. Tango and J.G. Robertson. IAU symposium No. **158** (1993) in press
Richichi, A., Salinari, P., Lisi, F.: Astrophys. J. **326** (1988) 791
Richichi, A., Lisi, F.: Astron. Astrophys. **230** (1990) 355
Richichi, A., Di Giacomo, A., Lisi, F.,Calamai, G.: Astron. Astrophys. **265** (1992) 535
Ridgway, S.T., in: High-resolution imaging by interferometry II, eds. J.M. Beckers and F. Merkle ESO Conference Proc. No.**39** (1992) 653
Ridgway, S.T., Joyce, R.R., White, N.M.,Wing, R.F.: Astrophys. J. **235** (1980) 126

Ridgway, S.T., Jacoby, G.H., Joyce, R.R., Siegel, M.J., Wells, D.C.: Astronom. J. **87** (1982) 1044

Ridgway, S.T., Keady, J.J.: Astrophys. J. **326** (1988) 843

Rigaut, F., Gendron, E.: Astron. Astrophys. **261** (1992a) 677

Rigaut, F., Léna, P., Madec, P.Y., Rousset, G., Gendron, E., Merkle, F., in: Progress in telescope and instrumentation technologies, ed. M.-H. Ulrich, ESO Conf. Proc. No. **42** (1992b) 399

Roddier, F.: Progress in Optics Vol.**XIX** (1981) 283

Roddier, F., in: Interferometric Imaging in Astronomy, J. Goad, ed., Tucson (1987) 1

Roddier, F.: Physics Reports **170** (1988) 97

Roddier, F., in: High-resolution imaging by interferometry II, eds. J.M. Beckers and F. Merkle ESO Conference Proc. No. **39** (1992) 571

Roddier, F., Léna, P.: J. Optics (Paris) **15** (1984) 171 and 363

Roddier, F, Northscott, M., Graves, J.E.: PASP **103** (1991) 131

Rohloff, R.-R., Leinert, CH.: Applied Opt. **30** (1991) 5031

Rousset, G., Fontanella, J.-C., Kern, P., Lńa, P., Gigan, P., Rigaut, F., Gaffard, J.-P., Boyer, C., Jagourel, P., Merkle, F.: SPIE Vol. **1237** (1990) 336

Rousset, G. and 13 coauthors, in: Progress in Telescope and instrumentation technologies, ed. M.-H. Ulrich ESO Conference Proc. No. **42**, M.H. Ulrich, ed. (1992) 403

Selby, M.J., Wade, R., Sanchez Magro, C.: Mon. Not. Royal Astron. Soc. **187** (1979) 553

Shaklan, St.B., Roddier, F.: Applied Opt. **26** (1987) 2159

Shaklan, St.B., Roddier, F.: Applied Opt. **27** (1988) 2334

Shao, M., Colavita, M.M., Hines, B.E., Staelin, D.H., Hutter, D.J., Johnston, K.J., Mozurkewich, D., Simon, R.S., Hershey, J.L., Hughes, J.A., Kaplan, G.H.: Astronom. Astrophys. **193** (1988a) 357

Shao, M., Colavita, M.M., Hines, B.E., Staelin, D.H., Hutter, D.J., Johnston, K.J., Mozurkewich, D., Simon, R.S., Hershey, J.L., Hughes, J.A., Kaplan, G.H.: Astrophys. J. **327** (1988b) 905

Shao, M., Colavita, M.M., Hines, B.E., Hershey, J.L., Hutter, D.J., Kaplan, G.H., Johnston, K.J., Mozurkewich, D., Simon, R.S., Pan, X.P.: Astronom. J. **100** (1990) 1701

Simon, M., Howell, R.R., Longmore, A.J., Wilking, B.A., Peterson, D.M., Chen, W.P.: Astrophys. J. **320** (1987) 344

Simon, M., Chen, W.P., Forrest, W.J., Garnett, J.D., Longmore, A.J., Gauer, T., Dixon, R.I.: Astrophys. J. **360** (1990) 95

Simon, M., Chen, W.P., Howell, R.R., Benson, J.A., Slowik, D.: Astrophys. J. **384** (1992a) 212

Simon, M., Leinert, Ch.: Sterne und Weltraum **31** (1992b) 380

Simon, M., Ghez, A.M., Leinert, Ch.: Astrophys. J. Letters **408** (1993) L33

Simons, D.A., Hodapp, K.-W., Becklin, E.E.: Astrophys. J. **360** (1990) 106

Sibille, F., Chelli, A., Léna, P.: Astron. Astrophys. **79** (1979) 315

Stecklum, B.: Die Sterne **61** (1985) 70

Sutton, E.C., in:High angular resolution stellar interferometry, eds. J. Davis and W.J. Tango, IAU Coll. No. **50** (1979) 16-1

Sutton, E.C., Storey, J.W.V., Betz, A.L., Townes, C.H.: Astrophys. J. Letters **217** (1977) L97

Tallon, M., Tallon-Bosc, I,: Astron. Astrophys. **253** (1992) 641

Toombs, R.I., Becklin, E.E., Frogel, J.A., Law, S.K., Porter, F.C., Westphal, J.A.: Astrophys. J. Letters **173** (1971) L71
Walker, J.G.: Optica Acta **28** (1981) 735
Wang, J.Y., Markey, J.K.: JOSA **68** (1978) 78
Warner, B.: High speed astronomical photometry. Cambridge University Press **1988**, Chapter 2.
White, N.M., Feierman, B.H.: Astronom. J. **94** (1987) 751
Weigelt, G.: Optics Comm. **21** (1977) 55
Weigelt, G., in: Evolution of galaxies. Astronomical observations, eds. I. Appenzeller, H.J. Habing, P. Léna, Lecture Notes in Physics No. **333** (1989) 283
Zappala, R.R., Becklin, E.E., Matthews, K.,Neugebauer, G.: Astrophys. J. **192** (1974) 109

Student Presentations

Abstracts of the presentations given by the pre-doctoral students at the Berlin School

ROSAT Survey Sources in Star Formation Regions

J.M. Alcalá[1,2], J. Krautter[1], R. Wichmann[1], J.H.M.M. Schmitt[3] and R.M. Wagner[4]

[1] Landessternwarte Königstuhl, D-6900 Heidelberg, FRG.

[2] Instituto Nacional de Astrofísica Optica y Electrónica, Tonantznintla–México.

[3] Max-Planck-Institut für Extraterrestrische Physik, D-8046 Garching, FRG.

[4] Department of Astronomy, Ohio State University, USA.

One important result of X-ray studies using EINSTEIN satellite is the discovery of a population of previously unrecognized PMS stars now called "*naked*" or weak-line T Tauri stars (WTTSs) (Walter 1986, Feigelson et al. 1987, Walter and Kuhi 1988). WTTSs are low-mass PMS stars which are found in star forming regions (SFR) and which display the Li I λ6707 absorption line (verifying their PMS nature), but lack both strong $H\alpha$ emission and strong IR excesses characteristic of classical T Tauri stars (CTTSs). WTTSs are believed to be PMS stars which lack dense surrounding matter, and therefore to be more easily detected by their strong solar-like coronal X-ray emission. There are indications that the WTTSs may outnumber the CTTSs by a factor of 2-10 (Walter and Kuhi 1988, Strom et al. 1990). This has strong consequences for estimates of the star formation efficiency and the initial mass function.

The ROSAT All Sky Survey offers for the first time the chance to study a spatially complete sample of X-ray sources in selected areas of the sky. In the course of a systematic study of ROSAT X-ray sources in star forming regions we have started spectroscopic observations in order to identify the optical counterparts. The main purpose of this study is the search for WTTSs. Here we give preliminary results of our research in the Orion, Chamaeleon, Lupus and Taurus-Auriga SFR. Part of the X-ray sources ($\sim$ 30%) could be identified with previously known sources from the SIMBAD catalogue, i.e., bright HR, HD, or SAO stars, CTTSs, and a few WTTSs known from previous Einstein observations. To identify the remaining X–ray sources we carried out optical low resolution spectroscopy at the European Southern Observatori (ESO), La Silla, at the Observatory of the Max-Planck Institut für Astronomie in Calar Alto, Spain and with the 1.8m telescope of the Ohio State University at the Lowell Observatory in Flagstaff, Arizona. Typically we observed 2–3 candidates on each ROSAT error box. These observations show that a significant fraction of the identified optical counterparts show indeed the characteristics of WTTSs, namely weak $H\alpha$ emission and Li absorption. There are, however, indications that the relative number of WTTSs differs from one SFR to the other (Krautter et al. 1992). It is too premature to give any definitive numbers for the ratio of WTTSs vs. CTTSs, but a preliminary comparison shows that the WTTSs outnumber the

CTTSs by at least a factor of three. However, this is a lower limit only, and higher numbers (up to 10) cannot be excluded yet. In the near future deep pointed ROSAT observations will be utilized to extrapolate the conclusions drown from the survey to more complete X-ray flux limits. Our results also show that WTTSs are distributed over a much larger area than the CTTSs, which tend to be concentrated near the cloud cores. This indicates that the WTTSs in fact may be used as tracers of the history of star formation. On the average WTTSs have significantly earlier spectral types than the CTTSs and those WTTSs with an observed $W_\lambda(H\alpha) \leq 0$ (filled in with emission or showing absorption features), tend to have the earliest spectral types among the WTTSs. Strong spectral line variability was also found among the WTTSs.

References

Feigelson, E.D., Jackson, J.M., Mathieu, R.D., Myers, P.C., Walter, F.M.: 1987, AJ, 94, 1251
Krautter et al. : 1992, Proc. Satellite Symposium 3, ESA ISY **3**, 187
Strom, K.M., Strom, S.E., Wilkin, F.P., Carrasco, L., Cruz-Gonzalez, I., Recillas, E., Serrano, A., Seaman, R.L., Stauffer, J.R., Dai, D., Sottile, J.: 1990, ApJ, 362, 168 Walter, F.M.: 1986, ApJ, 306, 573
Walter, F.M., Brown, A., Mathieu, R.D., Myers, P.C., Vrba, F.J.: 1988, AJ, 96, 297

Stellar Jets with Time-dependent Direction of Ejection

Susana Biro[1], Alejandro Raga[2] and Jorge Cantó[3]

[1] Astronomy Dept., The University, Manchester M13 9PL, U.K.
[2] Mathematics Dept., UMIST, P.O. Box 88, Manchester M60 1QD, U.K.
[3] Instituto de Astronomía, UNAM, Apdo. Postal 70-264, México D.F. 04510

Jets from young stars often show sinuous structures or apparent changes of direction in the plane of the sky. These effects can be interpreted as being a direct result of a time-dependence of the direction of ejection from the source.

To analyze this possible interpretation we numerically compute adiabatic 2D jets from sources with a time-dependent ejection direction. From these numerical simulations, we obtain a qualitative description of the general characteristics of such flows. We calculate a 2D, supersonic, adiabatic jet with periodic variations in the direction of ejection. Sideways shocks appear at the "elbows" of the sinuous jet. A pair of shocks (jet shock + bow shock) form at each of these elbows, in agreement with previously developed analytical models. These elbows periodically catch up with the head of the jet, forming a complex structure with several high density knots.

The Structure and Evolution of OB Associations

A.G.A. Brown

Sterrewacht Leiden, P.O. Box 9513, 2300 RA Leiden, The Netherlands

The study of OB associations is motivated primarily by the fact that they form the fossil record of star formation processes occurring in giant molecular clouds in the galactic plane. The stellar content and the internal kinematics of associations provide valuable information on: the initial mass function, the local star formation rate and efficiency, the velocity distribution of the young stars as a function of mass and position in the system, differential age effects between subgroups in an association, the characteristics of the binary population and the interaction between stars and gas.

Up to now the study of associations has been hampered by a lack of accurate knowledge of the stellar content as well as the internal kinematics. To remedy this problem project SPECTER was started at Leiden Observatory in 1982, prompted by the HIPPARCOS mission, which will measure proper motions and parallaxes of about 10000 candidate member stars in nearby (within 1 kpc) OB associations. The data will become available by 1995/96 and in anticipation thereof a number of studies are under way. These include Walraven photometry of all the southern associations, determination of accurate radial velocities for the stars in the Scorpio-Centaurus association, and a study of the interstellar medium surrounding the associations (cf. De Geus 1988).

The work presented at this school is based on Walraven photometry of the Orion OB1 association. This association is situated at a distance of ≈ 400 pc in the vicinity of the Orion Molecular Cloud complex which is itself a site of active star formation (cf. Genzel and Stutzki 1989). The most extensive study of the large scale stellar content of the Orion association was carried out by Warren and Hesser (1978) using $uvby\beta$ photometry. Their work goes down to A0 in spectral type. Our study is concentrated on a larger area on the sky and aims at obtaining information for spectral types down to late F.

The Walraven photometric system consists of 5 bands of intermediate width centred around 5441 (V), 4298 (B), 3837 (L), 3623 (U) and 3235 (W) Å (cf. Lub and Pel 1977). The (V-B) colour is similar to (B-V) in the Johnson system. With the aid of stellar atmosphere models a grid of synthetic Walraven colours can be constructed and used to derive log g and log T_{eff} for the stars. From these other physical parameters can be derived, including visual extinctions, absolute magnitudes and distance moduli. The derived parameters can subsequently be used to determine membership for the stars based on their distance moduli. The Orion OB1 association can be divided into four subgroups (Blaauw 1964) and knowledge of their stellar content enables us to calculate their distances and ages. For each subgroup we have derived the mass functions, the energy output of the stars (as compared to the surrounding gas) and an estimate can be made of the number of supernova events which have occurred in the past. This will allow us to constrain the scenario of sequential star formation in the Orion complex. Furthermore, the visual extinctions are compared to the gas/dust content in the Orion region with the aid of available data, such as IRAS $100\mu m$ maps, and the three-dimensional extent of the molecular clouds associated with Orion OB1 are constrained. The results of this photometric study will be described in a forthcoming paper (Brown et al. 1993).

References

Blaauw, A.: 1964, ARA&A, 2, 236

Brown, A.G.A., de Geus, E.J., de Zeeuw, P.T.: 1993, in preparation

Genzel, R., Stutzki, J., 1989, ARA&A, 27, 41
de Geus, E.J.: 1988, *Stars and Interstellar Matter in Scorpio Centaurus*, thesis, Leiden University
Lub, J., Pel, J.W.: 1977, A&A, 54, 137
Warren, W.H., Hesser, J.E.: 1978, ApJS, 34, 497

ROSAT X-ray Study of the Chamaeleon I Dark Cloud: The Stellar Population.

Sophie Casanova[1], Eric Feigelson[2], Thierry Montmerle[1] and Jean Guibert[3]

[1] Service d'Astrophysique, Centre d'Études de Saclay, 91191 Gif-sur- Yvette Cedex, France.

[2] Department of Astronomy & Astrophysics, Pennsylvania State University, University Park PA 16802.

[3] Observatoire de Paris and Centre d'Analyse des Images, 61 Avenue de l'Observatoire, F-75014, Paris, France.

Two soft X-ray images of the Chamaeleon I star forming cloud obtained with the *ROSAT* Position Sensitive Proportional Counter were presented. Seventy reliable, and 19 possible additional, X-ray sources are found. Eighty percent of these sources are certainly or probably identified with T Tauri stars formed in the cloud. Nineteen to 39 are proposed new "weak" T Tauri stars (WTTSs) which, when confirmed by optical spectroscopy, will significantly enlarge the known cloud population. Individual T Tauri X-ray luminosities range from $\sim 6 \times 10^{28}$ to $\sim 2 \times 10^{31}$ erg s^{-1} (0.4–2.5 keV), or $\sim 10^2 - 10^4$ times solar levels. The *ROSAT* images are an order of magnitude more sensitive, with 3-4 times more stellar identifications, that earlier *Einstein Observatory* images of the cloud.

A wide range of issues is addressed by these data. The spatial distribution and Hertzsprung-Russell diagram locations of the stars indicate that WTTSs and "classical" T Tauri stars (CTTSs) are coeval. Their X-ray luminosity functions are also essentially identical, suggesting that CTTSs have the same surface magnetic activity as WTTSs. The X-ray luminosities of well-studied Chamaeleon I cloud members are strongly correlated with a complex of four stellar properties: bolometric luminosity, mass, radius and effective temperature. The first relation can be expressed by the simple statement that low mass Chamaeleon I stars have $L_x/L_\star = 1.6 \times 10^{-4}$, within a factor of ± 2 (1σ) and the last relation by $F_x \propto R_\star$. There is thus no evidence of magnetic saturation of the stellar surfaces. We find no evidence for the absorption of soft X-rays in CTTS winds and/or boundary layers traced by the strength of the Hα emission. The mean X-ray luminosity for an unbiased optically selected T Tauri sample is 1.6×10^{29} erg s^{-1}, and we find evidence for temporal evolution of X-ray emission for stars within the pre-main sequence evolutionary phase. The total pre-main sequence population ($M_\star > 0.1\ M_\odot$) of the cloud is estimated to be $\geq$200 stars, with X-ray emitting WTTSs outnumbering CTTSs 2:1–3:1. The inferred star formation efficiency for the cloud cores is $\geq 20\%$.

Comparison of Molecular Line Data with IRAS and HI Data in High Latitude Clouds

Uwe Corneliussen, Andreas Heithausen

Physikalisches Institut, Zülpicher Str. 77, 5000 Köln 40, Germany

To explain how molecular clouds are formed from the interstellar medium and how the star forming process in such clouds will take place it is important to study the structure and physical properties (e.g. densities, temperatures, mass distribution, interstellar radiation field, chemical abundances) of molecular clouds. High Latitude Clouds (HLCs) are thought to be young, gravitationally unbound objects representing the earliest stages of molecular condensations from the interstellar medium (Magnani, Blitz and Mundy 1985). They are objects

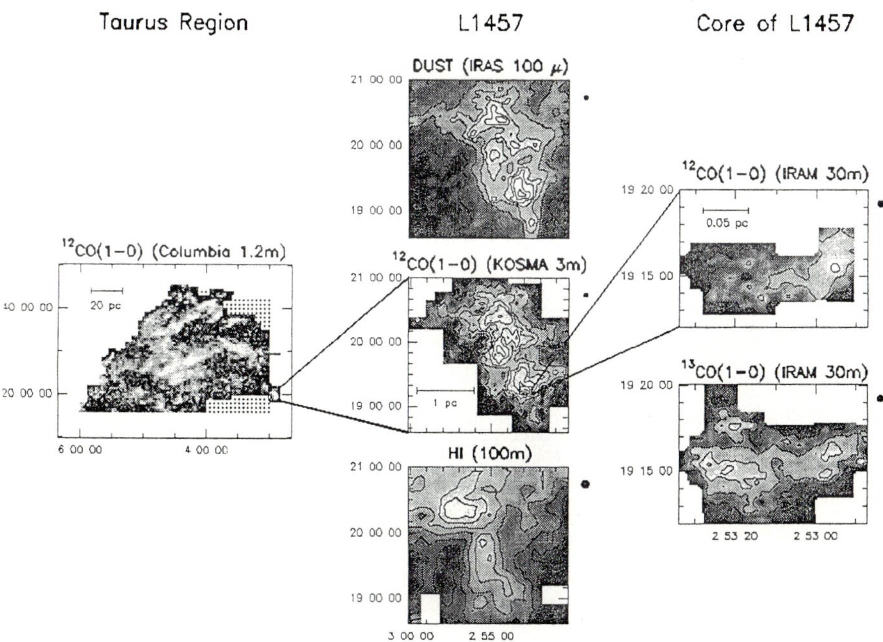

Fig. 1. Maps with different spatial resolutions of the high latitude cloud L1457 (MBM12).

with low visual extinctions and form the link between the classical diffuse clouds ($\leq$ 1 mag) and the dark clouds ($\geq$ 5 mag). This extinction range is of importance since the transformation from atomic to molecular hydrogen and from C^+ to CO occurs in this range (van Dishoek and Black 1988). As a practical advantage these objects can be studied with almost no confusing back- or foreground

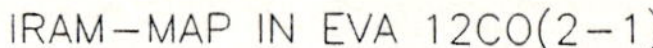

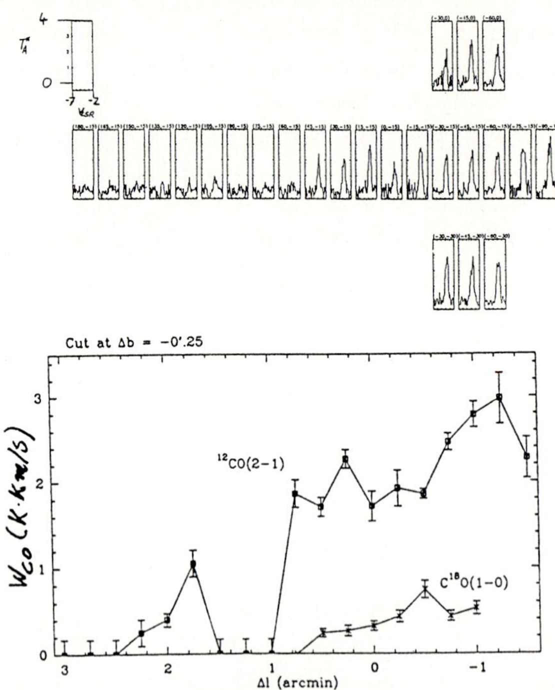

Fig. 2. Cut through the transition region of MCLD 126.6+24.5. **Top:** ^{12}CO(2-1) spectra along b=24°.53. **Bottom**: Integrated ^{12}CO(2-1) and C^{18}O(1-0) lines along b=24°.53. Values are derived from a single component Gaussian fit to the spectra.

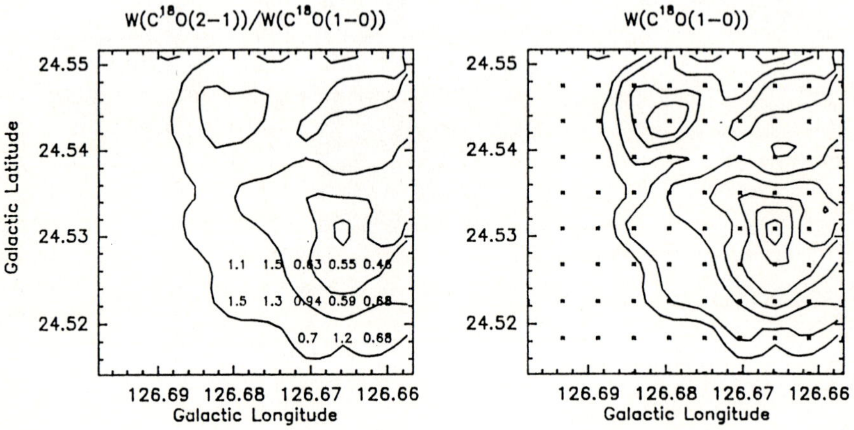

Fig. 3. High angular resolution map of MCLD 126.6+24.5. **Right**: The integrated C^{18}O(1-0) map, contours are every 0.1 kms^{-1}, starting at 0.1 kms^{-1}. **Left**: Ratio of the integrated C^{18}O(2-1) to (1-0) lines. The ratios are enhanced towards the edge of the cloud.

emission and with high spatial resolution due to their location at high latitude and small distances.

The nearest (65 pc) molecular cloud L1457 is a spatially isolated cloud in the Taurus region. Fig. 1 shows maps taken with the Columbia 1.2m telescope in ^{12}CO (Ungerechts and Thaddeus 1987), the KOSMA 3m telescope in ^{12}CO (Zimmermann and Ungerechts 1990) and the IRAM 30m telescope in ^{12}CO and $^{13}CO(1\text{-}0)$ (Zimmermann 1993) and maps of the atomic HI emission and IRAS IR emission. The investigation of the spatial and velocity structure of these molecular line data shows, that the cloud is clumpy on linear scales from 3 down to 0.003 pc. The molecular line data show several velocity components covering a range of 10 km s^{-1} and gives a clear hint to split up the whole cloud into several components. The different appearance of the ^{12}CO and ^{13}CO maps taken with the highest spatial resolution in Fig. 1 can be explained by optically thick ^{12}CO emission where only the surface of the core can be seen and optically thin ^{13}CO where the density distribution of this core can be seen. The good correlation between dust and CO allows one to use IRAS data to estimate independently a mass for the cloud which is about 100 $M_{\odot}$ in total. Since normally the IR emission originates from the molecular and the atomic part of the cloud, we expect a good correlation between the HI, IRAS and CO data. Fig. 1 shows that the HI emission peaks east of the CO emission with similar velocity gradients as the molecular material. This may be due to some kind of wind blowing away the less dense atomic gas.

The molecular cloud MCLD126.5+24.5 (Boden and Heithausen 1993) is an isolated dense core in the polaris flare. It enables a detailed study of the transition from the outer atomic part to the inner molecular part of the cloud. A steep intensity gradient which was unresolved in the 40″ beam of the 100m telescope in ammonia is seen in ^{12}CO, ^{13}CO and $C^{18}O$ towards the edge of this cloud. High resolution measurements with the IRAM 30m telescope show a sharp cut-off within 15″ corresponding to 1500 AU (Fig. 2). The measured ratio of the (2-1) and (1-0) integrated intensities of $C^{18}O$ increases in the transition region (Fig. 3). To examine whether this is due to increasing temperatures or densities further measurements, e.g. of the $NH_3(2,2)$ line, in the transition region are planned. The determined ^{13}CO and H_2CO abundances are higher and the calculated $N(H_2)/W(CO)$ ratio is lower than observed e.g. in the Taurus region whereas the NH_3 abundance is nearly the same. These abundances cannot be easily explained by steady state chemistry but one may have to invoke shock chemistry. Therefore we will search for further evidence of shock interference in this cloud and use this dataset of different resolutions and frequency ranges to model the cloud structure and investigate the physical conditions in the transition zone in detail in the future.

References

Boden K.-P., Heithausen A.: 1993, A&A, in press
Magnani L., Blitz L., Mundy L.: 1985, ApJ, 295, 402
Ungerechts H., Thaddeus P.: 1987, ApJS, 63, 645
van Dishoeck E.F., Black J.H.: 1988, ApJ, 334, 771

Zimmermann T.: 1993, PhD Thesis, in prep.
Zimmermann T,. Ungerechts H.: 1990, A&A, 238, 337

Photometric Study of Rotation in Low–mass PMS Stars

E. Covino
Osservatorio Astronomico di Capodimonte, Napoli–I

Since their discovery, T Tauri stars (TTS) and in general all young stars belonging to the so called Orion Population are well known for their, more or less prominent, erratic variability. This variability can be observed at practically all wavelengths and on very different time scales, making their interpretation very difficult. In spite of this, it had been suspected for a long time that TTS could also show sometimes periodic or quasi-periodic light variability on time scales of a few days.

As first studies have demonstrated (Rydgren and Vrba, 1983; Bouvier et al. 1986; Herbst et al. 1986, 1987), light-curve modulation with amplitudes of few tenths of a magnitude and periods of few days is often observed in young late-type stars. This modulation is explained in terms of the presence of inhomogeneously distributed spots on the photosphere of the rotating star. Thus, the observation of light curve modulation represents a powerful means for determining an important stellar parameter such as the axial rotation period of a star. Moreover, if the light-curve modulation is due to large stellar spots (as inferred, in most cases, from the colour dependence of the variations), these are also an indirect evidence of the presence of strong magnetic activity in those stars.

One striking argument in support of the interpretation of periodic light variations observed in low-mass PMS stars as modulation induced by the presence of photospheric inhomogeneities combined with axial rotation comes from the correlation between photometric periods and spectroscopic determinations of $v \sin i$ (Covino et al. 1989). This preliminary study of the sample of low–mass PMS stars with known rotation periods also suggests that the distribution of rotation periods probably presents a bimodal aspect.

Obviously, the phenomenon of light-curve modulation is easier to detect in moderately active TTS and in weak-line T Tauri stars (Rydgren and Vrba, 1983; Bouvier and Bertout, 1989), in which no large erratic variability is present. On the other hand, a large observational effort is required in order to reveal this kind of variability in classical TTS, as, for example, in the case of the star T Tauri itself (Herbst et al. 1986; Zaitseva 1990).

A photometric study of rotation in young stars in southern star forming regions has been undertaken recently by us using the ESO 0.5m telescope. As first results from our observations, new good candidates for rotational modulation have been found (Covino et al. 1992). Remarkably, all of these stars showing periodic behaviour have also been detected as strong X-ray emitters (Walter and Kuhi, 1984), three of them being weak–line T Tauri stars, discovered as a result of the X–ray survey of the Einstein satellite (Walter 1986).

Estimates of spot properties, e.g. spot temperature and coverage factor, have been derived for these stars through simple analysis of the amplitude of the light variations at different wavelengths. In two cases (namely Wa Oph 2 and Wa CrA 2) a bright–spot solution is found to satisfy the wavelength dependence of the amplitude of photometric variations.

A new sample of about thirty objects, mainly weak-line TTS discovered recently as a result of the ROSAT X–ray all–sky survey, is also at present under study in order to extend our search for rotation periods in low–mass PMS stars.

References

Bouvier, J., Bertout, C.: 1989, A&A, 211, 99
Bouvier, J., Bertout, C., Benz, W., Mayor, M.: 1986, A&A, 165, 110
Covino, E., Terranegra, L., Franchini, M., Chavarria, C., Stalio, R.: 1992, A&AS, 94, 273
Covino, E., Terranegra, L., Vittone, A.: 1989, Memorie SAIt **60**, No. 1-2, p111
Herbst, W., et al.: 1986, ApJ, 310, L71
Herbst, W., et al: 1987, AJ, 94, 137
Rydgren, A.E., Vrba, F.J.: 1983, AJ, 88, 1017
Walter, F.M.: 1986, ApJ, 306, 573
Walter, F.M., Kuhi,L.V.: 1984, ApJ, 284, 194
Zajtseva, G.V.: 1990, Proceedings of IAU Symp. No. 137, p173

EINSTEIN Observations of T Tauri Stars in Taurus-Auriga: Properties of X-Ray Emission and Relationships with Pre-Main-Sequence Activity

F. Damiani and G. Micela

Osservatorio Astronomico di Palermo, Palazzo dei Normanni, 90134 Palermo, Italy

We have performed a systematic re-analysis of the *Einstein* IPC X-ray data on the pre-main-sequence stars in the Taurus-Auriga star-formation region. We consider all catalogued stars in this region observed with the IPC; among them, 53 out of 69 can be identified with X-ray sources, and upper limits on the X-ray luminosity have been evaluated in the remaining cases. An X-ray luminosity function is derived, and we argue that the derived median L_x should indeed be representative of the true value, taking into account the incompleteness of the sample. As a result, it is confirmed that the known steady increase in the average level of X-ray emission towards younger ages can be extended even to the pre-main-sequence phase.

Our data suggest that X-ray emission might be reduced in binary systems. An anti-correlation between X-ray luminosity and rotational period is found, in qualitative agreement with earlier work; new ROSAT data, recently obtained by our group, give additional support to this result. We interpret this observed relation as a (further) indication of the magnetic origin of the T Tauri stars' X-ray emission. In order to explain a discrepant behaviour of some of our sample stars, however, this $L_x - P_{rot}$ relation should include a dependence on some additional parameter, for which age is the best candidate.

Taking into account the non-stellar origin of the T Tauri stars' Hα emission, we find much more meaningful physically to scale the Hα luminosity to the infrared (disk) emission, rather than to the stellar continuum (considering the line equivalent width). We find an anti-correlation between X-ray luminosity and this scaled Hα emission. This is interpreted as a competition between closed and open magnetic structures on the surface of these stars, respectively. In the stars without indications of winds and accretion disks, instead, a correlation between X-ray- and Hα luminosity is found, and interpreted as a further evidence of the chromospheric character of the line emission in these latter stars.

A Study on the Kinematics of the HII Regions of NGC 4449

Oriol Fuentes-Masip[1], Héctor O. Castañeda[1], Casiana Muñoz-Tuñón[1], Marcus V.F. Copetti[2], Roberto Terlevich[3]

[1] Instituto de Astrofísica de Canarias, 38200-La Laguna, Tenerife, Spain
[2] NEPAE, Universidade Federal de Santa Maria, Brazil
[3] Royal Greenwich Observatory, Madingley Road, Cambridge

We have used the TAURUS Fabry-Perot imaging spectrograph at the William Herschel Telescope of the Observatorio del Roque de los Muchachos to study the kinematics of the HII regions in the central zone (diameter of 80" or 2 Kpc, assuming a distance of 5 Mpc) of the giant, nearby, irregular galaxy NGC 4449. The Image Photon Counting System (IPCS) was used as our detector. Radial velocity, intensity and velocity dispersion maps have been obtained with the automatic fitting routines TAUFITS in the lines of Hα and [OIII]λ5007, allowing us to study the complex velocity structure of the ionized gas along the galaxy. In both emission lines the same general structures can be seen. We produced a complete set of maps fitted with gaussians and another set fitted with Cauchy functions, both profiles giving very similar results. For each one of the 91 HII regions detected (4 of which had not yet been identified) the two observed emission lines were fitted with gaussian profiles which allowed us to construct a table with the peak intensity, the radial velocity, the line width and the position with respect to the adopted center of the galaxy, of every HII region.

Star Formation in Dwarf Irregular Galaxies

Carme Gallart

Instituto de Astrofísica de Canarias, E-38200, La Laguna, Tenerife, Spain.

Dwarf Irregular Galaxies (DIG) have been revealed to be interesting laboratories where the process of star formation can be studied. Their characteristics and composition suggest that we are dealing with young systems, in the sense that they are in an early stage in the processing of gas into stars: they are simple, low metallicity, relatively gas rich galaxies with a young stellar population of bright, blue stars and HII regions. Their simple structure, lacking spiral arms, bars, or high density gradients, makes it easier to achieve complete and unbiased information on its composition.

We are working with a couple of tools that provide interesting information on the young stellar population and, consequently, on the related processes of recent star formation. On the one hand, the nearby DIG offer the possibility of resolving individual stars up to a relatively high absolute magnitude, and the construction of the HR diagram. It is a useful tool for the kind of study mentioned above, in particular to derive the luminosity function and the initial mass function, at least at their bright end. On the other hand, the H_α luminosities of the HII regions, provide an estimate of the total number of massive ionizing stars. Since they live a short time, the present day star formation rate can also be derived.

In this context, we present preliminary results from our UBVRI CCD photometry of the stellar content of NGC 6822, one of the first DIG discovered in the Local Group, and a very nearby system. Our photometry improves on previous data since we reached a deep limiting magnitude for resolved stars of about 24 in the B, V, R and I passbands, and it also provides information in a larger set of colours. The global colour-magnitude diagram shows an overall richness in stars of the upper main sequence of NGC 6822, indicating a high rate of recent star formation.

Centimeter Continuum Emission from IRAS 16293-2422

José M. Girart

Departament d'Astronomia i Meteorologia, Universitat de Barcelona, Av. Diagonal 647, E-08028, Barcelona, Spain

IRAS 16293-2422 is a deeply embedded young stellar system located in the eastern streamer region of the Rho Oph molecular cloud complex. The source has been the subject of numerous far-infrared, millimeter and centimeter wavelength observations, which have revealed a complex structure within this source. IRAS 16293-2422 is a cold and intense far-infrared source and nearly all of its luminosity (about 30 $L_\odot$) is radiated in this spectral region. The infrared and millimeter spectrum is well fitted by a $\nu^{1.5\pm0.5}$ emissivity law, and a dust temperature of 41 ± 2 K (Mundy, Wilking and Myers 1986). The mass of the gas and dust is about two solar masses. IRAS 16293-2422 is associated with an unusual quadrupolar outflow (e.g. Walker et al. 1988) and water masers emission (e.g. Wilking and Claussen 1987). High resolution observations at centimeter and millimeter wavelengths show that IRAS 16293-2422 is resolved into two sources, components A (southeast) and B (northwest), separated by $5''$ along a northwest-southeast axis. Component A has been resolved into two subcomponents separated by $0.3''$ (Wootten 1989).

In star forming regions the continuum emission in the centimeter range is usually due to the ionized gas surrounding the young stellar object. In the case of IRAS 16293-2422 the ionized gas around components A and B has different physical properties. For component A the continuum emission between 1.3 and 6 cm is well fitted by a power law with spectral index $\alpha = 0.6 \pm 0.1$ (Estalella et al. 1991). This value is expected for a partially optically thick HII region or an ionized, isothermal stellar wind. For component B, the spectral index between 2 and 6 cm ($\alpha = 2.1 \pm 0.2$) is consistent with an optically thick HII region. However, the spectral index between 2 and 1.3 cm is $\alpha = 3.2 \pm 0.5$, a value too high for an HII region. That is, the flux measured at 1.3 cm (9.6 ± 1.0 mJy) is higher than the value that would correspond to the emission arising for the optically thick HII region associated with component B (≈ 6.1 mJy). The excess of about 3.5 ± 1.0 mJy at 1.3 cm could be due to the dust emission. Observations of IRAS 16293-2422 at far-infrared and millimeter wavelengths indicate that the dust emission dominates in this spectral range. At 2.7 mm the dust emission arises from the two components (Mundy et al. 1992). Then, if we assume that the dust emission from component B can be extrapolated to the centimeter range with a ν^1 emissivity law, we expect a dust emission at 1.3 cm of about 2 mJy. This value is consistent with the excess observed at 1.3 cm. Mundy et al. (1992) observed the source at 1.3 cm with higher resolution ($0.2''$). The flux measured for component B presents an excess of only 1.8 ± 0.8 mJy over the emission from ionized hydrogen expected at this wavelength. This can be interpreted as dust emission coming from a region with a size $\gg 0.2''$, and partially resolved by the observations of Mundy et al. (1992). Moreover the observations made by Estalella et al. (1991) do not resolve the component B at 1.3 cm, so the dust emission arises from a region with a size less than $2''$. These limits for the size of the dust emission are compatible with the size of the compact component obtained from multi-component analysis of the dust emission spectrum from IRAS 16293-2422 (Mezger et al. 1990).

We have again observed IRAS 16293-2422 at 2 and 1.3 cm with the VLA in the C configuration. The reduction of the data is currently being done. The aim of these observations is to obtain a better determination of the continuum emission of the component B at these wavelength in order to confirm the detection of dust emission at centimeter wavelength in IRAS 16293-2422.

References

Estalella, R., Anglada, G., Rodríguez, L.F., and Garay, G.: 1991, ApJ, 371, 626
Mezger, P.G., Sievers, A., Zilka, R.: 1990, in IAU Symp. 147, *Fragmentation of Molecular Clouds and Star Formation*, eds. E. Falgarone, F. Boulanger, G. Duvert (Dordrecht:Reidel), p245
Mundy, L.G., Wilking, B.A., Myers, S.T.: 1986, ApJ, 311, L75
Mundy, L.G., Wootten, A., Wilking, B.A., Blake, G.A., Sargent, A.I.: 1992, ApJ, 385, 306
Walker, C.K., Lada, C.J., Young, E.T., Margulis, M.: 1988, ApJ, 332, 335
Wilking, B.A., Claussen, M.J.: 1987, ApJ, 320, L133
Wootten, A.: 1989, ApJ, 337, 858

N(CO)/N(H_2)-ratio in the Local Interstellar Medium

P. Harjunpää and K. Mattila
Observatory, P.O. Box 14, SF-00014 University of Helsinki, Finland

CO has become the “canonical” tracer of the generally unobservable molecular hydrogen, because of its large abundance and the strong spectral lines at 115GHz (^{12}CO) and 110GHz (^{13}CO). The CO to H_2 column density ratio as well as the ratio of directly observable line area I(^{12}CO) or I(^{13}CO) to N(H_2) are important scaling factors in studies of the molecular gas distribution in our Galaxy and in other galaxies. It has become customary that masses of H_2 in individual molecular clouds, the Galaxy, and other galaxies are derived using one canonical N(H_2) to N(CO) ratio (see e.g. Sanders, Solomon and Scoville 1984).

However, Williams (1985) has argued that current theories of interstellar chemistry do not justify the use of a uniform $N(H_2)$ to N(CO) ratio in dark clouds. He finds instead that the ratio may vary from cloud to cloud and even within a cloud. In regions where the star forming activity is high the $N(CO)/N(H_2)$ ratio can be higher (even by a factor of 5) as compared to regions without star formation.

We have investigated the N(CO) to A_V (and thus N(CO) to $N(H_2)$) ratio in the direction of three different local clouds: Coalsack, Chamaeleon I and R Coronae Australis. For these dark clouds a uniform set of extinction values towards highly reddened background stars (A_V up to $\sim 25^m$) is available from the near-infrared work by Hyland, Jones and Mitchell (1982), Jones et al. (1980, 1985), Strom et al. (1989) and Wilking et al. (1986). Using the SEST-telescope we have observed the ^{12}CO, ^{13}CO and $C^{18}O$ (1-0) emission lines towards these background stars. Our study supports the theory that $N(CO)/N(H_2)$ ratio is higher in active star forming regions (Cha I and R Coronae Australis) than in more quiescent regions without star formation (Coalsack). According to our preliminary results the dependence of $N(^{13}CO)$ on A_V can be fitted by $N(^{13}CO) \sim 1.7\times10^{15}(A_V\text{-}1.0)$ molecules cm^{-2} for Cha I (see Fig. 1a) and by $N(^{13}CO) \sim$

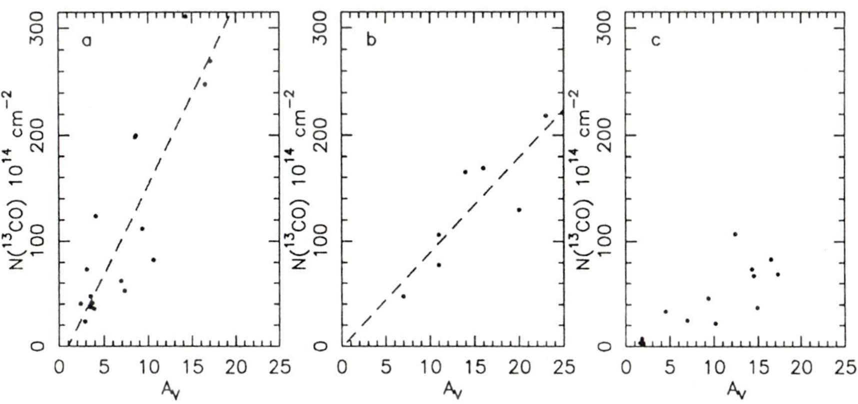

Fig. 1. Correlation between $N(^{13}CO)$ and A_V *from left to right* in the direction of (a) Cha I, (b) R Cor Aust, and (c) Coalsack.

$9.0\times10^{14}(A_V\text{-}0.1)$ molecules CM^{-2} for R Coronae Australis (see Fig. 1b). In the Case of Coalsack the problem of the extinction caused by dust lying behind the cloud is under investigation at present time. Anyway, at the moment our view is that for Coalsack (see Fig. 1c) the $N(^{13}CO)/A_V$ ratio is about half of that for R Coronae Australis.

References

Hyland, A.R., Jones, T.J., Mitchell, R.M.: 1982, MNRAS, 201, 1095

Jones, T.J., Hyland, A.R., Harvey, P.M., Wilking, B.A., Joy, M.: 1985, AJ, 90, 1191
Jones, T.J., Hyland, A.R., Robinson, G., Smith, R., Thomas, J.: 1980, ApJ, 242, 132
Sanders, D.B., Solomon, P.M., Scoville, N.Z.: 1984, ApJ, 276, 182
Strom, K.M., Newton, G., Strom, S.E., Seaman, R.L., Carrasco, L., Cruz - Gonzalez, I., Serrano, A.: 1989, ApJS, 71, 183 Wilking, B.A., Taylor, K.N.R., Storey, J.W.V.: 1986, AJ, 92, 103 Williams, D.A.: 1985, QJRAS, 26, 463

Study of the Physical and Chemical Conditions towards the W3 Region

F.P. Helmich[1,2], D.J. Jansen[1], Th. de Graauw[2] and E.F. van Dishoeck[1]
[1] Leiden Observatory, P.O. Box 9513, 2300 RA Leiden, The Netherlands
[2] SRON- Groningen, P.O. Box 800, 9700 AV Groningen, The Netherlands

The interstellar and circumstellar gas associated with high–mass young stellar objects (YSOs) has become amenable to detailed investigations with the advent of large ground–based submillimeter telescopes. The best studied case is the Orion/KL region (e.g. Sutton et al. 1985, Blake et al. 1987) for which complete, uniformly calibrated spectral scans have been made in the 230 GHz window. The data provide information on the physical and chemical behaviour of the warm gas and its interaction with the dust. Recent developments in infrared technology have made it possible to study the same gas by molecular absorption lines against the (mid–)infrared continuum of the embedded or background YSO (Lacy et al. 1989a, Evans et al. 1991, Mitchell et al. 1990). Due to the very small apertures used, this is a very powerful technique, even though it is limited to the line of sight to the YSO. Not only the gas–phase molecules, but also the solid–state composition can be probed in this way. Also, the full rotational population distribution of the gas–phase molecules can be obtained from a single infrared spectrum, thereby obviating the problems of differing beam sizes and calibration associated with the submillimeter observations. We are currently in the process of applying both techniques to the study of the W3 region.

The W3 region is a Giant Molecular Cloud complex at a distance of about 2.4 kpc. We have chosen 3 high–mass star–forming clumps for study: W3(OH), W3 IRS5 and W3 IRS4. W3(OH) is a well–known ultracompact HII region, which lies about 16′ away from the submillimeter and infrared core of the GMC containing the deeply embedded YSOs W3 IRS4 and IRS5. The three sources have, based on their $C^{18}O$ line strengths, comparable H_2 column densities of about 2×10^{23} cm^{-2}, corresponding to more than 200 magnitudes of visual extinction. The first part of the project consists of a complete spectral scan in the 345 GHz submillimeter window and selected parts of the 230 GHz window using the James Clerk Maxwell Telescope on Mauna Kea, Hawaii (Helmich et al. 1993). Initial observations show hundreds of molecular lines, especially towards W3(OH). There appear to be large differences in excitation and abundances between the three sources: IRS4 is the coldest region with temperatures around 50 K, whereas W3(OH) is much warmer with temperatures up to 200 K. Molecules such as CH_3OH are very prominent towards W3(OH), but are hardly detectable towards IRS5. As described by e.g. Blake et al. (1987) and Williams and Hartquist (1991), this may be caused by the interplay between gas and dust during the star–formation process. Initially, the chemistry is dominated by standard gas–phase ion–molecule reactions, but subsequently, during the collapse phase, it will turn into predominantly grain surface chemistry. When the radiation of the newly–formed star becomes strong enough, (processed) molecules such as CH_3OH will evaporate from the grains. The observed differences in chemical abundances in W3 therefore may reflect different evolutionary stadia. Work on more detailed models is in progress.

The second part of the project consists of high–resolution infrared absorption line observations of gas–phase molecules toward W3 IRS5, performed at the IRTF with IRSHELL (Lacy et al. 1989b). Numerous absorption lines of CO, ^{13}CO and C_2H_2 have already been detected (Lacy et al. 1989; Mitchell et al. 1990), but searches for other species such as HCO^+ and H_2S are proving to be more difficult. In the future, observations of molecules such as CO_2, which have lines in parts of the spectrum that are inaccessible from the ground, will be performed with the *Infrared Space Observatory*.

References

Blake, G.A., Sutton, E.C., Masson, C.R., Phillips, T.G.: 1987, ApJ, 315, 621

Evans II, N.J., Lacy, J.H., Carr, J.S.: 1991, ApJ, 383, 674

Helmich, F.P., Jansen, D.J., De Graauw, Th., van Dishoeck, E.F.: 1993, *in preparation.*

Lacy, J.H., Evans II, N.J., Achtermann, J.M., Bruce, D.E., Arens, J.F., Carr, J.S.: 1989a, ApJ, 342, L43

Lacy, J.H., Achtermann, J.H., Bruce, D.E., Lester, D.F., Arens, J.F., Peck, M.C., Gaalema, S.D.: 1989b, PASP 101, 1166

Mitchell, G.F., Maillard, J.P., Allen, M., Beer, R., Belcourt, K.: 1990, ApJ, 363, 554

Sutton, E.C., Blake, Geoffrey A., Masson, C.R., Phillips, T.G.: 1985, ApJS, 58, 341

Williams, D.A., Hartquist, T.W.: 1991, MNRAS, 251, 351

Spatial and Kinematic Properties of Winds from T-Tauri-Stars

Gerhard A. Hirth

MPI für Astronomie, Königstuhl 17, W-6900 Heidelberg, Germany

Metallic forbidden emission lines are thought to probe the outer (r>1AU) low-density ($n_e \approx 10^{4\cdots7}cm^{-3}$) regions of mass outflows originating from Young Stellar Objects (YSO). Making use of long-slit spectroscopy with high spectral ($\approx$ 20 kms^{-1}) and spatial ($\approx 1''$) resolution and combining the information gathered from various slit positions centered on selected T-Tauri-Stars (TTS) it is possible to reconstruct the projected spatio-kinematic structure of the YSO's near environment.

Detailed investigations of the [OI]$\lambda\lambda$6300,6363 and [SII]$\lambda\lambda$6717,6731 line intensities, line profiles and line ratios as a function of space give useful information on the underlying wind geometry, collimation, density stratification and excitation condition of the emitting region. With this method it is in principle possible to detect yet unresolved protostellar disks and jets. At least for seven nearby TTS the forbidden emission emission line region appears to be extended (typically 1-3″). In most cases the centroid of the extended emission region is shifted with respect to the TTS by 0.2-1.0″ suggesting the presence of occulting material (disk) which makes only one side of the bipolar outflow visible. This effect could be also due to intrinsic variations in the emissivity of the region. The offset is smaller for [OI] than for [SII] implying that [OI] is formed in the denser regions of the wind near the TTS. Due to the occultation of the receding part of the flow the observed profiles are mostly blueshifted. Additionally they are in most cases double-peaked showing a high velocity component of $\approx$ -150 kms^{-1} and a low velocity component near the stellar rest velocity. It is intriguing that profiles of the same object and species at different wavelengths can be both single and double-peaked. If double-peaked both components display a distinct spatial variation in intensity. These observational facts could be due to a spatial separated line formation region for each component or because of different physical formation mechanisms for the two components.

Since the forbidden lines are also detectable in apparently more evolved TTS, studies of the forbidden emission line region in principle allow to investigate evolutionary effects of the wind properties of TTS.

CO Deficiency in Galaxies of the Fornax Cluster?

Cathy Horellou[1], Fabienne Casoli[1,2] and Christophe Dupraz[1,2]

[1]DEMIRM, Observatoire de Paris, Section de Meudon, F-92195 Meudon Cedex, France

[2]Ecole Normale Supérieure, 24 rue Lhomond, F-75231 Paris Cedex 05, France

There is ample observational evidence that cluster galaxies are different from those in the field. Interaction with the hot intracluster medium affects the morphology of the galaxies, their gaseous content and possibly their star-formation activity. Tidal encounters between galaxies also play an important role.

The atomic hydrogen component has been investigated in detail for several clusters, among them our neighbor Virgo. A large fraction of the bright Virgo spirals have about one order of magnitude less HI than isolated counterparts of same optical size and type (Giovanelli and Haynes 1983). *What about the molecular phase of the interstellar medium in cluster galaxies ?* Molecular clouds are the raw material for star formation. CO surveys of Virgo bright galaxies (Stark et al. 1986; Kenney and Young 1989) do not show any evidence of H_2 deficiency : the mechanisms tearing out the low-density atomic gas leave intact the denser molecular clouds. A similar result has recently been found for the much denser and more distant Coma cluster (Casoli et al. 1991). To date, CO observations are available for those two rich clusters only.

With the Swedish-ESO 15 m telescope, we have observed in the $^{12}CO(1-0)$ transition the 23 brightest spirals and lenticulars of the Fornax cluster. With the Nancay radiotelescope, we have measured their HI emission. Fornax is the nearest southern cluster of galaxies, at the same distance as Virgo (D= 17 Mpc adopted, consistent with H_o=75 kms^{-1} Mpc^{-1}). Fornax and Virgo have a similar galaxy population (same ratio of ellipticals to spirals, of giants to dwarfs). Fornax contains 7 times less galaxies than Virgo, but its galactic density is 2.5 higher in the central region. The x-ray luminosity of Fornax is two orders of magnitude lower than that of Virgo.

Our study of Fornax galaxies has lead to a surprising result: in CO, we detected only 9 galaxies out of 23 despite long integration times and the low noise level (we could detect H_2 masses as low as 5$\times$ 10^7 $M_\odot$). *The average CO emission of Fornax galaxies is 9 times lower than in a comparison sample* based on the FCRAO Extragalactic Survey (Young et al. 1989).

On the other hand, *the HI emission of Fornax galaxies is normal*, as derived from a comparison with the HI content of the large sample of isolated galaxies observed by Haynes and Giovanelli (1984). This result is consistent with the low x-ray emission of Fornax. However, the weak CO emission combined with a normal HI content is hard to explain, because all of the mechanisms which can be invoked (ram pressure stripping, evaporation, collisions between galaxies) affect the diffuse and extended atomic gas much more efficiently than the denser molecular gas. For instance the non detection of CO in the prominent barred and ringed Fornax galaxy NGC1350 and in the Sb galaxy NGC1425 is astonishing, the more so that these objects have a high HI emission (see Fig. 1).

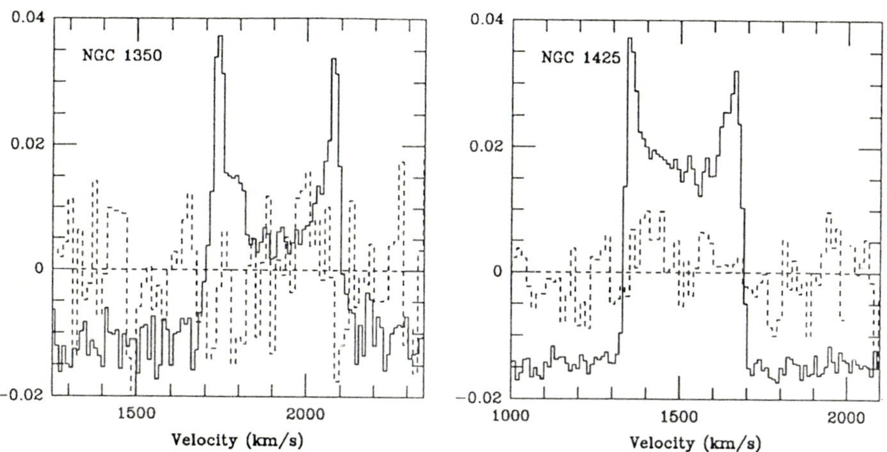

Fig. 1. HI and $^{12}CO(1-0)$ (dashed line) profiles obtained toward the center of Fornax spiral galaxies NGC1350 and NGC1425. The HI spectra are in arbitrary scale.

Both galaxies exhibit wide and symmetric HI lines whose double-horn shape is typical for rotating disks. NGC1350 (resp. NGC1425) has at least 25 (resp. 100) times less H_2 than HI gas!

Are the Fornax galaxies genuinely deficient in molecular gas, or must we suspect the conversion factor from CO emissivities to H_2 column densities? As for the latter, we used the value $N(H_2)/I(CO) = 2.3\ 10^{20} cm^2 (K kms^{-1})^{-1}$ derived for the Milky Way by Strong et al. (1988). The molecular masses may be underestimated if the Fornax galaxies have a low metallicity, but certainly not by a factor as large as 9. Indeed, even in the metal-poor Magellanic Clouds, the correction factor does not exceed 2 (Johansson 1991). Furthermore, our Fornax sample galaxies are more luminous than the LMC, thus more massive and metallic, and other indicators such as the color indices U-B and B-V or the FIR luminosity are consistent with a low star-formation activity. The low CO emission can be partly due to the small size of the Fornax galaxies with respect to the template sample, but this effect alone cannot account for the observed deficiency.

In conclusion, we suggest that the reference sample, as selected from the infrared, is strongly biased towards CO-rich galaxies. Could the Fornax galaxies be more representative of "normal" galaxies ? Be what it may, our result stresses the need for CO observations of optically-selected isolated galaxies, in order to determine the molecular content of "normal" galaxies.

References

Giovanelli, R., Haynes, M.P.: 1983, AJ, 88, 881

Stark, A. Knapp, G., Bally, J. Wilson, R.W., Penzias, A., Rowe, H.: 1986, ApJ, 310, 660

Kenney, J., Young, J.: 1989, ApJS, 144, 171

Casoli, F. Boisse, P., Combes, F., Dupraz, C., Pagani, L.: 1988, A&A, 249, 359

Young, J., Xie, S. Keney, J. Rice, W.: 1989, ApJS, 70, 699

Haynes, M.P., Giovanelli, R.: 1984, AJ, 89, 758

Strong, A., Bloemen, J., Dame, T., Grenier, I., Hermsen, W. Lebrun, F., Nyman, L.-A., Pollock, A., Thaddeus, P.: 1988, A&A, 207, 1

Johannson: 1991, IAU Symposium No. 146, p1

Properties and Distribution of Gas and Dust in the Thumbprint Nebula

K. Lehtinen[1], K. Mattila[1], G. Schnur[2], and T. Prusti[3]

[1] Helsinki University Observatory, Finland

[2] Astronomisches Institut der Ruhr-Universität Bochum, Germany

[3] Astrophysics Division, ESA, ESTEC, The Netherlands

The Thumbprint Nebula (TPN) is an opaque, roundish, bright-rimmed dark globule in Chamaeleon (Fitzgerald, Stephen and Witt 1976). Because of its isolated position and regular shape it is ideally suited for a comparison of observations and models for dense molecular clouds.

Molecular line observations. The globule has been observed with the SEST* in $^{12}CO(1-0)$ and $(2-1)$, $^{13}CO(1-0)$ and $(2-1)$, $C^{18}O$ $(1-0)$, $C^{17}O(1-0)$

and CS(2 – 1) transitions. The derived excitation temperature is $T_{ex} = 6.6K$.

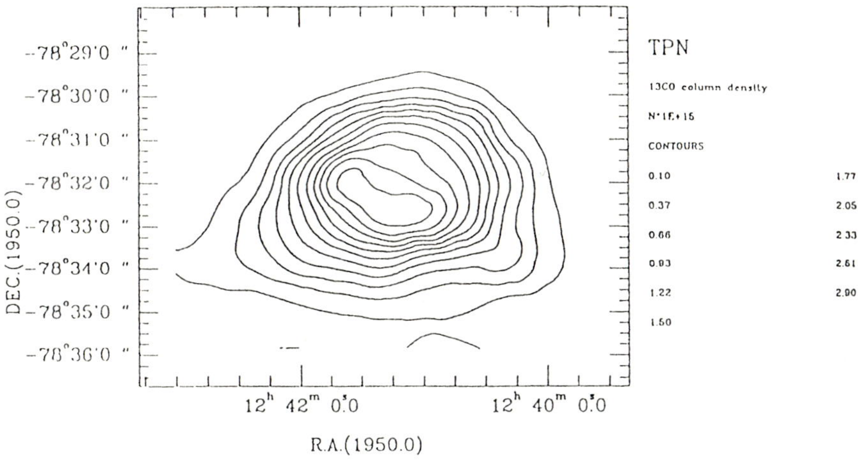

Fig. 1. A contour map of ^{13}CO column density (see text).

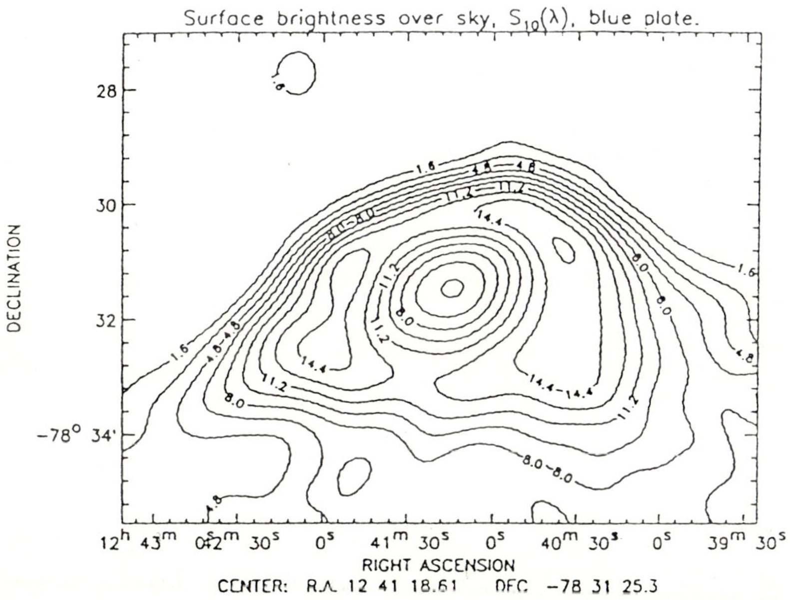

Fig. 2. A contour map of the surface brightness of the blue plate (see text) in units of $S_{10}(\lambda)$

Values of column density and optical depth at the center of the globule are $N = 3.0 \times 10^{15}\mathrm{cm}^{-2}$ and $\tau = 1.7$ for ^{13}CO, and $N = 5.9 \times 10^{14}\mathrm{cm}^{-2}$ and $\tau = 0.4$ for $C^{18}O$. A contour map of ^{13}CO column density is presented in Fig. 1a. To calculate the mass of TPN we have used a $N(^{13}CO)/A_V$ conversion factor determined by Harjunpää (these proceedings) for the Coalsack II globule. CO molecules are supposed to coagulate on grain surface in the cold environment of globules, thus making the $N(^{13}CO)/A_V$ ratio lower than in typical dark clouds. The mass of the molecular gas is found to be about 10 $M_\odot$ at a distance of 200 pc. There is indication of a velocity gradient $\Delta v = 0.2$km/s between the cloud center and the edge. The ratio between turbulent, thermal and rotational energy is $E_{turb} : E_{therm} : E_{rot}(\sin i)^{-2} = 1 : 0.7 : 0.2(\sin i)^{-2}$, where i is the angle between rotational axis and line of sight. As a result of a very low kinetic temperature the turbulent energy is the most important supporting force. Because the ratio of gravitational potential energy and kinetic energy $|E_{pot}|/E_{kin} \approx 4$, TPN seems to be in a state of gravitational collapse, if there are no other supporting forces, like magnetic field.

Optical surface brightness observations. The surface brightness of the globule was measured from blue and red glasscopies of the ESO/SRC-survey with a microdensitometer at ESO/Garching. TPN shows a dark core and a bright rim, which is caused by anisotropic scattering of interstellar photons inside the cloud. A contour map of the surface brightness of the blue plate in units of $S_{10}(\lambda)$ is presented in Fig. 2. The nebula shows a dark core and a bright rim surrounding the core. The edge of the globule facing the galactic plane is found to be very sharp, whereas the opposite side is more diffuse. The reason for this structure may be the radiation pressure from the galactic disk. The position of the surface brightness minimum, which corresponds to the dust density maximum, is found to be about 40"-50" north of the ^{13}CO and $C^{18}O$ column density maximum. Our interpretation is that the dust is protecting molecules against UV-radiation coming from the galactic disk, causing the molecular column density maximum to shift towards the south.

Infrared observations. TPN contains no IRAS infrared point source. We have made infrared surface brightness maps from the IRAS survey at 25, 60 and 100 μm and extracted individual IRAS scans crossing the TPN. However the angular resolution of the surface brightness maps seems to be insufficient to resolve the TPN. The globule is visible only at 100 and 60 μm; the dust is apparently too cold to be visible at shorter wavelengths. Both the maps and individual scans show that the coldest dust is situated at the center of TPN.

References

Fitzgerald, M.P., Stephens,T.C., Witt, A.N.: 1976, ApJ, 208, 709

A Disk Around The Young Stellar Object Z Canis Majoris?

F. Malbet

Laboratoire d'Astrophysique, Observatoire de Grenoble, BP 53X, F-38041 Grenoble Cedex, France

Z Canis Majoris is a young stellar system of intermediate mass (2-4 $M_\odot$)

located at 1150 pc. It has been recognized as an FU Orionis object with a high accretion rate (about $10^{-3} M_{\odot}/yr$) and is associated with a collimated outflow at P.A. 60° traced by both several Herbig-Haro objects and an optical jet. The linear polarization (1.5%) is orthogonal to the outflow. Recently, speckle observations in the near-infrared revealed that Z CMa is a double star: the south-eastern component emits mostly in the visible and is assumed to be the FU Ori disk; the north-western component becomes more luminous in the infrared and is interpreted as a dust photosphere.

We observed Z CMa in January 1991 with COME-ON, the ESO adaptive optics system at the ESO 3.6-m telescope, at 3.87 μm (L′) and 4.75 μm (M). The images are limited only by the diffraction of the telescope. The data processing consisted of the standard procedure for infrared images and a deconvolution performed in Fourier space. A fit by χ^2 minimization has been performed which shows that, in addition to the double star, we detect an elongated structure (0.4" length, 0.15" width) surrounding the binary and perpendicular to the jet.

We interpret these observations as follows: the south-eastern object is the FU Ori disk from which the outflow emanate; the north-western object is a dust photosphere, the color temperature of which is about 700 K; we think that the FU Ori disk is prolongated by a flattened dust reservoir which scatters the light of the infrared component located above the plane of the disk. That is the elongated structure that we detect in the near-infrared.

References

Malbet, F., Léna, P., Bertout, C.: 1991, The ESO Messenger, **66**, 32
Malbet, F., Rigaut, F., Bertout, C., Léna, P.: 1993, A&A, in press

A CS ($J=1 \rightarrow 0$) Study of Regions Previously Mapped in Ammonia

Oscar Morata

Departament d'Astronomia i Meteorologia, Universitat de Barcelona, Av. Diagonal 647, E-08028 Barcelona, Spain.

It is well known that star-formation takes place in the densest regions of molecular clouds. Thus, in order to understand the phenomena associated with the early stages of stellar evolution, it is important to study the morphology and kinematics of this high density gas. The interstellar gas is traced using the emission of the CO molecule, but if we intend to study the high density gas we need to use molecules that only have significant emission at high enough densities. Good examples of such high density gas tracers are NH_3, CS, and HCO^+.

The energy levels of the CS molecule are more complex than those of an individual atom. The force that acts over the electrons is not central-like and both nuclei have quantized rotational and vibrational movements. In molecular clouds, where the temperature is lower than 100 K, the molecular electrons are in their fundamental electronic and vibrational energy levels. Thus, the observed emission arises from energy transitions between adjacent rotational levels. When temperature is as low as in molecular clouds, typical frequencies for the emission of a variety of molecules are in the infrared and radio regions of the spectrum, particularly the latter. The thermalisation critical density for the inversion transition (J,K)=(1,1) of NH_3, with a frequency of 23.69 GHz, and for the rotational transition (J=1 $\rightarrow$ 0) of CS, with a frequency of 48.99 GHz, is $\approx 10^4$ cm^{-3}. However, the observed emission of these two molecules has been found to be very different in some sources, in spite of having very similar critical densities. Therefore, if we wish to find the actual distribution of the high density gas, we must try to clarify the intrinsic differences between the emission of these two tracers.

In 1987, our group began a program of CS (J=1 $\rightarrow$ 0) observations of sources with signs of molecular or optical outflows already mapped in the (J,K)=(1,1) inversion transition of NH_3. The goal of this program was to compare the emission of these two molecules. In order to avoid morphological discrepancies in the maps due to the use of instruments with different angular resolutions, observations of both molecules were carried out with radiotelescopes with similar beamwidths. The ammonia observations were made with the 37 m Haystack (USA) radiotelescope, which has a beamwidth of 1.′4. The CS observations were carried out with the 14 m telescope of Yebes (Spain), with a beamwidth of 1.′9.

A first set of CS observations was made between 1987 and 1990. During this period six sources were mapped (Pastor et al. 1991). In 1992, six new regions have been mapped : AFGL6366S, HH32a, HH38, HH43, L1251A and B and RNO109 (Morata et al. 1993). In both observational runs, additional $C^{34}S$ (J=1 $\rightarrow$ 0) observations have been made for calculating physical parameters for each source. The positions observed in the sky were on the same grid that was used in the NH_3 observations made at the Haystack Observatory, so as to make the comparison between the emission of both molecules more meaningful. Thus, if the emission is intrinsically similar, details like emission peak position would be at the same point. Integrated line intensity, line center velocity, line width and peak antenna temperature maps have been made for each source. Also, a bibliographical research has been made looking for manifestations of the outflow phenomenon, e.g. Herbig-Haro objects or optical jets, and of the power sources, e.g. IR or radio continuum sources.

We found CS (J=1 $\rightarrow$ 0) emission in all of the regions observed. The results of the first set of observations show that there is a general correlation between the emission of NH_3 and CS. In all of the sources, the emission peaks of CS and ammonia are located very close to centers of star-forming activity, such as IRAS sources, but usually the position of the peaks of the NH_3 and CS maps do not coincide. It was found that the CS emission tends to be more extended than the NH_3 emission. Preliminary results from the second set of observations seem to be along the same trends of earlier observations. This is specially apparent in some regions where the CS emission is very extended and appears as a background connecting several tracers of star formation like ammonia clumps and IRAS sources : HHL73 (22.′8 $\times$ 7.′3), NGC2068(HH19-27) (13.′0 $\times$ 7.′8) (Pastor et al 1991), HH38, HH43 (21′ $\times$ 8′) and RNO109 (16.′8 $\times$ 8.′4) (Morata et al. 1993). The CS emission in the HH38, HH43 region extends as far as the Haro 4-255 FIR region.

References
Morata, O., López, R., Sepúlveda, I., Estalella, R., Anglada, G., Planesas, P.: 1993, in preparation
Pastor, J., Estalella, R., López, R., Anglada, G., Planesas, P., Buj, J.: 1991, A&A, 252, 320

Interacting H_2O Masers in Star-forming Regions

Konstantinos G. Pavlakis and Nikolaos D. Kylafis
Univ. of Crete, Physics Dept., 714 09 Heraklion, Crete, Greece

We have studied the interaction of H_2O masers in star forming regions (and through it the enhancement of the observed luminosity) as a physical mechanism for the explanation of the very strong H_2O maser sources. For each individual maser we have taken a collisional pump model. Such a model can explain the low and medium luminosity masers but not the very powerful ones. We have carried out detailed numerical calculations for both saturated and unsaturated masers and have derived approximate analytic expressions for the expected brightness temperature from interacting masers. We have found that the interaction of two low or medium power H_2O masers can in principle lead to the appearance of a very strong one. Extremely strong OH masers have not been observed yet, but this could be a result of interstellar scattering.

Surface Adjustment of the KOSMA 3m Telescope Using Phase Retrieval "Holography"

J. Staguhn[1], W. Fuhr[1], A. Schulz[1], R.E. Hills[2], A.N. Lasenby[2], J. Lasenby[2], M. Miller[1], R. Schieder[1], J. Stutzki[1], and G. Winnewisser[1]
[1] I. Physikalisches Institut der Universität zu Köln, Zülpicher Straße 77, 5000 Köln 1, Germany
[2] Cavendish Laboratory, Mullard Radio Astronomy Observatory, Madingley Road, Cambridge, CB3 OHE, England

The beam efficiency of a telescope depends strongly on the roughness of its reflecting surfaces. The power collecting efficiency η of a surface with a Gaussian distribution of surface errors is given by $\eta = \eta_0 e^{-(4\pi\sigma/\lambda)^2}$ (Ruze 1966), where σ is the surface rms and λ the wavelength. A σ of $\lambda/20$ corresponding to a gain loss of 30%, is usually accepted as being tolerable. The KOSMA 3m telescope (Winnewisser et al. 1986) is located on the 3125m high Gornergrat near Zermatt, and is suitable for sub-mm observations up to 900 GHz in terms of the atmosphere (Kramer and Stutzki 1991). Though initially adjusted mechanically to 30μm rms by the manufacturer Dornier, 345 GHz observations in Winter 90/91 showed strong sidelobes in the beampatterns, indicating a systematic large scale deformation. It was thus necessary to apply a method for measuring and adjusting the telescope's surface in situ. A practical way for this measurement is the determination of the phase distribution in the aperture plane, either by measuring both the phase and amplitude of the beampattern with an interferometry system, or by using phase retrieval as described by Morris (1984) and Fienup (1982). In the latter, only amplitudes, respectively intensities, of beampatterns in several focus positions are measured. The phase is then reconstructed from these, taking into account the Fourier Transform relationship between the beam pattern (outside the near-field region) and the field in the aperture plane. Such methods are commonly referred to as "holographic". In phase retrieval methods, beam maps are usually measured at two (or more) focus positions. One then tries to estimate the amplitude and phase in the aperture plane which best fits the measured patterns in the different focus positions.

The phase retrieval algorithm we used, was first applied at the James Clerk Maxwell Telescope (JCMT) by Hills and Lasenby (1988, 1992). In that approach the problem is treated as one of obtaining a model of the aperture field which gives a least squares fit to the data. This software was recently modified for easy use at other telescopes than the JCMT. Adapting it to the special requirements at KOSMA was straightforward. Only a few new computations were required, like the evaluation of the specific diffraction pattern of the KOSMA telescope's secondary. In order to achieve the high dynamic range necessary for the power beam pattern measurements, we used as our power detector a combination of a high resolution acousto-optical deflector and a photomultiplier for the detection the deflected light.

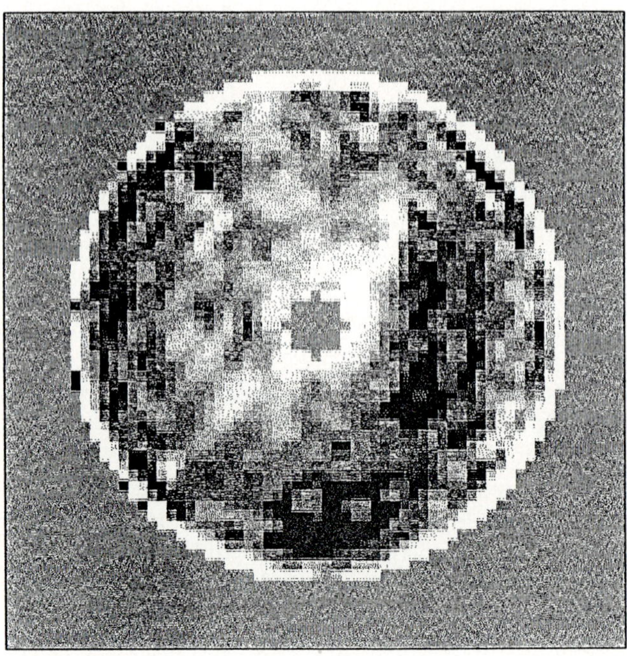

Fig. 1. KOSMA 3m Telescope primary surface before the alignment. Grey scale: black = +150μm and white = -150μm.

The aperture field distribution is evaluated in 2 steps: in the first iteration large scale phase deviations are fitted, using a set of specified functions. These include terms describing the effects of pointing errors and defocus. Additionally, we fitted a set of Zernike polynomials to the phase distribution in the aperture. In the second step, the measured beam patterns at different foci are fitted by varying independently the amplitude and phase at each grid point in the aperture plane, giving the best estimate of amplitude and phase in the aperture plane.

The holographic dish map before the alignment (Fig. 1) shows a large deformation in the lower right section of the KOSMA primary. We were able to correct this deformation iteratively in 19 steps of adjustment. The surface rms achieved after the final adjustment of the primary is shown in Fig. 2.

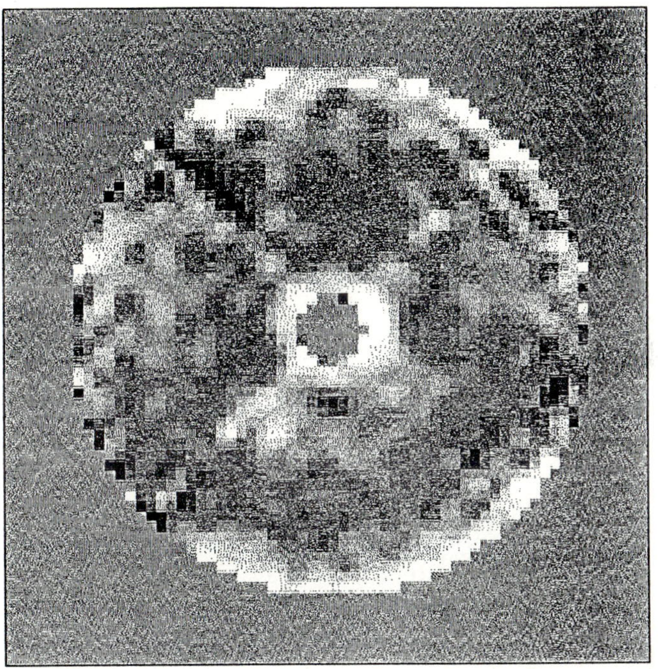

Fig. 2. KOSMA 3m Telescope primary surface after alignment. Grey scale: black = +150μm and white = -150μm.

There is essentially no difference between two large scale maps measured subsequently under the same conditions (the difference maps have a rms of about 6μm). Including the small scale fits, however the maps show a difference with pure noise at a level of 40μm, corresponding to a noise of $(40\mu m/\sqrt{2})$ rms in an individual map, consistent with the observed power drifts due to atmospheric scintillations. Correcting for this measured statistical uncertainty, blocking out the phase disturbances around the central blockage which are produced by the forward scattering around the edge of the secondary and weighting the average with a 14 dB Gaussian edge taper as used in astronomical measurements, yields a roughness rms of 40μm. This value is consistent with that derived independently from aperture efficiency measurements at different frequencies after the alignment. These data closely follow the Ruze formula and a least square fit gives a value of 42μm rms.

The improvement of the surface accuracy together with the good atmospheric conditions opened the sub-mm window for the KOSMA 3m Telescope. We have already successfully measured the fine structure line of [CI] at 492 GHz (Hernichel et al. 1992). The main beam efficiency at this frequency was measured to be 29%. The predicted value of 19% beam efficiency for 660 GHz (calculated from the fit results of the aperture field distribution) implies that the telescope can be used for measurements up to this frequency with its present surface.

References

Fienup, J.R.: 1982, App. Optics, 21, 2758

Fuhr, W., Staguhn, J., Schulz, A., Hills, R.E., Lasenby, A.N., Lasenby, J., Miller, M., Schieder, R., Stutzki, J., Vowinkel, B., Winnewisser, G.: 1993, submitted to A&A

Hernichel, J., Krause, D., Röhrig, R., Stutzki, J., Winnewisser, G.: 1992, A&A, 259, L77

Hills, R.E.: 1986, Proposal for Holographic Measurement of the JCMT, MRAO, Cambridge

Hills, R.E., Lasenby, A.N.: 1993, Millimetre-Wave Metrology of the James Clerk Maxwell Telescope, in preparation

Kramer, C., Stutzki, J.: 1991, Atmospheric transparency at Gornergrat, Technical Memorandum

Misell, D.L.: 1973, J.Phys.D: Appl.Phys., Vol. 6

Morris, D.: 1984, Proc. U.R.S.I. Int. Symp. Mm. and Submm. Wave Radio Astr., Granada, p29

Rusch, W.V.Y.: 1978, Reflector Antennas, A. Love, IEEE Press

Ruze, J.: 1966, Proc. IEEE, Vol. 54

Winnewisser, G., Bester, M., Ewald, R., Hilberath, W., Jacobs, K., Krotz-Vogel, W., Miller, M., Olberg, M., Pauls, T., Pofahl, T., Rau, G., Schieder, R., Stubbusch, H., Vowinkel, B., Wieners, C., Zensen, W.: 1986, A&A, 167, 207

Ammonia Observations of Dense Cores in Molecular Clouds

Inma Sepúlveda

Departament d'Astronomia i Meteorologia, Universitat de Barcelona, Av. Diagonal 647, E-08028 Barcelona, Spain

The dense cores of molecular clouds play an important role in the star forming process. It is well known, that in these regions of high density ($> 10^4$ cm^{-3}) low-mass star formation takes place. Thus the study of the morphology and kinematics of the dense cores is very useful in order to understand the star forming process. There are several molecules which have significant emission at high densities, in particular NH_3 and CS have their thermalisation critical density at 10^3-10^4 cm^{-3}, so these molecules are used as tracers of the high density gas.

The NH_3 molecule is a symmetric top, and each rotational level is characterized by 2 quantum numbers (J,K). All the (J,K) rotational levels are split into inversion doublets (except for K=0) and the transition between the two inversion states is the inversion line. Each inversion transition has hyperfine interactions, the most important is the electric quadrupolar interaction, this splits the line in five components, the main line, two internal satellites and two external satellites. For the (J,K)=(1,1) transition at 23.69 GHz, the satellite hyperfine components are strong and thus, detectable; so, we observe in a single spectrum the five lines. This particular structure allows a direct determination of the optical depth of the transition, due to the ratio of the antenna temperature of the main line and the internal satellite becomes a function of optical depth alone. In this way, observing only one transition, we can obtain the optical depth; in contrast to other molecules like CO and CS, where two different isotopes must be observed to obtain the optical depth. Observations of the (J,K)=(2,2) transition, at 23.72 GHz, allow one to obtain more information, the frequency of the two transitions is so close that they can be observed with the same radio telescope and with the same angular resolution. From the ratio between the (2,2) line and the (1,1) line, the kinetic temperature of the region can be derived. From these results, the ammonia column density can be calculated. Then assuming that the ratio between the abundances of NH_3 and H_2 is known (e.g., $[NH_3/H_2]=10^{-8}$, Herbst and Klemperer, 1973), the column density of H_2 of the observed region can be obtained, and also, the mass of the region. Also, using a two-level excitation model for the NH_3 molecule (Ho, 1977), the volumetric density of H_2 can be derived independently of the NH_3 abundance. In this sense, NH_3 is a very useful molecule determining the physical conditions in the region.

Outflows are a common phenomenon in star forming regions and dense cores are intimately related with this phenomenon. The high density gas is usually associated with the central region of the outflows. Most of the outflows show a bipolar morphology and many bipolar outflows have a high density gas structure elongated perpendicularly to the outflow axis. Probably, high density structures play an important role in the collimation of the outflows (Torrelles et al. 1983). Another important aspect in the study of the outflow phenomenon is the identification of the powering source. The exciting source is frequently embedded in the high density gas, traced by the ammonia emission, and is usually located near the position of the ammonia peak. Due to this fact, ammonia observations have been used to identify the exciting source (e.g., Anglada et al. 1989, Verdes-Montenegro et al. 1989).

In order to study the physical parameters of dense cores in star forming regions and its role in the outflow phenomenon, we have carried out ammonia observations of a sample of regions, using the 37 m radio telescope at Haystack Observatory (Sepúlveda et al. 1993). This sample includes regions with molecular outflows and optical outflows (jets and Herbig-Haro objects). The observed regions are: L1228, L1251A, L1251B, RNO 109, L1641N, HH 83, HH 84, HH 86-88, V571, L483, L100, L1262, RNO 43, HHL 73, IRAS 20188+3928, L1048.

We have detected and mapped ammonia emission in all of these regions, except in L100 and L1048. In most of the sources mapped there is an IRAS source near the position of the ammonia peak. There are two sources, L1641N and RNO 109, where this association is specially clear. In L1641N we have found two ammonia peaks and two IRAS sources, one in each ammonia peak, and in RNO 109 we have found three ammonia peaks, and also, one IRAS source coincident with each peak. In L483, IRAS 20188+3928, L1262, L1228, L1251B, an IRAS source coincides spatially with the maximum of ammonia emission, and with the proposed exciting source of the outflow, giving support to the idea that the exciting source is embedded in the high density gas, at the position of the ammonia peak. There is one case, L1251A, where the proposed exciting source is displaced 2′ from the position of the ammonia peak. This suggest that an undetected source can be embedded at this position. Further infrared and optical observations of this region could reveal this object.

References

Anglada, G., Rodríguez, L.F., Torrelles, J.M.,Estalella, R., Ho, P.T.P., Cantó, J., López, R., Verdes-Montenegro, L.: 1989, ApJ, 341, 208
Herbst, E., Kemplerer, W.: 1973, ApJ, 185, 505
Ho, T.P.T.: 1977, Ph.D., Massachusetts Institute of Technology.
Sepúlveda, I., Anglada, G., Estalella, R., López, R., Gómez, J.F., Torrelles, J.M., Curiel, S., Rodríguez, L.F., Ho, P.T.P.: 1993 (in preparation)
Torrelles, J.M., Rodríguez, L.F., Cantó, J., Carral, P., Marcaide, J., Moran, J.M., Ho, P.T.P.: 1983, ApJ, 274, 214
Verdes-Montenegro, L., Torrelles, J.M., Rodríguez, L.F., Anglada, G., López, R., Estalella, R., Cantó, J., Ho, P.T.P.: 1989, ApJ, 346, 193

Tidally-induced Warps in T Tauri Disks: First-Order Perturbation Theory

Caroline Terquem and Claude Bertout

Lab. d'Astrophysique, Obs. de Grenoble, BP 53X, 38041 Grenoble Cedex, France

We study the perturbation caused by a stellar companion on the potential of a T Tauri disk in the first-order approximation. We compute the resulting disk warp as a function of stellar and disk parameters in both cases of self-gravitating and accretion disks. We then calculate the radiative flux emitted by the warped disk, allowing for heating by both viscous accretion and reprocessing of stellar light. We predict that T Tauri stars with tidally warped circumstellar disks may display far-infrared and submillimetric radiative flux well in excess of that expected from flat circumstellar disks.

Near Infrared Images of Galactic Water Masers

L. Testi[1], M. Felli[2], P. Persi[3] and M. Roth[4]

[1]Dipartimento di Astronomia, University of Florence, Largo E. Fermi 5, I-50125 Firenze, Italy

[2]Osservatorio Astrofisico di Arcetri, Largo E. Fermi 5, I-50125, Firenze, Italy

[3]Istituto di Astrofisica Spaziale, CNR, CP 67, 00044 Frascati, Italy

[4]Las Campanas Observatory, Carnegie Institution of Washington, Casilla 6011, La Serena, Chile

H_2O masers in star forming regions are usually very powerful and the energy source for the pumping mechanism is still very uncertain. Recently it has been proposed (Elitzur, Hollenbach and McKee 1989) that strong hydrodynamic shocks in the dense medium around newly formed stars can produce the conditions for maser events.

Early studies at infrared wavelengths of galactic water maser regions have found a correlation between infrared and radio sources with a typical angular resolution of $\approx 30''$. Modern near infrared detectors can be successfully used to make high resolution $\leq 1''$ images of these sources.

In order to probe the processes that take place around newly formed massive stars and to test the shock model for maser pumping, we decided to observe water masers for which high resolution VLA data were available (Forster and Caswell 1989).

The observations were made during June 1991 and July-August 1992 in the three standard J, H and K filters using the near-infrared camera at the Las Campanas Observatory of the Carnegie Institution of Washington. Data reduction and analysis are still in progress at the moment.

References

Elitzur, M., Hollenbach, D.J., McKee, C.F.: 1989, ApJ, 346, 983
Forster, J.R., Caswell, J.L.: 1989, A&A, 213, 339

Multiwavelength Study of Star Formation Related Objects

L. Viktor Tóth[1,2]

[1] Helsinki University Observatory, Tähtitorninmäki, SF-00420, Finland

[2] L. Eötvös University Dept. of Astronomy, Budapest, Ludovika tér 2. H-1083, Hungary

Two dimensional (i.e. imaging) observations have an important role in modern astronomy, in particular in the field of star formation when one investigates extended objects. Image type data are practically available in all wavelength ranges from X-ray (e.g. ROSAT) through optical CCD images to HI (21cm) and CO (J=1-0) surveys.

A simultaneous study of the multiwavelength data base is required to learn the structure and physical processes of the objects. The observed reality is a superposition of physically different structures. The aim of the multiwavelength observations is to identify the acting physical effects, and separate the substructures. Multidimensional statistics (MDS) is a powerful tool when the number of measured variables (e.g. intensities at several wavelengths), and the number of observed "objects" (e.g. pixels or gridpoints of a map) are large. In thc optimum case there is a linear combination behind the observed "mixture", in which case MDS may provide not only a clear, unbiased view on the data, but may help to find the background physical variables.

We may expect answers to thc following questions using MDS: (1) Is the number of hidden variables smaller then the measured? (2) Could the number of observed objects be reduced? in our investigations of star forming regions we successfully performed cluster analysis, principal component analysis (PCA) and factor analysis (FA) on IRAS and radio line data (see Murtagh and Heck 1987 for a discussion of MDS methods).

Example 1: Applying PCA on IRAS maps we recognised and separated two main components corresponding to the Zodiacal Light and the galactic dust as the media responsible for the observed fluxes in a Taurus region (Balázs and Tóth 1992).

Example 2: We carried out an analysis of HI 21cm spectra and IRAS fluxes from 891 positions of a Cep-Cas region. First the HI spectra were replaced by the 3 new variables (the first three factors of PCA), then a cluster analysis was performed on the 7 dimensional parameter space. The major clouds have been identified by this procedure (Tóth et al. 1992).

Example 3: The spectra of the 21cm HI radiation from the direction of L1780, a small high galactic latitude dark/molecular cloud have been analysed. FA has been found to be effective in separating small spectral features, and was able to differentiate among very similarly looking input spectra. The number of variables has been reduced from the number of spectrometer channels to the number of main factors which explained the data as a superposition of radiation from an extended HI layer and from L1780 (Tóth et al. 1993).

L.V.T. acknowledges financial support from the EADN and the NSRC Finland. The above mentioned research have been carried out in collaboration with the Konkoly Observatory Budapest, under the supervision of Prof. L. G. Balázs and Prof. K. Mattila.

References

Balázs, L. G., Tóth, L. V.: 1991, in "Physics and Composition of interstellar Matter", eds. J. Krelowski and J. Papaj, Torun University.

Murtagh. F., Heck, A.: 1987, "Multivariate Data Analysis", Ap. Sp. Sc. Lib., Dordrecht Riedel, Holland

Tóth, L. V., Balázs, L. G., Ábrahám, P.: 1992, in "ADASS'92" eds. Worall, D. M., Biemesderfer, C., Barnes, J., ASP Conf. Ser. Vol. 25, p251

Tóth. L. V., Mattila, K., Haikala, L., Balázs, L. G.: 1993, in "ADASS'93" ASP Conf. Ser. in press